Risk Engineering

Series editor

Dirk Proske, Vienna, Austria

For further volumes:
http://www.springer.com/series/11582

The Springer Book Series *Risk Engineering* can be considered as a starting point, looking from different views at Risks in Science, Engineering and Society. The book series publishes intense and detailed discussions of the various types of risks, causalities and risk assessment procedures.

Although the book series is rooted in engineering, it goes beyond the thematic limitation, since decisions related to risks are never based on technical information alone. Therefore issues of "perceived safety and security" or "risk judgment" are compulsory when discussing technical risks, natural hazards, (environmental) health and social risks. One may argue that social risks are not related to technical risks, however it is well known that social risks are the highest risks for humans and are therefore immanent in all risk trade-offs. The book series tries to cover the discussion of all aspects of risks, hereby crossing the borders of scientific areas.

Junbo Jia

Essentials of Applied Dynamic Analysis

Springer

Junbo Jia
Aker Solutions
Bergen
Norway

ISSN 2195-433X ISSN 2195-4348 (electronic)
ISBN 978-3-642-37002-1 ISBN 978-3-642-37003-8 (eBook)
DOI 10.1007/978-3-642-37003-8
Springer Heidelberg New York Dordrecht London

Library of Congress Control Number: 2013955237

© Springer-Verlag Berlin Heidelberg 2014
This work is subject to copyright. All rights are reserved by the Publisher, whether the whole or part of the material is concerned, specifically the rights of translation, reprinting, reuse of illustrations, recitation, broadcasting, reproduction on microfilms or in any other physical way, and transmission or information storage and retrieval, electronic adaptation, computer software, or by similar or dissimilar methodology now known or hereafter developed. Exempted from this legal reservation are brief excerpts in connection with reviews or scholarly analysis or material supplied specifically for the purpose of being entered and executed on a computer system, for exclusive use by the purchaser of the work. Duplication of this publication or parts thereof is permitted only under the provisions of the Copyright Law of the Publisher's location, in its current version, and permission for use must always be obtained from Springer. Permissions for use may be obtained through RightsLink at the Copyright Clearance Center. Violations are liable to prosecution under the respective Copyright Law. The use of general descriptive names, registered names, trademarks, service marks, etc. in this publication does not imply, even in the absence of a specific statement, that such names are exempt from the relevant protective laws and regulations and therefore free for general use.
While the advice and information in this book are believed to be true and accurate at the date of publication, neither the authors nor the editors nor the publisher can accept any legal responsibility for any errors or omissions that may be made. The publisher makes no warranty, express or implied, with respect to the material contained herein.

Printed on acid-free paper

Springer is part of Springer Science+Business Media (www.springer.com)

To Jing and Danning, who make life a gift

Preface

The subject of dynamics originated from Sir Isaac Newton's monograph *Philosophiæ Naturalis Principia Mathematica*, and Lord Rayleigh paved the way for its further development with his *Theory of Sound*. These provided the basis for the unique position of the field of dynamics in mechanics. Since then, many scientists and engineers have applied and furthered this knowledge in various fields of applied science and technology.

With the enormous investment made in civil, mechanical, and aerospace engineering during the twentieth century, designs were pushed to the limits of their performance capacity, with the trend being toward high-speed operations, adverse environment capability, light weight, etc. With the requirement of functionality in an unpredictable, highly uncertain environment, practitioner engineers encountered more and more problems with regard to dynamics. Pure mathematics is sometimes satisfied with showing that the non-existence of a solution implies a logical contradiction, while engineers might consider numerical results the desirable goal. Although dynamics as a scientific topic is by no means fully understood (and perhaps never will be), the great amount of activity in this field during the last century has made it possible to form a practical subject in a fairly systematic, coherent, and quantitative manner. All these factors have pushed applied dynamics into a greater complexity than it has ever had before, and also promoted the subject into one of the essential tools in current engineering.

Thanks to the rapid development of computer technology, more portable and accurate testing equipment and techniques, as well as a few breakthroughs in computation algorithms, during the last 50 years applied dynamics has found efficient and unique ways of developing. This raised a vast amount of challenges in implementing designs in reality, while also putting ever higher demands on engineers, requiring a thorough understanding of the subject. In spite of increased engineering knowledge, the practical problems regarding dynamics and vibrations are in some cases handled without success despite large expenditures of money. Moreover, even if engineers can perform sophisticated computer-based dynamic analysis tasks, many of them lack an actual understanding of the essential principles of dynamics, and hence of the links between theory and application. This leads to an insurmountable barrier when they are requested to validate/verify and provide insightful explanations of analysis results, or to further improve the engineering designs with regard to vibrations, which poses a significant safety hazard and can also result in significant economic loss. These considerations motivated the author to write this book.

With the objective of providing up-to-date knowledge of dynamic analysis, which is of great importance from the point of view of engineering, in the preparation of the book, the author tried to link the general principles of dynamics with their applications from various angles in order to make it possible for readers from various backgrounds to appreciate their significance. The book aims to be as elegant as is possible given this wide-ranging treatment of the subject.

The book is intended to serve as an introduction to the subject and also as a reference book with advanced topics. A balance between the theoretical and practical aspects is sought. All the chapters are addressed to practitioner engineers who are looking for answers to their daily engineering problems, and to students and researchers who are looking for links between theoretical and practical aspects, and between phenomena and analytical explanations. It should also be of use to other science and engineering professionals and students with an interest in general dynamic analysis.

The book is written in such a way that it can be followed by anyone with a basic knowledge of structural analysis. The mathematical background assumed for reading this book is a working knowledge of differential equations, matrix manipulation, and an elementary knowledge of statistics/probability. In addition, readers are also assumed to have basic knowledge on the strength of materials.

The book covers topics on the concepts, principles, and solutions of dynamics and vibrations. These are essential for engineers and researchers to further explore any type of dynamic analysis, such as mechanical vibrations or dynamic structural responses due to environmental loads such as wave, wind, earthquake, and ice loading, etc. The core knowledge of linear and nonlinear dynamics, damping effects, random vibrations, and modal analysis is elaborated. The various solution schemes and selection criteria for a given problem are discussed. The modeling and measuring of damping are also elaborated. Special topics on seismic responses, fatigue assessment, human body vibrations, and vehicle-structure interactions are discussed. The engineering applications, relevant codes and practice, and their links with theory are also provided in relevant chapters.

The first three chapters present and discuss the phenomena, concepts, and principles of dynamic analysis with discussions on their applications.

Chapter 1 gives an introduction to dynamics in the physical world, distinguishes its essential differences from its static counterpart, and briefly summarizes general methods for treating a dynamic problem. Chapter 2 elaborates the basic formulation of governing equations of motions, which include the formulation of and relationships among Newton's second law of motions, Hamilton's principle, and Lagrange's equation, the three pillars of classical dynamics. This chapter also provides preparatory work for solving both free (Chap. 3) and forced (Chap. 11) vibration problem. Between chapters on free and forced vibrations, important topics focusing on eigenfrequencies and mode shapes are examined in Chap. 4 (for presenting eigenanalysis for discrete systems and a brief introduction on vibration-based structural health monitoring), Chap. 5 (eigenproblem for continuous system), Chap. 6

Preface   ix

(vibration under axial load), and Chap. 7 (eigenproblem for nonuniform beams and foundations).

Note that explicit and concise equations to describe a dynamic system and its responses, like a deterministic one such as Newton's, is seldom able to reflect real-world phenomena, which are complex, noisy, high-dimensional, etc., and for which the instantaneous value cannot be explicitly predicted at any time instant or reproduced. These can be treated by statistical description and characterizing randomness (probability distribution) of loads and responses, which promoted the research and application of stochastic dynamics. Therefore, Chaps. 8 and 9 systematically examine the deterministic and stochastic loads and responses from a statistical point of view. The essential concepts of Fourier and power spectrum as well as the relationship between a spectrum and its statistical properties are discussed. These form the pillar for stochastic dynamics, which is in parallel to and promotes a wider application of Newton's equations.

In Chap. 10, concepts of short and long-term probability distribution and number of occurrence are introduced. They pave the way for a reasonable understanding of load level at a given return period and for a further extension to reliability and risk assessment. This is also a part of background knowledge for assessing fatigue damage due to dynamic loading (Chap. 17).

With the understanding of spectrum analysis and power spectrum (Chap. 9), the power spectrum densities due to specific environment loads with wind, wave, ice, and earthquake loadings are presented in Chap. 12.

As Chaps. 8, 9 and 10 provide a broad overview of loads and responses, they enable efficient solutions for forced vibration problems as elaborated in Chap. 11. When reading Chap. 11, readers need to bear in mind that if the excitations are of a deterministic nature, a direct solving of equations of motions is preferred. However, if excitations are of strong stochastic nature, a random vibration approach is more efficient.

In Chap. 13, the solution to the dynamic responses is extended from a single-degrees-of-freedom to a multi-degrees-of-freedom system. In addition, the most popular numerical methods (i.e., the direct/exact method, modal superposition method, and the direct integration methods) are discussed with an emphasis on their applicabilities.

As the estimation and modeling of the damping are rather difficult tasks for both engineering and research purposes, and in the meantime the resulting uncertainties with damping pose a great challenge to reach a reasonable accuracy for the calculated dynamic responses (a phenomenon more apparent for dynamic sensitive structures), Chap. 14 is therefore dedicated to an elaboration of the effects, modeling, and measuring of various types of damping.

As almost all applied processes exhibit nonlinearities in various forms and extents, it is of particular importance to study nonlinear dynamics and vibrations. Therefore, Chap. 15 elaborates this topic by distinguishing them from their linear counterpart, summarizing their causes and sources, and by presenting the relevant numerical solution strategies used in engineering practice.

For dynamic analysis with any extent of difficulty for a real system or a structure, the numerical challenges generally arise from three aspects: space and time discretization and various types of nonlinearities. In the last 60 years, these have attracted extensive research efforts and become almost matured for engineering applications by finite element analysis (for space discretization), finite difference (Newmark's type) method (for time discretization), and linear iteration (Newton's type) method (for solving nonlinearities). These three methods form the cornerstones of current applied dynamic analysis. The finite element method can be studied in many available literatures, and the finite difference and linear iteration methods are elaborated in Chaps. 11 and 15, respectively.

After digesting the first 15 chapters, readers should have the capability to find solutions of dynamic responses in their specific fields of applications. In Chaps. 16 to 19, the essential knowledge presented in the first 15 chapters is extended to a few of their application areas, with discussions on seismic responses (Chap. 16), fatigue assessment (Chap. 17), human body vibrations (Chap. 18), and vehicle-deck dynamic interactions (Chap. 19).

While the book does not seek to promote any specific "school of thought," it inevitably reflects this author's "best practice" and "working habit." This is particularly apparent in the topics selected and level of detail devoted to each of them, their sequences, and the choices of many mathematical treatments and symbol notations, etc. The author hopes that this does not deter the readers from seeking to find their own "best practice" and dive into the vast knowledge basin of modern dynamics, which is extremely enjoyable as readers go deeper and wider.

Most of the chapters in this book can be covered in a two-day industry course in a brief manner, a one-week intensive course for either industry or university, or a two-semester course in an elaborated form for graduate students. The first four chapters together with Chaps. 11, 13, and 14 can also form a one-semester undergraduate course on structural dynamics or mechanical vibrations.

I am indebted to many individuals and organizations for assistance of various kinds, such as participation in book reviews, technical discussions, research co-operation, contributing illustrations, and copyright clearance. These include: Gunnar Bremer, Håkon Sylta, Tore Holmås, Olav Helset Lien, Zhibin Jia, Rikard Mikalsen, Peng Zheng, Vicky McNiff (Aker Solutions), Wai-Fah Chen (University of Hawaii), Andy Ruina (Cornell University), Wengang Mao, Jonas Ringsberg, and Igor Rychlik (Chalmers University of Technology), Douglas Stock (Digital Structures, Inc. Berkeley), Stefan Herion (Karlsruhe Institute of Technology), Anders Ulfvarson (The Royal Swedish Academy of Engineering Sciences), Alaa Mansour (University of California at Berkeley), Salvador Ivorra Chorro (University of Alicante), Christopher Stubbs and Philip Wilmott (Colebrand International Limited), Weicheng Cui (Shanghai Jiaotong University), Tadashi Shibue (Kinki University), Matthew S. Allen (University of Wisconsin-Madison), Derek A. Skolnik (Kinemetrics, Inc), Lance Manuel (University of Texas at Austin), Rune Elleffsen, Terje Nybø, Tor Inge Fossan, and Odd Jan Andersen (Statoil), Ketil Aas-Jakobsen (Dr. Ing. A. Aas-Jakobsen AS), Preben Terndrup

Pedersen (Technical University of Denmark), Flemming Jacobsen, Martin J. Sterndorff and Lyngberg Kim (Dong Energy), Jeffrey Wang (North America Wave Spectrum Science and Trade Inc), International Organization for Standardization (ISO), International Society of Offshore and Polar Engineers (ISOPE), Springer, Cambridge University Press, and Elsevier. Moreover, there are numerous others not named to whom I extend my sincere thanks.

This book has an extensive list of references reflecting both the historical and recent developments on the subject. I would like to thank all the authors in the references for their contribution to the area.

I wish to thank all colleagues at Aker Solutions Bergen, especially to those at Structural and Marine Department for providing a technically and socially inspiring working environment.

I would also like to acknowledge the support from Concept and Technology at Aker Solutions MMO, especially from Daniel Cazòn, Nils-Christian Hellevig, and Kristian Risdal.

Most importantly, I dedicate this book to those who live with me every day, and who brought me into existence. I conclude this preface with an expression of deep gratitude to them.

Contents

1 Introduction 1
   1.1 Experiencing Dynamics 1
   1.2 Utilize Dynamics.................................. 14
   1.3 Dynamics Versus Statics 19
   1.4 Solving Dynamic Problem 24
   1.5 Pioneers of Dynamic Analysis 29

2 Governing Equation of Motions 31
   2.1 Dynamic Equilibrium............................... 31
   2.2 Principle of Virtual Displacements 33
   2.3 Hamilton's Principle Through Lagrange's Equations 34
   2.4 Momentum Equilibrium 38
   2.5 Validity of Classical Dynamics........................ 39

3 Free Vibrations for a Single-Degree- of-Freedom (SDOF)
System–Translational Oscillations.......................... 41
   3.1 Definition of Harmonic Oscillations.................... 41
   3.2 Undamped Free Vibrations of a SDOF System 42
   3.3 Damped Free Vibrations of an SDOF 47

4 Practical Eigenanalysis and Structural Health Monitoring....... 55
   4.1 Eigenpairs, Global-, Local- and Rigid-Body Vibrations 55
   4.2 Hand Calculation of Natural Frequency for Systems
       with Distributed Masses............................. 58
       4.2.1 Classical Method for Exact Solutions 58
       4.2.2 Equivalent System Analysis for Approximate Solutions. . 61
       4.2.3 Natural Frequency with Distributed Masses:
            Dunkerley Method for Approximate Solutions 68
   4.3 Using Symmetry and Anti-Symmetry in Eigenanalysis........ 72
   4.4 Vibration-Based Structural Health Monitoring.............. 74

5 Solving Eigenproblem for Continuous Systems: Rayleigh Energy Method 79

6 Vibration and Buckling Under Axial Loading 87
    6.1 Vibration Versus Buckling 87
    6.2 Vibration and Buckling Under Harmonic Axial Loads 88
    6.3 Eigenvalues Under the Influence of Axial Loads 89

7 Eigenfrequencies of Non-uniform Beams, Shallow and Deep Foundations 95
    7.1 Non-uniform Beams 95
    7.2 Shallow and Deep Foundations 97

8 Deterministic and Stochastic Motions 99
    8.1 Category of Motions 99
    8.2 Deterministic Motions 100
    8.3 Random/Stochastic Process 102

9 Time Domain to Frequency Domain: Spectrum Analysis 109
    9.1 Fourier Spectrum 109
    9.2 Power Spectrum Density 114

10 Statistics of Motions and Loads 119
    10.1 Narrow- and Wide-Banded Process 119
    10.2 Gaussian Distribution 121
    10.3 Short-Term Distribution for Continuous Random Process: Rayleigh Distribution 125
    10.4 Long-Term Distribution for Continuous Random Process: Weibull distribution 129
    10.5 Number of Occurrence Within a Fixed Time or Space Interval: Poisson Distribution 132
    10.6 Joint Probability Distribution 134
    10.7 Long-Term Prediction 137
    10.8 Environmental Contour Line Method 138

11 Forced Vibrations 141
    11.1 Forced Vibrations Under Harmonic Excitations 141
        11.1.1 Responses to Harmonic Force 141
        11.1.2 Responses to Harmonic Base Excitations 152
    11.2 Forced Vibrations Under Complex Periodical Excitations 155
    11.3 Forced Vibrations Under Non-periodical Excitations 156
        11.3.1 Transient Responses to Force Excitation with Short Duration 157

Contents xv

11.3.2 Responses Due to Arbitrary Base Excitations
Using Convolution Integral 161
11.3.3 Responses to Non-Periodical Excitations
with Fourier Integral 162
11.4 Forced Vibrations Under Random Excitations............... 164
11.4.1 Method 164
11.4.2 White Noise Approximation..................... 169
11.5 Cross-Covariance, Cross-Spectra Density Function
and Coherence Function............................... 170
11.5.1 Cross-Covariance in Time Domain 170
11.5.2 Cross-Spectra Density in the Frequency Domain....... 171
11.5.3 Coherence Function in the Frequency Domain 171

12 Calculation of Environmental Loading Based on Power Spectra

12 **Calculation of Environmental Loading Based on
Power Spectra** 173
12.1 Wave Loads 173
12.1.1 Calculation of Hydrodynamic Wave Loads 173
12.1.2 Power Spectrum Density for Ocean Wave Kinematics... 175
12.2 Wind Loads 183
12.2.1 Calculation of Aerodynamic Wind Load 183
12.2.2 Power Spectrum Density for Wind Velocity Fields 184
12.3 Ice Loads on Narrow Conical Structures 196
12.4 Earthquake Ground Motions........................... 198
12.4.1 Power Spectrum of Seismic Ground Motions 198
12.4.2 Spatial Variation of Ground Motions
by Coherence Function 200

13 **Vibration of Multi-Degrees-of-Freedom Systems** 203
13.1 Equations of Motions................................ 203
13.2 Free Vibrations of the Two-Degrees-of-Freedom System:
Direct/Exact Method 208
13.3 Forced Vibrations of Two Degrees-of-Freedom Systems:
Direct Method..................................... 210
13.4 Forced Vibrations of MDOF: Modal Superposition Method 211
13.5 Forced Vibrations of MDOF: Direct Time Integration
Method... 220
13.5.1 Introduction to the Method..................... 220
13.5.2 Explicit Integration Method 224
13.5.3 Implicit Integration Method 226
13.5.4 Comparison between Modal Superposition
and Direct Time Integration Method 229
13.6 Lumped and Consistent Mass 230

xvi

14 Damping 233
  14.1 Types of Damping and Its Effects 233
  14.2 Damping Modeling 234
    14.2.1 Pure Viscous Damping 235
    14.2.2 Friction/Coulomb Damping 236
    14.2.3 Frequency-Dependent Hysteretic Damping 238
    14.2.4 Frequency-Independent Hysteretic Damping 240
    14.2.5 Fluid (Hydrodynamic or Aerodynamic) Damping 240
    14.2.6 Equivalent Viscous Damping 241
    14.2.7 Equivalent Viscous Damping with Coulomb
         Damping 243
    14.2.8 Equivalent Viscous Damping with Frequency
         Dependent Hysteretic Damping 243
    14.2.9 Practical Damping Modeling for Dynamic Analysis 244
  14.3 Measuring Damping 248
    14.3.1 Free Decay Method 250
    14.3.2 Step Response Method 252
    14.3.3 Hysteresis loop method 253
    14.3.4 Amplification-factor Method from Forced Vibrations . . . 253
    14.3.5 Half-power/Bandwidth Method from Forced
         Vibrations 254
  14.4 Relationship Among Various Expressions of Damping 255
  14.5 Damping for Engineering Structures 256
    14.5.1 Material Damping 256
    14.5.2 Structural/Slip Damping 257
    14.5.3 System Damping 257
    14.5.4 Hydro- and Aerodynamic Damping 258
    14.5.5 Typical Damping Levels 258
  14.6 Comparison of Cyclic Responses Among Structures
    Made of Elastic, Viscous and Hysteretic
    (Viscoelastic) Materials 259

15 Nonlinear Dynamics 261
  15.1 From Linear to Nonlinear 261
  15.2 Sources of Nonlinearities 263
    15.2.1 Material Nonlinearity 263
    15.2.2 Geometrical Nonlinearity 275
    15.2.3 Buckling 278
    15.2.4 Displacement Boundary Nonlinearity 279
    15.2.5 Force Boundary Nonlinearities 279
    15.2.6 Nonlinearities Due to Temperature Effects 279
  15.3 Load Sequence Effects 280
  15.4 Eigenfrequencies Influenced by Nonlinearities 283
    15.4.1 Material Nonlinearity 283
    15.4.2 Geometrical Nonlinearity 285

15.4.3 P-Delta (P-Δ) Effects.	288
15.5 Numerical Solutions for Nonlinear Problem	291
15.5.1 Characteristics of Nonlinear Responses	291
15.5.2 Load Control (Newton-Type) Methods	295
15.5.3 Displacement Control Methods.	299
15.5.4 Load-Displacement Control Method—Arc-Length Method (ALM).	300

16 Structural Responses Due to Seismic Excitations............. 303
16.1 Seismic Ground Motions 303
    16.1.1 Transmission of Seismic Wave from Bedrock to Ground 303
    16.1.2 Resonance Period of Soil—Site Period 304
    16.1.3 The Amplitude and Duration of Bedrock Motions...... 305
    16.1.4 Spatial Variation of Earthquake Ground Motions 307
16.2 Seismic Response Spectrum 309
    16.2.1 Introduction 309
    16.2.2 Construction of Response Spectrum 310
    16.2.3 Modal Combination Techniques for Response Spectrum Analysis 312
16.3 Characteristics of Seismic Responses Varying with Frequencies.................................. 314
16.4 Influences from Structures' Orientations and Ice Covering..... 316
16.5 Whipping Effects................................. 317
16.6 Seismic Analysis Methods 319

17 Fatigue Assessment 321
17.1 Failure of Structural Components 321
17.2 Fatigue Damage Assessment......................... 323
    17.2.1 Classification of Fatigue Assessment Approaches 323
    17.2.2 Stress-Based Approach 324
    17.2.3 Strain-Based Approach 341
    17.2.4 Fracture Mechanics Approach 341
    17.2.5 Cumulative Damage 347
17.3 Dynamic Analysis Methods for Calculating Fatigue Damage ... 349
    17.3.1 Deterministic Fatigue Analysis Method 349
    17.3.2 Simplified Fatigue Analysis Approach. 351
    17.3.3 Stochastic Fatigue Analysis Method with Narrow-Banded Responses...................... 353
    17.3.4 Deterministic vs Stochastic Fatigue Analysis for Structures Subjected to Wave Loads 360
    17.3.5 Fatigue Analysis Methods Accounting for Bandwidth, Multi-modal Frequency and Nonlinearities. 362

18 Human Body Vibrations 373
    18.1 General 373
    18.2 Criteria Related to Human Body Vibrations 374

19 Vehicle-Structure Interactions 379
    19.1 Introduction to the Topic 379
    19.2 Physical Modeling 384
        19.2.1 General 384
        19.2.2 Vehicle, Lashing and Tire Models 384
        19.2.3 Modeling of Supporting Structures 391
        19.2.4 Interaction Models for Vehicle and Supporting
               Structures. 393
    19.3 Finite Element Simulations. 394
    19.4 Analysis of Vehicle Securing 396

References 399

Index 419

Chapter 1
Introduction

1.1 Experiencing Dynamics

Every day, we are surrounded by environments full of dynamics. The ringing of the alarm clock in the morning, the voice of your beloved, sound from radio and TV, noise due to traffic, the swaying of masts and trees in the wind, and even the heart beating.

A fundamental example of dynamics is a playground swing. In order to increase the amplitude of a swing (as shown in Fig. 1.1), either the rider or the external excitor must excite the swing in phase with the movement of the swing, i.e., the period of excitation is close or identical to the natural period of the swing, leading to a resonance condition. For the swing shown in Fig. 1.1, if its amplitude is small, the natural period T_n, defined as the time that the swing spends in its arc back and forth once, is constant:

$$T_n \approx 2\pi \sqrt{\frac{L}{g}} \tag{1.1}$$

where L is the length of a single hanging rope in meter and g is the acceleration of Earth's gravity.

It is noted that, in the calculation above, the rope of the swing is assumed to be weightless and the child on the swing is idealized as a point mass. For a swing with a rope of 1.5 m in length, its natural period is 2.5 s. This equation is often referred to as the law of pendulum, which was originally discovered in 1583 by Galileo Galilei at the age of 19, when he noticed that a lamp was swinging overhead with a constant period while he was sitting in Pisa Cathedral (Fig. 1.2).

We do not even think about many dynamic phenomena because they are so common in our daily life, even though we naturally utilize them. However, if we think them through a little bit more, we may find that our life can be safer, cheaper, more enjoyable, convenient, and environmentally friendly.

In the engineering world, the design or maintenance of many engineering structures such as high-rise buildings, bridges, ships, offshore structures, aircrafts, land-based and space vehicles, mechanical equipment, tiny electronic components

J. Jia, *Essentials of Applied Dynamic Analysis*, Risk Engineering,
DOI: 10.1007/978-3-642-37003-8_1, © Springer-Verlag Berlin Heidelberg 2014

Fig. 1.1 A child on a playground swing

and even human bodies require dedicated considerations of their dynamic responses.

A sense of the importance of dynamics can be conveyed through the illustration of a few representative accidents, all of which were due to improper accounting for dynamics.

In 1985, an 8.1 magnitude of earthquake occurred in Mexico City. 412 buildings collapsed and another 324 were seriously damaged (Fig. 1.3). A large percentage of the damaged buildings in the downtown area were between 8 and 18 stories high. Not surprisingly, those buildings had a resonance vibration period of around 2.0 s, indicating general resonance with the soft soils under the city's ground, through which the seismic waves were transmitted to the ground with a dominant period of around 2.0 s [1]. The 1994 Northridge earthquake also witnessed similar structural damage due to the resonance of the upper structure with the seismic ground excitations [2]. During seismic ground shaking, through a process called soil liquefaction the loose sandy soils may behave more like a liquid than a solid. This resembles the situation in which, if a person stands statically on a layer of wet sand, the sand will easily support him or her, but if this person starts to jump or shakes his/her body dynamically, the sand will flow as a result of liquefaction, and finally the person's feet will sink into the sand. This is exactly what

1.1 Experiencing Dynamics

Fig. 1.2 The lamp hanging in Pisa Cathedral (the photo was taken in 2013, it may not represent the exact status of the lamp in 1583)

happened in many earthquake events such as the one occurred during the Niigata earthquake of 1964, shown in Fig. 1.4.

Resonance is responsible for many failures of engineering structures. For example, many leaks in pipes are caused by cracks due to vibrations or acoustical resonance with excessive excitation forces. Even if the resonance may not cause an immediate structural failure, it can be responsible for significant deflection or acceleration on a structure, leading to objects falling, termination or instability of mechanical and electronic equipment installed on the structure, or human discomfort, injury or casualty. Figure 1.5 shows the chaotic situation on board an offshore platform that was caused by the excessive movements of the platform due to the resonance vibrations caused by a large storm. The excess of rolling (rotate around the longitudinal axis of the ship) can be attributed to the resonance of ship roll motions with the sea waves. It has been reported that half of all serious accidents on board ships are caused by vibrations, either directly through structural failures or indirectly through symptoms of fatigue among the crew [5]. The human body, especially the abdomen, head and neck, are also sensitive to vibrations. When subject to vibrations with a frequency range of 1–30 Hz, people experience difficulty in maintaining correct posture and balance [2]. Even if it may be difficult

Fig. 1.3 An eight-story building was broken into two during the 8.1 magnitude of earthquake occurred in Mexico City in 1985

Fig. 1.4 Appearance of soil liquefaction after the 7.4 magnitude Niigata earthquake, Japan June 16, 1964 (courtesy of National Geophysical Data Center)

to relate some vibration effects on the human body to specific frequencies, the resonance of human organs is an important contribution to motion sickness. Motion sickness can also occur in animals due to the resonance of transportation vehicles with the animals' organs. It has been observed that, during transportation, healthy chickens can become ill and are unable to stand [4].

The swill of red wine shown in Fig. 1.6 is known as sloshing, and involves the dynamic responses of liquids under excitations, which can be amplified if the motions of a liquid container (glass) has a period close to the period of liquid (wine) sloshing. Sloshing must be considered for almost any moving vehicle or for

1.1 Experiencing Dynamics

Fig. 1.5 The chaotic situation in the office (*top*) and archive room (*bottom*) of an offshore platform after a significant storm (courtesy of Aker Solutions)

structures containing a liquid with a free surface [6]. This means that, for example, if similar phenomena occur for a chemical tank on a running truck, the liquids' sloshing motions inside the tank (Fig. 1.7) can exert significant impact on the tank, making the truck unstable or even causing a rollover. It has been reported that 4 % of heavy-truck road accidents are directly caused by the sloshing of the liquid cargo within the tank trucks [7]. For onshore tanks, earthquakes may also induce tank liquid sloshing, causing structural damage. An accident of this kind was the damage caused to seven large oil storage that occurred during the Tokachi-oki earthquake in 2003. Forensic investigations have found that, at the tanks' sites, the earthquake generated ground motions with long peak period of 4–8 s, which is in the range of the tanks' sloshing period of 5–12 s [8]. For LNG (Liquefied Natural Gas) ships, the coupling between liquid cargo sloshing and LNG ship motions can be significant at certain partial filling levels of the liquid cargo, as the large liquid movement creates highly localized impact pressure on tank walls, causing problems with regard to structural integrity and stability of the LNG ships. On various cruise ships, sloshing of the water in the swimming pool on the sun deck occurs frequently, often on even a monthly basis. The reader may refer to the movie online [9] for an illustration. This is mainly caused by the surge (heading) and pitch (rotate around the transverse axis of the ship) motions of the ship that

Fig. 1.6 The sloshing of wine in a glass (photo by Stefan Krause)

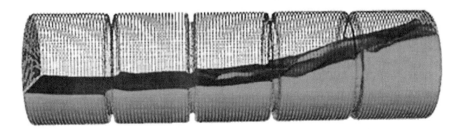

Fig. 1.7 Sloshing of the liquid inside a tank

coincide with the natural sloshing frequency of the pool. In addition, during a strong earthquake or storm event, the seismic wave and wind blowing can also excite waters in a lake or semi-enclosed sea, causing sloshing with high water surges, known as seiches. Harbors, bays, and estuaries are often prone to small seiches with amplitudes of a few centimeters and periods of a few minutes. The North Sea often experiences a lengthwise seiche with a period of about 36 h. Geological evidence indicates that the shores of Lake Tahoe may have been hit by seiches and tsunamis as large as 10 m (33 feet) high in prehistoric times, and local researchers have called for the risk to be factored into emergency plans for the region [11]. On the 26th of June, 1954, eight fishermen at Lake Michigan were

1.1 Experiencing Dynamics

Fig. 1.8 **a** Mild seiche at Canal Park in Duluth, MN, and **b** the situation just minutes before the seiche took place (courtesy of Minnesota Sea Grant)

swept away and drowned by a seiche more than 3 m high. Figure 1.8 shows two photos of storm-induced seiche at Canal Park in Duluth, Lake Superior, Minnesota. The two photos were taken just minutes apart.

It is well known that structures loaded repeatedly tend to fail at a lower load level than expected, a phenomenon known as fatigue failure. This type of failure is responsible for most of the material failures in engineering structures. Furthermore, the fatigue damages are often accompanied by unfavorable dynamic excitations relevant to resonance, high frequency loading or repeated significant loadings. There have been a number of accidents due to fatigue failure that have become well known. On the 3rd of June, 1998, a high-speed intercity train traveling from Hannover to Hamburg derailed in the village Eschede at a speed of 200 km/h, and the train crashed into a road bridge after derailment (Fig. 1.9), leading to a loss of 102 lives and 88 injuries. The forensic study followed showed that the accident was

Fig. 1.9 The destruction of rear passenger cars that were pushed into each other and crashed into a road bridge (Eschede accident, 3rd of June, 1998, photo by Nils Fretwurst)

caused by a broken wheel tire on the first middle car, in which an undiscovered crack grew to an unacceptable size under repeated fatigue loading.

Environmental loading such as wind or sea waves hit structures repeatedly, and the fatigue is a typical problem posing risks for structural safety. For example, the sinking of the *MS Estonia* on the 28th of September, 1994, was simply caused by the repeated wave impact on her bow door, leading to fatigue failure of the bow visor locking devices and the formation of opening moments about the deck hinges. This simple failure led to 852 casualties [12]. Other well known accidents leading to structural failure are the *Ranger I* jackup (Gulf of Mexico, 84 fatalities) and the collapse of the *Alexander Kielland* semi-submersible (North Sea, 123 fatalities), in 1979 and 1980 respectively. The sequence of failure in the *Alexander Kielland* platform accident (lower figure in Fig. 1.10) was [13]: fatigue failure of one brace (shown in Fig. 1.10); overload failure of five other braces; loss of column; flooding into deck; and capsizing. For *Ocean Ranger* the accident sequence was [13]: flooding through a broken window into the ballast control room; closed electrical circuit; disabled ballast pumps; erroneous ballast operation; flooding through chain lockers; and capsizing. Except for the fatigue cracks that caused the failure, it is noticed that both of these structures are statically determinate platforms with a lack of redundancy. Figure 1.11 shows a member in an offshore jacket structure breaking due to repeated wave loading. The wave induced ship vibrations are nowadays regarded as an important source contributing to fatigue damage. The vibrations are normally referred to as hull girder vibrations including whipping and springing, in which the vibrations at resonance period, the so called 2-node mode vibrations, are typically dominant. The hull girder vibrations often occur on ships. In many cases, people on board a ship can easily feel them when the wave height reaches only a few

1.1 Experiencing Dynamics

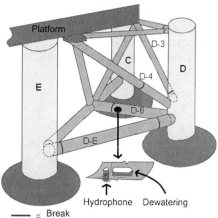

Fig. 1.10 Capsizing of *Alexander Kielland* semi-submersible (*upper*, photo courtesy of Norwegian Petroleum Museum) initiated by a fatigue crack in one brace (*lower*)

meters. Moreover, due to the low damping and large size of the blunt ships, they may experience more vibrations than slender ships.

Tacoma Narrows Bridge, shown in Fig. 1.12, was opened to traffic on the 1st of July, 1940. It spanned over a mile, the third longest suspension span in the world at

Fig. 1.11 A primary structural member of an offshore jacket structure breaks due to the repeated wave loading. The photo was taken after the jacket structure was decommissioned and transported onshore

Fig. 1.12 The collapse of Tacoma Narrows Bridge due to wind flutter

that time, with a combination of a cable-supported suspension structure and steel plate girder supporting the deck. The slenderness of the suspended deck represented a distinct departure from earlier suspension bridge designs, but because of this the bridge had shown vibratory tendencies even during construction. From the beginning of its service, it received many complaints from users because even with a light wind, the bridge behaved like a ship riding the waves with pronounced vertical oscillations, causing the "seasickness" of many passengers in cars [14], thus earning it the name "Galloping Gertie." On 7th of November, 1940, in a wind of 64 km/h, the bridge twisted so much that the left side of road descended significantly, with the right side rising, and this motion alternating rapidly. The twisting vibrations became more and more significant, finally leading to the total collapse of the bridge, as shown in Fig. 1.12. From an aerodynamic point of view, such violent vibrations are caused by the aero-elastic fluttering due to the feeding of energy when the bridge was subjected to alternative unstable oscillations in strong wind. From a structural engineering point of view, this is a type of self-

1.1 Experiencing Dynamics

Fig. 1.13 A large deflection amplitude of a bridge deck can be clearly observed due to the self-excited vibrations: the *left* and *right* figures show the relative position along vertical direction at two time instants [16]

Fig. 1.14 Proposals to avoid wind flutter for Tacoma Bridges (courtesy of University of Washington Library)

excited vibrations, which are due to the sustained alternating excitations that induce the instability of a system at its own natural or critical frequency (note that this is different from a typical resonance phenomenon, as will be explained later on in Sect. 3.3). The entire bridge-wind system therefore behaved as if it had an effective negative damping, leading to exponentially growing responses.

Fig. 1.15 The Golden Gate Bridge (photo by Rich Niewiroski Jr.)

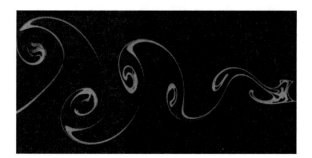

Fig. 1.16 Vortex produced by fluid passing through a cylinder

Figure 1.13 shows another example of self-excited vibrations of a bridge. While they did not cause a collapse, the large deflection amplitude of the bridge deck can clearly be observed. After the accident on Tacoma Narrows Suspension Bridge, engineers had proposed various mitigation measures to prevent similar accidents from occurring again, such as cutting holes in the web of the underdeck girder or installing curved outriggers to divert the wind (Fig. 1.14), making the wind pass through the holes and thus avoiding the wind fluttering. The Tacoma disaster

1.1 Experiencing Dynamics

Fig. 1.17 Cracks (*marked with circle*) found on a tubular joint due to the wind-induced VIV on a high-rise flare boom in the North Sea (Courtesy of Aker Solutions)

Fig. 1.18 A spiral strake installed on the *upper* part of a chimney to diminish VIV (photo by Jing Dong)

provided a great impetus to research in the field of aerodynamic stability and structural dynamics, which led to the modifications of the Golden Gate Bridge (Fig. 1.15) and several other significant suspension bridges [15].

Note that in the accident on Tacoma Narrows Suspension Bridge, the aeroelastic torsional fluttering is strongly coupled with the vortex-induced vibrations (VIV), causing a higher onset flutter velocity. For bridge decks with twin- and multi-box girders, VIV seems to be the most important problem. VIV is normally induced on members interacting with external fluid flows, and this produces periodical irregularities (vortices) in the flow, as shown in Fig. 1.16. When the vortices are not symmetrical around the body (with respect to its mid plane), lift forces will be applied on each side of the member, leading to members vibrating perpendicular to the fluid flow. Cylinder members, such as a subsea pipes or chimneys, are most susceptible to VIV. Figure 1.17 shows the cracks due to fatigue on a tubular member's end (joint). From a forensic investigation, it is found that the typical wind load on the members cannot cause fatigue cracking on this joint [17]. Therefore, VIV is most likely to be the reason for the development of these particular fatigue cracks. Another example of VIV-induced vibrations is the "loud singing" of external hand railings of ships during storms or hurricanes. To diminish VIV, it is common to put obstacles around free spans of cylinders. Figure 1.18 shows spiral strakes installed on the upper part of a chimney to diminish VIV.

1.2 Utilize Dynamics

While avoiding resonance disasters is an important concern in the engineering world, vibrations and resonance can also be put to use. Resonant systems can be used to generate vibrations at a specific frequency (e.g. musical instruments), or pick out specific frequencies from a complex vibration containing many frequencies (e.g. filters). For example, many clocks keep time by mechanical resonance in a balance wheel, pendulum, or quartz crystal.

Another example is the vibration plate (power vibration plate), which was originally used by Russian scientists to stop the reduction of bone density and muscle atrophy in cosmonauts, and is now used as a piece of fitness equipment to strengthen muscle and reduce weight. As shown in Fig. 1.19, it is essentially a flat base that vibrates. An exerciser stands on the plate while the plate is vibrating with a frequency ranging from 0.4 to 2 Hz. This forces the exerciser's entire body to react to the relatively high frequency vibrations, causing the muscles to contract and stretch in order to maintain balance. By tuning the machine to an appropriate vibration frequency, most of the muscles in the body can be effectively tightened, thus strengthening muscles and reducing weight.

As will be elaborated in a Chap. 4, the natural period is an inherent property of the system. Therefore, if people can find a convenient way to measure it, several essential characteristics of the system can be obtained. To demonstrate this, we first go to the London Eye observation wheel, as shown in Fig. 1.20. It has a

1.2 Utilize Dynamics

Fig. 1.19 A power vibration plate as a piece of fitness equipment

structural system similar to a bicycle wheel, with its rim stiffened by 16 rotation cables and 64 spoke cables. It is obvious that part of the tension load in each cable will be lost within the lifetime of the structure, requiring a re-tensing in order to maintain the cable tension in accordance with the design requirement. However, it is rather challenging to directly measure the tension force in each cable. Therefore, engineers used a much more convenient alternative to measure the natural frequency due to the transverse vibration of each cable, from which the tension load can be calculated using the relationship between the natural period and the tension load (see Sect. 6.3).

The measurement of important mechanical properties of materials, such as Young's modulus, has traditionally been carried out through a series of mechanical tests by placing the specimen on costly traction-torsion machines. Engineers nowadays have found a less costly and more convenient way to obtain part of the basic mechanical properties: simply hitting the material sample and inducing vibrations on it, as shown in Fig. 1.21. A high precision microphone close to the sample captures vibration signals and transfers them to computers; the signals are then analyzed, and the natural period and internal friction of the sample can be obtained. Thereafter, the resonant frequency, together with the dimensions and the weight of the sample, are used to calculate the elastic properties (Young's modulus, Shear modulus and Poisson ratio).

As a counter-measure, masses with their supporting stiffness can be installed to absorb energy (through momentum exchange) of another structure at their resonant frequency and further dissipate the absorbed energy through the damping of the system. Therefore, the resonance of the structure can be canceled or greatly decreased. This is normally referred to as a dynamic absorber [18]. As shown in

Fig. 1.20 The London Eye observation wheel (*upper*) with the rim supported by tensioned steel cables (*lower*, photo by Christine Matthews). The wheel works like a huge spoked bicycle wheel

Fig. 1.22, in a dynamic absorber, the mass and stiffness of it m_a and k_a are tuned such that the absorber's natural frequency coincides with the resonance frequency of the main structure.

1.2 Utilize Dynamics

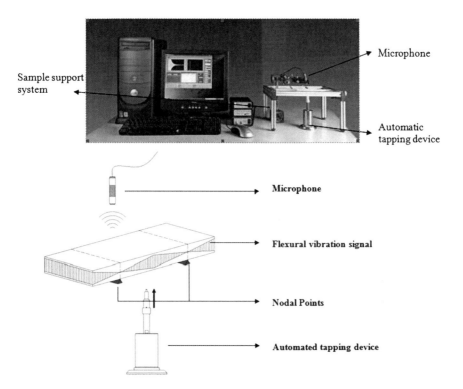

Fig. 1.21 Measurement of flexural and torsional eigenfrequencies to determine the Young's modulus and shear modulus, together with a measurement of damping through the free decay of the sample's vibrations (courtesy of IMCE Belgium)

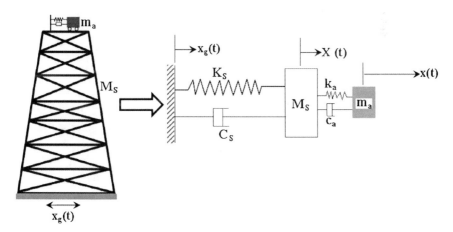

Fig. 1.22 Mechanism of a dynamic absorber with mass m_a, stiffness k_a and viscous damping c_a. It is used to mitigate the dynamic responses of the main structure with mass M_s

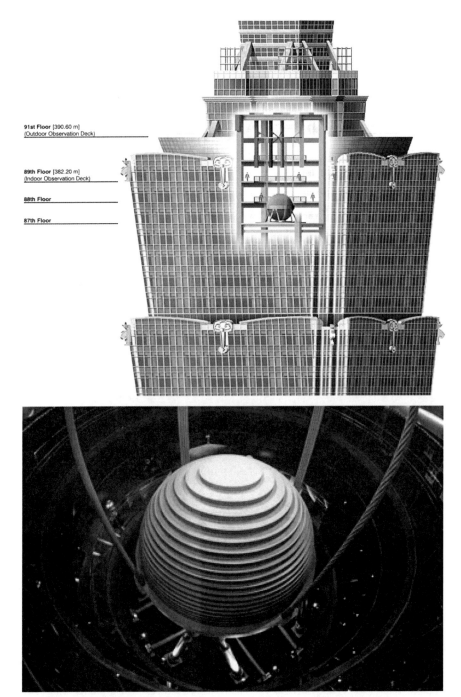

Fig. 1.23 A 660 ton pendulum tuned mass damper system installed in the Taipei World Financial Center (with a height of 509.2 m) to mitigate the wind- and earthquake-induced responses of the building. The TMD hangs from the 92nd to the 88th floor

1.2 Utilize Dynamics

Similar to the mechanism of a children's swing in Figs. 1.1 and 1.23 shows a type of dynamic absorber—the tuned mass damper (TMD) installed in the Taipei World Financial Center with a height of 509.2 m. The TMD weighs 660 tons and is suspended by 8 steel cables, arranged in 4 pairs, from the frame on the 92nd floor as a pendulum system. By adjusting the free cable length, the mass in this pendulum system moves with the building at similar natural period of 6.8 s. Eight primary hydraulic viscous dampers situated beneath the TMD automatically dissipate energy from vibration impacts. A bumper system of eight hydraulic viscous dampers beneath the TMD absorbs vibration impacts, particularly in major typhoons or earthquakes where movements exceed 1.5 m [19]. This can greatly decrease dynamic responses due to seismic and wind loadings.

Similar to a TMD, a tuned liquid damper (TLD), another type of dynamic absorber, is also a passive damping system in which the damping effects are provided by the motion of liquid in tanks. The moving liquid has a function similar to the moving mass of a TMD, in that gravity is harnessed as a restoring force. Energy is mainly dissipated by using damping baffles to create turbulence in the liquid, as well as through the wave breaking and the impact of liquid on the tank wall. The geometry of the tank that holds the water is determined theoretically to give the desired natural frequency of water motions in accordance with the space in which the tank is to be located. Liquid tanks used as a TLD are typically rectangular or circular, with the former being able to be tuned to two different frequencies in two perpendicular directions. An engineering example of TLD is the water tanks installed on the top of the skyscraper One Rincon Hill in San Francisco, which can hold up to 190 tons of water. The water level in the tanks is adjusted to achieve a tank sloshing natural frequency close to that of the building structure. Baffles are installed inside the water tanks in order to increase the damping when the water is in motion. In addition to the function as a TLD to mitigate wind- and earthquake-induced responses, the water tanks were also built to hold water for fire fighters. It is noticed that, compared to a TMD, the TLD has the advantage of low manufacturing and maintenance costs, and it can also serve the purpose of liquid (water, fuel, crude oil or mud, etc.) storage [20] for emergency, industry, or everyday purposes if fresh water is used [21, 22]. Furthermore, without adversely affecting the functional use of tanks, TLD tanks can be designed with proper dimensions or reconfigured with internal partitions of existing tanks, which is helpful to cope with physical and architectural requirements [23].

1.3 Dynamics Versus Statics

Over history, the safety and serviceability of structures have basically been measured on the basis of their static behavior, which required adequate stiffness and strength. This was perhaps because the necessary knowledge of dynamics was less accessible to engineers than their static counterpart. Nowadays, it is common

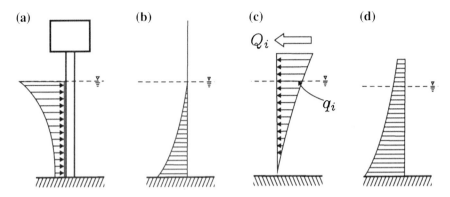

Fig. 1.24 Wave induced static versus instantaneous dynamic forces and moments in a bottom-fixed cantilevered tower [24]

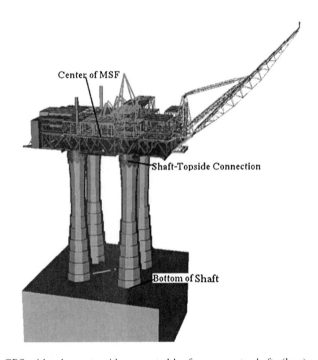

Fig. 1.25 A GBS with a heavy topside supported by four concrete shafts (legs)

knowledge that all bodies possessing stiffness and mass are capable of exhibiting dynamic behavior.

The major difference between dynamic and static responses is that dynamics involves the inertia forces associated with the accelerations at different parts of a structure throughout its motion. If one ignores the inertia force, the predicted responses can be erroneous. As an example, let's consider a bottom fixed cantilevered

1.3 Dynamics Versus Statics

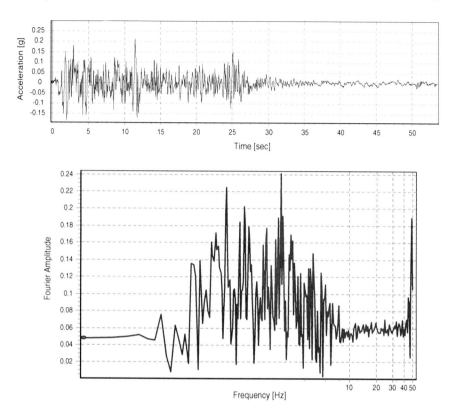

Fig. 1.26 Ground motions EW component (*upper*) recorded during El Centro earthquake and its Fourier amplitude (*lower*)

tower subjected to sea wave loadings as shown in Fig. 1.24 [24]. In addition to the static bending moment due to wave loadings applied on the structure, as shown in Fig. 1.24b, the stiffness and mass of the structure will react to the wave loadings and generate internal forces on both the top mass block (Q_i) and the tower (q_i), shown in Fig. 1.24c. Rather than a single function of mass, the amplitudes of the inertia forces are related to a ratio between stiffness and mass (eigenfrequency), mass, as well as damping, thus resulting in additional dynamic bending action (Fig. 1.24d).

As another example, consider a gravity-based structure (GBS), shown in Fig. 1.25, that is subjected to the ground motions recorded during El Centro earthquake, which have a high energy content at the vibration period above 0.2 s (below 5 Hz in Fourier amplitude shown in Fig. 1.26, which will be explained in Sects. 9.2 and 12.4). The dynamic responses of the platform are investigated by varying the thickness of four shafts from half of the reference thickness, to the reference thickness, to twice the reference thickness. It is obvious that the GBS becomes stiff by increasing the shafts' thickness. If a static analysis is performed, under the same seismic excitations the stiffer structure would have lower responses. However, the seismic responses involving dynamic effects may not

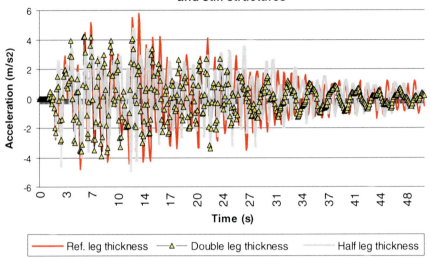

Fig. 1.27 Acceleration at the shaft-topside connection with various leg/shaft stiffness (peak acceleration: 4.7 m/s^2 for double leg thickness, 5.8 m/s^2 for reference leg thickness, 4.2 m/s^2 for double leg thickness)

obey this rule. Figure 1.27 shows the acceleration at the shaft-topside connection. It is clearly shown that the peak acceleration for the reference shaft thickness case is higher than that of the half-thickness case. However, the trend of peak acceleration response variation with the change of stiffness cannot be identified, as the peak acceleration for the double shaft thickness (the stiffest one) is lower than that for other cases with lower stiffness. This indicates the effects of inertia, which are more complex than their static counterpart. As will be discussed in Sect. 12.4 and Chap. 16, the response variation trend can be identified by relating the seismic responses to the dynamic characteristics of both structures and excitations.

Even for dynamic insensitive structures with low periods of resonance compared to that of the dynamic loading, dynamics does include the inertia effects due to loading that varies with time, even if this load variation may be quite slow. The inertia effects could lead to the fatigue failure of the materials at stress conditions well below the breaking strength of the materials (Chap. 17). They may also be responsible for the discomfort of human beings (Chap. 18). Figure 1.28 shows an offshore jacket structure subjected to two consequent sea waves; the jacket has a resonance period of 2.5 s. Figure 1.29 compares the calculated axial force time history at a leg C1 with and without accounting for the dynamic inertia effects. When the dynamic effects are ignored (right figure), the axial forces history entirely follows the variation of the wave and has a period of wave loading (15.6 s) well above the structure's resonance period (2.5 s). However, when the dynamic effects are accounted for, fluctuations (left figure) of the axial force can be clearly

1.3 Dynamics Versus Statics

First wave crest hit Leg C1   Second wave crest hit Leg C1

Fig. 1.28 An offshore jacket structure subjected to a wave with a wave height of 31.5 m and a wave peak period 15.6 s (*Courtesy of Aker Solutions*)

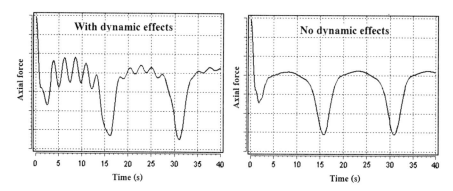

Fig. 1.29 Axial force time history on the lower part of leg C1 of the offshore jacket with and without dynamic inertia effects. (The exact magnitude of axial forces are omitted to protect the interests of the relevant parties)

observed as a background noise with the resonance period of the structure (2.5 s). Depending on the magnitude of this background noise, it may influence the integrity of the structure with regard to fatigue damage.

From another angle, the dynamic loading often has a different orientation than the static one. For example, the static loading of a structure under the gravity of the Earth is strictly toward the Earth. However, when the structure is subjected to dynamic loading due to, for example, wind, earthquake or sea waves, the direction of resultant loadings change from downward to the one that is more toward a horizontal orientation, this can result in an entirely different pattern regarding the load level and load path, and this obviously influences the structural design. Therefore, structural engineers are required to have a complete picture of load path and level, and structures designed must have corresponding load resisting systems that form a continuous load path between different parts of the structures and the foundation. The structure shown in Fig. 1.28 represents a typical configuration of the jacket structure and a clear path for load transferring, i.e., the gravity and acceleration loads from topside, the wave load applied on the upper part of the jacket, and the jacket gravity and acceleration loads are all transferred through legs and braces down to the pile foundation at the bottom.

Before concluding this section, it is of great importance to emphasize that dynamics is a rather more complex process than its static counterpart. The natural frequency of a structure can change when a change in its stiffness, mass or damping occurs. What makes dynamics even more complicated is that, strictly speaking, regular harmonic loadings or responses, with a sine or cosine form at a single frequency, do not exist in the real world, even if they can be a good simplification when the dynamics at a single frequency is dominating. This implies that one should always assess whether the vibrations in various frequencies need to be accounted for or not.

1.4 Solving Dynamic Problem

With the presence of inertia effects as discussed in Sect.1.3, the dynamic analysis is generally much trickier to solve than their static counterpart. This is mainly due to the fact that when the inertia term appears in the equilibrium equation (Eq. 1.2) as will be elaborated in Sect. 1.2, in order to uniquely determine the solution, apart from the boundary conditions, initial values are also needed, earning the dynamic analysis problem the name "initial boundary value problem." Furthermore, rather than a linear equation, as a static equilibrium has (Eq. 1.3), the additional inertia and damping terms make the equilibrium equation an ordinary second-order differential equation and time dependent, which requires more in-depth knowledge to examine. In addition, if time series responses are needed, a decent time stepping procedure has to be employed, making the dynamic problems even more demanding than their static counterpart.

$$m\,\ddot{x}(t) + c\,\dot{x}(t) + kx(t) = F(t) \qquad (1.2)$$

1.4 Solving Dynamic Problem

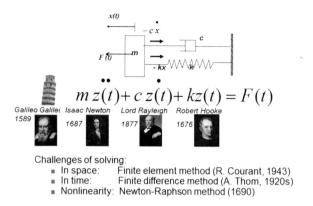

Fig. 1.30 Essentials of applied dynamic analysis (from a presentation by the author at the 11th International Conference on Recent Advances in Structural Dynamics, Pisa, 2013.)

$$kx = F \tag{1.3}$$

where F, k, m and x are the external force on a body, linear stiffness (between the body and the fixed ground), mass, and displacement of the body, respectively; t is the time; the dot over the symbol represents differentiation with respect to t. See Fig. 1.30 for an illustration.

Therefore, if responses can be calculated based on a static analysis, dynamic calculation should always be avoided. However, this is unfortunately not the case for many problems in engineering. The general rule is that if the excitations (loads) have a dominant frequency close to the natural frequency of structures, dynamic analysis has to be adopted. However, even if the load frequency is far from the natural frequency of the structure, the inertia effects may still be important and responsible for certain types of integrity problem (e.g. fatigue), such as the one shown in Fig. 1.29. In addition, transient loads can also cause dynamic responses, such as explosion, car collision, etc., in which the inertia effects of the relevant structures can be rather significant. In these situations, dynamic analysis is also normally required. It is sometimes convenient to use an amplification factor to simulate the dynamic effects, as will be elaborated in Sect. 11.1.1, so that only static analysis is needed and can be scaled with a dynamic amplification factor to predict the dynamic response level. However, this method lacks a solid theoretical background and has its limitations, and may become seriously erroneous under certain situations, which will be discussed at the end of Sect. 11.1.1.

Before solving a dynamic problem, one needs to classify the vibration problem in terms of whether external excitations are presented or not (forced and free vibrations), whether the excitations are of a deterministic or stochastic type, whether the damping is presented or not (damped or undamped), whether the system can be modeled as a discrete or continuous one, and whether the responses present linear or nonlinear characteristics. The relevant knowledge will be elaborated throughout this book.

For performing a dynamic analysis, analysts should fully understand the essential dynamic characteristics of a system or a structure: eigenfrequencies, mode shapes and damping. This part of the field is explained in detail from Chaps. 3 and 7.

In order to find the solutions of vibration responses, the system or the structure must be represented by an idealized model. This model can be either discrete or continuous. The former one can be modeled by limited degrees-of-freedoms, while the later theoretically has infinite degrees-of-freedoms.

For a discrete model, one needs to first construct the governing equations of motions, which can be described in terms of a second-order differential equation with constant coefficients (as elaborated in Chap. 2). The information with regard to displacements or rotations (essential boundary conditions) and external forces excitations (natural boundary conditions) need to be clarified. If time is involved, the initial conditions (boundary conditions in time) need to be known as well. After gathering sufficient information on these boundary conditions, the solutions of the equilibrium equations are then unique [25]. One can then solve the equations using decent mathematical treatment.

The system under study can be undamped or damped and with or without external excitations. We pay particular attention to the solutions for forced vibrations, which typically consist of a steady-state term that oscillates and gradually becomes dominant at the forcing frequency, and a transient term at the system's natural frequency that may be important initially but gradually dies out and eventually becomes insignificant due to the presence of damping. Under certain conditions the dominant forced vibrations become rather significant, indicating the occurrence of resonance.

Engineers sometimes need to choose the type of dynamic analysis method to be adopted. Each method has its unique characteristics, merits and limitations, and the various methods also fit different situations with regard to structural and load characteristics, design requirements, limitations of computation tools, and even the skills of analysts, in addition to other factors. Understanding all these factors is essential for choosing the right method: on the one hand, this can increase the accuracy; on the other hand, it may also simplify the computation efforts to a certain extent without degrading reliability level. In certain cases, the trade-off may be difficult to judge even for experienced researchers and engineers.

For structures or systems with single or very few degrees-of-freedoms, depending on the types of excitations and responses (duration, shape, deterministic or stochastic) and their eigenpairs (eigenfrequencies and mode shapes) in comparison with the excitations, based on the pure mathematical formulation of the stiffness, mass and damping of the structures, various types of analytical methods are available for solving the dynamic responses, all of which result in exact solutions.

However, for a structure or a system with multiple or many degrees-of-freedoms, it is almost impossible to perform a dynamic analysis by the classical analytical methods. Therefore, two approximation methods can be adopted. The first involves approximating the solutions using either a series of solutions or an

1.4 Solving Dynamic Problem

energy criterion to minimize the error, such as the Rayleigh energy method (Chap. 5). The second one is essentially a discretization of structures into many sub-domains (elements), and an assembly of these elements expressed in a matrix form for solving, which practically promotes the application of the finite element method.

As illustrated in Fig. 1.30, practically, in complex dynamic analyzes for engineering structures, the finite element method, the finite difference method (Sect. 13.5) or the modal superposition method (Sect. 13.4), and linear iteration method (Sect. 15.5) are the three most commonly used numerical methods in computational solid mechanics [26], solving problems associated with space, time and nonlinearities, respectively, but normally in a combined manner. They are important not only because of their efficiency and generality of application, but also due to the simplicity of their computer implementation [24].

In the modal superposition method, the coupled equations of motions are transformed into a series of uncoupled/independent equations. Each of these equations is analogous to the equation of motions for a single-degree-of-freedom system, and can be solved in the same manner. The responses are calculated as the linear sum of product between the eigenvectors (constant with time) and the generalized/modal coordinates (varied with time) for each eigenmode. Note that the number of uncoupled equations needing to be solved is equal to the number of eigenmodes to be accounted for. For structures with the dynamic responses dominated by the first few eigenmodes, the modal superposition method leads to high computation efficiency. This is more obvious if the structure has a large number of degrees-of-freedoms.

In linear dynamic analysis, the responses of a system/structure are proportional to the loads/excitations to which it is subjected. This enables the utilization of superposition, which brings significant convenience with regard to mathematical treatment, and, in most cases, also ensures the calculation accuracy. However, when nonlinearities appear in the system/structure, the stiffness and/or load are dependent on the deformation, and the responses of a system are generally not amenable to any analytical method that can provide exact solutions. A general method for obtaining the exact solution of nonlinear differential equations is not available, and most of the analytical methods that have been developed only yield approximate solutions. Further, the available techniques vary greatly according to the type of nonlinear equation [18].

Despite the significant efficiency of modal analysis, it generally applies only to linear dynamic problems. Therefore, the nonlinearities involved in a dynamic analysis are theoretically and practically treated with the support of the finite difference method and linear iteration method discussed in Sect. 13.5 and Chap. 15. The former (typically referred to as the Newmark method) is a step-by-step time integration of the equations of motions, and it can solve for example the transient phenomena such as nonlinear vibrations or shock wave propagation. The latter one is a generalization of Newton–Raphson method, which is essentially the application of a linearization in a locally approaching curve between load and deformation, and which is able to overcome numerical challenges introduced by

Fig. 1.31 The pioneers in history who contributed to dynamic analysis (from *upper left* to the *lower right*: Galileo Galilei, Newton Isaac, Robert Hooke, the third Baron Rayleigh, Joseph Louis Lagrange, and William Rowan Hamilton)

the geometric (e.g. buckling), material (e.g. plasticity), boundary (e.g. contact) and force (e.g. follower forces with change of geometry or hydrodynamic drag load) nonlinearities.

Damping exists in all types of real world structures or systems. It mainly provides a dissipation of energy. In most cases, it is beneficial to decrease dynamic responses. Most types of damping are most effective at or close to a structure/system's eigenfrequencies. An elaboration of damping effects and their modeling is presented in Chap. 14.

Dynamic loadings and responses induced by wave, wind, ice and earthquakes are often the governing environmental loads for designing an engineering structure, and are therefore discussed in separate sections in Chap. 12 using the power spectra to represent the loadings. Furthermore, special interest topics on seismic responses (Chap. 16), dynamics regarding fatigue (Chap. 17), human body vibrations (Chap. 18), and vehicle-structure dynamic interactions (Chap. 19) are elaborated in separate chapters.

1.5 Pioneers of Dynamic Analysis

Several great scientists (Fig. 1.31) in history need to be mentioned here, as without them and many others, classical dynamics might still today be called "modern" dynamics: Galileo Galilei (1564–1642), who showed that the acceleration due to gravity is independent of mass; Isaac Newton (1642–1727), who disclosed the three laws of motions (specifically the second law of motion); Robert Hooke (1635–1703) who developed the law of elasticity; the third Baron Rayleigh (1842–1919), who introduced the concept of modal analysis and viscous damping; Joseph Louis Lagrange (1736–1813), who presented the Lagrange multipliers; and William Rowan Hamilton (1805–1865), who illustrated the Hamiltonian formulation of dynamics (which is essentially the reformulation of Newtonian dynamics).

We should also acknowledge the more recent contributions: Goldstein [27], Whittaker and Synege [28], Timoshenko and Young [29], Den Hartog [30], Griffith [31], Nayfeh [32], Crandall and Mark [33], Robson [34], Zienkiewicz [35] and many others have contributed to the development of dynamic analysis in the last century, and have thus made the solving of rather sophisticated dynamic analysis and real engineering vibration problems possible.

Before leaving off the general information on dynamics, it should also be noted that the dynamic analysis elaborated in this book is for a real-time causal system, in which the present responses depend only on the past and present inputs, and not on the future inputs. It is assumed that non-causal systems do not exist in nature.

Chapter 2
Governing Equation of Motions

Although all dynamic modelings are essentially "lies," in this book I try to present those that may prove useful about a dynamic system in the real world.

The first step for performing a dynamic analysis is to set up the equations of motions. We start with Newton's second law, which is followed by the virtual work principle (D'Alembert's principle), Hamilton's principles, and Lagrange's equations. We shall also see the important role of energy in the study of dynamic analysis. It is noted that each of the formulations basically represents the same dynamic equilibrium but in a unique form of expression.

2.1 Dynamic Equilibrium

We start with Newton's second law of motions from *Philosophiæ Naturalis Principia Mathematica* (Fig. 2.1), which is the most powerful of Newton's three laws of motions. It states that "the rate of change of momentum of a mass equals the force acting on it." This is illustrated in Fig. 2.2 and expressed as:

$$\frac{\mathrm{d}}{\mathrm{d}t} m \left(\frac{\mathrm{d}x}{\mathrm{d}t} \right) = F(t) \tag{2.1}$$

where F, m and a are the external force on a body and the mass and acceleration of the body, respectively, and t is the time.

Here, we describe the position of mass in a Cartesian coordinate, which is referred to as an inertia frame in three dimensions X, Y, and Z (in the equation above, only one dimension X is used). Essentially, no inertia frames exist—a conclusion derived from the debates going back to the late nineteenth. However, it is usually convenient to find a frame for the purpose of a particular situation so that the dynamic analysis agrees with the observations. In this sense, the Earth is normally taken to be an inertial frame even if it rotates forever. However, the Earth is not an appropriate inertia frame for large scale motions such as those of the atmosphere and oceans [36]. For example, it is common knowledge that winds

J. Jia, *Essentials of Applied Dynamic Analysis*, Risk Engineering,
DOI: 10.1007/978-3-642-37003-8_2, © Springer-Verlag Berlin Heidelberg 2014

32   2 Governing Equation of Motions

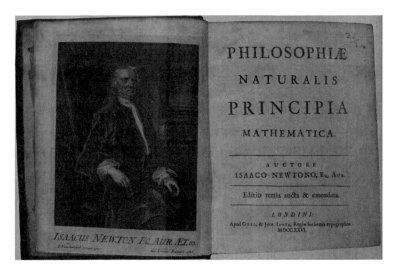

Fig. 2.1 *Philosophiæ Naturalis Principia Mathematica*, which laid the foundation for dynamic equilibrium

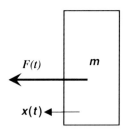

Fig. 2.2 A single mass under external force $F(t)$

blow more frequently along an east–west direction, but they would blow north–south if the Earth was not rotating.

Newton's second law of motion was a breakthrough in the understanding of dynamics, and it confirms that force only causes a change of velocity, correcting the previous view, proposed by Aristotle (384–322 BC), that force maintains the velocity.

If one attaches a spring and a damper to the mass, as shown in Fig. 2.3, by assuming that the spring obeys Hooke's law and the damping is of viscous type (the damping force is proportional to the velocity of the mass), the equilibrium is formulated by adding the terms of spring- and damping-induced forces:

$$m\ddot{x}(t) + c\dot{x}(t) + kx(t) = F(t) \tag{2.2}$$

where the dot over the symbol represents differentiation with respect to time t.

In the equilibrium equation above, all the motion parameters (displacement, velocity and acceleration) are lower-order differentiations of displacement with

2.1 Dynamic Equilibrium

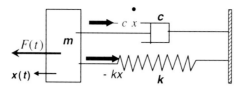

Fig. 2.3 A SDOF spring-mass-damper system

respect to time, and are sensible by human beings. Higher-order differentiations do not appear in the equilibrium equation because scientists do not find their link to equilibrium of force, but they are already used in many engineering applications. For example, the time-derivative of acceleration is called jerk/jolt, i.e., the change of acceleration with time. The jerk can be illustrated by the example of a person gradually pressing the surface of a wall: one's hand can feel that the force is increasing (a change in the force) until it reaches a constant force pressure. The jerk can also be sensed when, moving quickly on a bicycle, one suddenly brakes hard. If the bicycle were to slide off a paved track onto wet grass, even though the friction between the bicycle tire and grass is still present due to sliding, it will decrease and the bicycle will slow down less rapidly, thus undergoing a positive jerk. The rider would feel pressure on his/her hip, as if the bicycle were speeding up, even if it does not really go any faster. As its name suggests, jerk is important when evaluating the destructive effect of motions on a mechanism or the discomfort caused to passengers onboard vehicles. Movement-sensitive instruments need to be kept within specified limits of jerk as well as acceleration to avoid damage. For passenger comfort, a train in operation will typically be required to keep jerk below 2 m/s^3. In the aerospace industry, a type of instrument called a jerk-meter is used for measuring jerk.

A system that has a small number of degrees-of-freedoms can be evaluated efficiently by directly using the equation of the motions above. However, when the degrees-of-freedom become too large for an analyst to handle using this simplified direct method, other methods for formulating and solving the equations of motions have to be used instead. Examples of the former type are Hamilton's principle or the finite element method, while the latter type include the modal superposition method (Sect. 13.4), and direct time integration method (Sect. 13.5) etc. In addition, for continuous systems with non-uniform mass or stiffness distribution, the Rayleigh energy method (Chap. 5) can also be used to obtain a quick but approximate characterization of the dynamic characteristics.

2.2 Principle of Virtual Displacements

Newton's second law can be more practically expressed as D'Alembert's principle: the condition for dynamic equilibrium is that the total force is in equilibrium with the inertia forces.

For a complex structure or a system, it is not practical to describe the forces acting on each mass point as a vectorial addition of all those forces. In this situation, the principle of virtual displacement is particularly appealing. It solves the dynamic equilibrium by indirectly formulating the equations of motions, which is essentially an energy approach, i.e., the total virtual work done by effective forces applied on a system through virtual displacements, which is compatible with the system constraints, will be zero. By saying "effective forces," they contain both the normal and inertia forces.

2.3 Hamilton's Principle Through Lagrange's Equations

Note that even if D'Alembert's principle eliminates the problem of force addition in a vectorial context, the virtual work itself is still a product of force vector and virtual displacement vector, and the equations of motions are formulated in terms of position coordinates that may not all be independent. This problem can be solved by the more powerful Hamilton's principle through Lagrange's equations.

Many principles describe a system by minimizing certain physical quantities, as does Hamilton's principle, which states that (as shown in Fig. 2.4), for a conservative system, of all the possible paths along which a dynamical system may travel from one point to another within a specified time interval (consistent with any constraints), the actual path (called the true, Newtonian or dynamical path) followed is that which minimizes the time integral of the difference between the kinetic and potential energy:

$$\delta I = \int_{t_1}^{t_2} \delta(T - V)dt = \int_{t_1}^{t_2} \delta L dt = 0 \qquad (2.3)$$

where δ is the first variation and L is called the Lagrangian function, which represents the difference between kinetic (T) and potential (V) energy of the system; the former is a function of particle velocity, the latter is a function of position. Hamilton's principle represents the most condensed description of motion for a given system. For a single-degree-of-freedom system, as shown in Fig. 2.3, it can be expressed as:

$$T = \frac{1}{2}m\dot{x}(t)^2 \qquad (2.4)$$

$$V = \frac{1}{2}kx(t)^2 \qquad (2.5)$$

Surprisingly, in certain senses, Hamilton's principle coincides with the statements of the Chinese philosopher Laozi (Fig. 2.5), who expressed in his *Daodejing* (No 48, 《道德经》) around 500 BC that "non-action is all action" ("无为而无不为").

2.3 Hamilton's Principle Through Lagrange's Equations

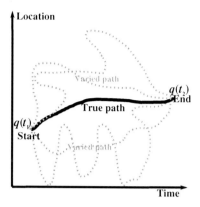

Fig. 2.4 Many paths, among which the true path (*solid line*) is that which follows Newton's law (minimize the action); varied paths are not possible [38]

Fig. 2.5 Chinese philosopher Laozi (老子) (painting by Jing Yu)

The convenience of Hamilton's principle lies in the fact that all the system differential equations of motions can be derived from two scalar functions, the kinetic energy and the potential energy, with the virtual work corresponding to

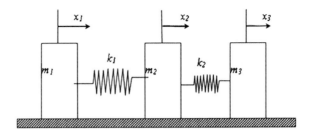

Fig. 2.6 *Three* masses resting on the horizontal ground are free to move horizontally, no friction is assumed between the mass and the ground surface

non-conservative forces [37]. Describing a system by applying Hamilton's principle allows people to determine the equations of motions for a system for which we would not be able to derive these equations easily from Newton's laws. However, one nevertheless needs to bear in mind that Hamilton's principle is not a new law, but simply provides a new description of Newton's laws.

The extension from Newton's second law to Hamilton's principle also makes it possible to handle dynamic problems for deformable bodies by using continuum mechanics. This paves the way for the development of finite element discretization of deformable bodies [24].

In a more general case with a system comprising non-conservative forces (such as the ones caused by frictions), one may express Hamilton's principle via Lagrange's equations:

$$\frac{d}{dt}\left(\frac{\partial T}{\partial \dot{q}_j}\right) - \frac{\partial T}{\partial q_j} + \frac{\partial V}{\partial q_j} + \frac{\partial D}{\partial \dot{q}_j} = Q_j, j = 1, 2, \ldots, n \tag{2.6}$$

where D is the dissipation function. For a single degree-of-freedom system shown in Fig. 2.3, $D = \frac{1}{2} c \dot{x}(t)^2 \cdot Q_j$ represents the non-conservative forces and q_j is the generalized degree-of-freedom (coordinate or path), which is not unique and related to the physical coordinate. The dot over the symbols represents differentiation with respect to time. It is noted that the forces are not direct knowns; instead, their information is contained in the kinetic and potential energy terms.

Since $\frac{\partial T}{\partial q_j}$ is zero, the equation above is finally written as:

$$\frac{d}{dt}\left(\frac{\partial T}{\partial \dot{q}_j}\right) + \frac{\partial D}{\partial \dot{q}_j} + \frac{\partial V}{\partial q_j} = Q_j, j = 1, 2, \ldots, n \tag{2.7}$$

Each item in this equation exactly corresponds to the equations of motions (Eq. (2.2)).

2.3 Hamilton's Principle Through Lagrange's Equations

Example [39]: Consider a system with three masses connected by springs as shown in Fig. 2.6. Establish the equations of motions using Lagrange's equations and summarize them in a matrix form.

Solution: The Lagrange's equations are written as:

$$\frac{d}{dt}\left(\frac{\partial T}{\partial \dot{q}_j}\right) - \frac{\partial T}{\partial q_j} + \frac{\partial V}{\partial q_j} + \frac{\partial D}{\partial \dot{q}_j} = Q_j, j = 1, 2, \ldots, n$$

Since there is neither friction nor external forces, $D = 0$ and $Q_j = 0$. Let q_1, q_2 and q_3 be the generalized degree-of-freedom. We have

$$T = \frac{1}{2}m_1\,\dot{q}_1^2 + \frac{1}{2}m_2\,\dot{q}_2^2 + \frac{1}{2}m_3\,\dot{q}_3^2$$

$$V = \frac{1}{2}k_1(q_2 - q_1)^2 + \frac{1}{2}k_2(q_3 - q_2)^2$$

$$= \frac{1}{2}k_1\left(q_2^2 - 2q_1q_2 + q_1^2\right) + \frac{1}{2}k_2\left(q_3^2 - 2q_3q_2 + q_2^2\right)$$

With regard to q_1:

$$\frac{d}{dt}\left(\frac{\partial T}{\partial \dot{q}_1}\right) - \frac{d}{dt}\left(\frac{\partial T}{\partial q_1}\right) = \frac{d}{dt}\left(m_1\,\dot{q}_1\right) - 0 = m_1\,\ddot{q}_1$$

$$\frac{\partial V}{\partial q_1} = \frac{1}{2}k_1(-2q_2 + 2q_1) = -k_1q_2 + k_1q_1$$

With regard to q_2:

$$\frac{d}{dt}\left(\frac{\partial T}{\partial \dot{q}_2}\right) - \frac{d}{dt}\left(\frac{\partial T}{\partial q_2}\right) = \frac{d}{dt}\left(m_2\,\dot{q}_2\right) - 0 = m_2\,\ddot{q}_2$$

$$\frac{\partial V}{\partial q_2} = k_1q_2 - k_1q_1 - k_2q_3 + k_2q_2$$

With regard to q_3:

$$\frac{d}{dt}\left(\frac{\partial T}{\partial \dot{q}_3}\right) - \frac{d}{dt}\left(\frac{\partial T}{\partial q_3}\right) = \frac{d}{dt}\left(m_3\,\dot{q}_3\right) - 0 = m_3\,\ddot{q}_3$$

$$\frac{\partial V}{\partial q_3} = k_2q_3 - k_2q_2$$

Insert the equations above into the Lagrange's equations:

$$m_1\,\ddot{q}_1 = k_1q_2 - k_1q_1$$

$$m_2\,\ddot{q}_2 = -k_1q_2 + k_1q_1 + k_2q_3 - k_2q_2$$

$$m_3 \ddot{q}_3 = -k_2 q_3 + k_2 q_2$$

Sum up the three equations above in a matrix form:a

$$\begin{bmatrix} m_1 & 0 & 0 \\ 0 & m_2 & 0 \\ 0 & 0 & m_3 \end{bmatrix} \begin{Bmatrix} \ddot{q}_1 \\ \ddot{q}_2 \\ \ddot{q}_3 \end{Bmatrix} + \begin{bmatrix} k_1 & -k_1 & 0 \\ -k_1 & k_1 + k_2 & -k_2 \\ 0 & -k_2 & k_2 \end{bmatrix} \begin{Bmatrix} q_1 \\ q_2 \\ q_3 \end{Bmatrix} = 0$$

2.4 Momentum Equilibrium

Once again, we recall Newton's second law: if there are no forces acting on the mass, the momentum will be constant. For systems comprising more than one body mass, each one has an individual momentum, but their sum will be constant if there are no external forces acting on the system. The sum of the momentum is expressed as:

$$Momentum\,sum = m_1 v_1 + m_2 v_2 + \ldots + m_n v_n \tag{2.8}$$

where the lower indices serve to identify the mass.

The momentum equilibrium has abundant applications in the engineering world. One example of its application is the calculation of the speed of two cars before and after their collision. In structural engineering, we can also find their applications in the design of various types of dynamic absorbers [23].

In civil engineering, this momentum equilibrium can be utilized in designing, for example, a tuned mass damper (TMD) or an impact damper, both of which comprise a secondary mass attached to (in the case of TMD) or constrained by (in the case of impact damper) a vibrating structure (main structure). This mass has dynamic characteristics that relate closely to that of the main structure. By varying the ratio of the mass to the primary body (main structure), the frequency ratio between the two masses, and the damping ratio associated to the secondary mass, the momentum exchange can control the maximum responses of the main structure.

Figure 2.7 shows a single TMD fitted to the underside of a concrete deck at the Infinity Bridge in the north-east of England. The installation of more such TMDs is planned for when the issue of maintenance arises. Today, many TMDs are installed on high-rise buildings and bridge structures to mitigate the dynamic responses due to dynamic loadings induced by wind, earthquake, impact and mechanical vibrations [23]. Representative examples are the Taipei World Financial Center (Fig. 1.23), Washington National Airport Tower, Sydney Tower, Citicorp Center (New York), the John Hancock Building (Boston), and the Crystal Tower (Osaka, Japan).

Figure 2.8 shows an impact damper, which comprises a small rigid mass placed inside a container mounted on the side of the structure. There is a small optimal

2.4 Momentum equilibrium

Fig. 2.7 TMDs installed under the bridge deck of the Infinity Bridge (photo by John Yeadon)

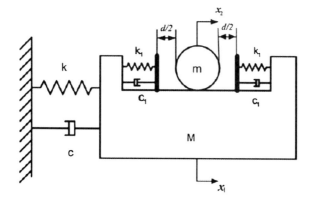

Fig. 2.8 An impact damper model [40]

clearance between the small mass and the container wall, thus allowing collisions between the mass and the container wall to occur when the displacement along the clearance direction exceeds the optimal clearance. The collision achieves both momentum exchange and energy dissipation, the latter of which is mainly produced on the contacting surface between the mass and the wall. A schematic diagram of an impact damper is shown in Fig. 2.8.

2.5 Validity of Classical Dynamics

Before ending this chapter, this author would like to quote a famous conversation that is reported to have taken place between Napoleon Bonaparte, Laplace and Lagrange (Fig. 2.9) [41]:

Fig. 2.9 Laplace (*left*), Napoleon (*middle*) and Lagrange (*right*)

Napoleon: How is it that, although you say so much about the Universe in this huge book, you say nothing about its Creator?

Laplace: No, Sire, I had no need of that hypothesis. [This is an example of his arrogance at that time].

Lagrange: Ah, but it is such a good hypothesis, it explains so many things!

Laplace: Indeed, Sire, Monsieur Lagrange has, with his usual sagacity, put his finger on the precise difficulty with the hypothesis, it explains everything, but predicts nothing.

Laplace was confident that he could predict the motions of everything, but now we know that the equations of motions introduced in this chapter are only valid for the mechanical universe [38] and do not apply for particles at rather small distances or with extremely high velocities. Nevertheless, they remain valid for the scientific research in the fields of civil and mechanical engineering.

Arrogant and humble attitudes are both necessary for scientific research. In the author's opinion, however, it is most important to be humble when investigating problems that are beyond our knowledge.

Chapter 3
Free Vibrations for a Single-Degree-of-Freedom (SDOF) System–Translational Oscillations

> *The grasping of truth is not possible without an empirical basis. However, the deeper we penetrate and the more extensive and embracing our theories become, the less empirical knowledge is needed to determine those theories.*

> Albert Einstein, December 1952

3.1 Definition of Harmonic Oscillations

Even though a realistic system or a structure possesses many degrees-of-freedom, it can be analyzed in terms of its separate modes, each of which in the case of light damping can be considered a single-degree-of-freedom (SDOF) system.

Before we go into a detailed study of free vibrations, we shall go through a few basic definitions of harmonic oscillations, as shown in Fig. 3.1. Those oscillations may represent force, stress, strain, displacements, velocity, or accelerations. They can be expressed as a function of time:

$$x(t) = X \cos(\omega t - \theta) \tag{3.1}$$

where ω and t are angular frequency and time, X is the maximum peak of the oscillations, $(\omega t - \theta)$ is the phase angle (rad), and θ is the initial phase angle.

The time required for the oscillations to go through a complete cycle is called period $T = \frac{2\pi}{\omega}$, as shown in Fig. 3.1.

Harmonic oscillations (sinusoidal and/or co-sinusoidal) can normally be represented in one form or another by using Euler's equation, which states that a complex number may be written in exponential form:

$$e^{i\omega t} = \cos(\omega t) + i\sin(\omega t) \tag{3.2}$$

$$\sin(\omega t) = \frac{1}{2i}(e^{i\omega t} - e^{-i\omega t}) \tag{3.3}$$

$$\cos(\omega t) = \frac{1}{2}(e^{i\omega t} - e^{-i\omega t}) \tag{3.4}$$

where i is the imaginary unit.

J. Jia, *Essentials of Applied Dynamic Analysis*, Risk Engineering,
DOI: 10.1007/978-3-642-37003-8_3, © Springer-Verlag Berlin Heidelberg 2014

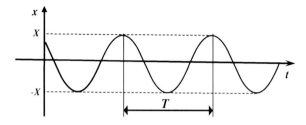

Fig. 3.1 Definition of period in harmonic oscillations

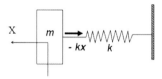

Fig. 3.2 An SDOF spring-mass system

Harmonic oscillations can generally be expressed in four forms [181]:

$$x(t) = A\cos(\omega t) + B\sin(\omega t) = X\cos(\omega t - \theta) \qquad (3.5)$$

$$x(t) = Ce^{-i\omega t} = Xe^{-i(\omega t - \theta)} \qquad (3.6)$$

where $C = Xe^{i\theta} = A + iB$ is the complex amplitude containing both magnitude X and phase $\theta = B/A$.

It is obvious that:

$$|C| = X = \sqrt{A^2 + B^2} \qquad (3.7)$$

$$\sin\theta = B/\sqrt{A^2 + B^2} = B/X \qquad (3.8)$$

3.2 Undamped Free Vibrations of a SDOF System

First consider the simplest form of an oscillator—a SDOF spring-mass system shown in Fig. 3.2. By giving an initial disturbance to the system (i.e., the mass m is released at a distance of X_0 from the neutral position), an initial velocity produced by an impact, or a combination of the two, the resulting motions, unaffected by any external force, are called free vibrations. Now apply Newton's second law; the governing linear equilibrium (differential) equation of motions for this system is:

$$m\ddot{x}(t) + kx(t) = 0 \qquad (3.9)$$

3.2 Undamped Free Vibrations of a SDOF System

By dividing the equation above by m, one has:

$$\ddot{x}(t) + \frac{k}{m}x(t) = 0 \tag{3.10}$$

The equation above indicates simple harmonic vibrations, i.e., x varies sinusoidally or cosusoidally with time t. The motions are repetitive if no damping is presented in the system. The general form of the solution is given as:

$$x(t) = A\cos(\sqrt{\frac{k}{m}}t) + B\sin(\sqrt{\frac{k}{m}}t) \tag{3.11}$$

Here we introduce a parameter $\omega_n = \sqrt{\frac{k}{m}}$, which is at this stage called the undamped angular frequency of the motions with the unit of rad/s. It is normally referred to as the natural frequency of the oscillatory system, which is defined as the number of cycles per unit time.

By substituting Eq. (3.11) into (3.10), one has:

$$-\omega_n^2(A\cos\omega_n t + B\sin\omega_n t) + \frac{k}{m}(A\cos\omega_n t + B\sin\omega_n t) = 0 \tag{3.12}$$

where A and B are constants depending on the initial conditions at time $t = 0$.

In case there are motions in the system, it is obvious that $A\cos\omega_n t + B\sin\omega_n t \neq 0$. Therefore, the natural frequency is calculated as:

$$\omega_n = \sqrt{\frac{k}{m}} \tag{3.13}$$

It is noticed that Eq. (3.10) is of the second order, and two constants are required to obtain the solution. For the condition in which the initial displacement and velocity are X_0 and V_0, respectively, from Eq. (3.5) one has $A = X_0$ and $B = V_0/\omega_n$. Substituting these expressions for the A and B constants into Eq. (3.11), one obtains:

$$x(t) = X_0\cos(\omega_n t) + \frac{V_0}{\omega}\sin(\omega_n t) \tag{3.14}$$

The solution of the equation above represents oscillation responses with constant natural frequency $\omega_n = \sqrt{\frac{k}{m}}$, hence giving the system the name "harmonic oscillator." The displacement motions $x(t)$ for the equation above are shown in Fig. 3.3. The mode shape corresponding to the natural frequency ω_n is defined as special initial deflections that cause the entire system to vibrate harmonically, i.e., $x(t = 0) = X_0$.

Note that a complete harmonic cycle occurs for each angular increment 2π, i.e., $x(t) = x(t + \frac{2\pi}{\omega_n})$. The natural period, T_n, which is the time required for the system to go through a complete cycle, is therefore expressed as:

$$T_n = \frac{2\pi}{\omega_n} \tag{3.15}$$

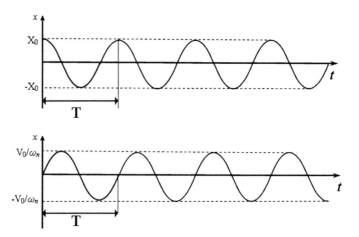

Fig. 3.3 Free vibration displacement response x varies with time *t*, for cases of initial displacement X_0 (*upper*) and initial velocity V_0 (*lower*), respectively

It is obvious that the natural frequency is the inverse of the natural period:

$$f_n = \frac{1}{T_n} \quad (3.16)$$

A vertically or rotationally vibrating system can be analyzed in a similar manner.

From an engineering point of view, the natural frequency can be explained as the frequency at which a structure or a system can very easily be excited to vibrate, even with a rather small excitation. This will cause resonance, as shown in examples in Sects. 1.1 and 1.2. Figure 3.4 shows the collapse of a building during a strong earthquake, due to the fact that the building's natural period is similar to that of the earthquake ground motions.

From the point of view of energy, for an undamped system, rather than a force-equilibrium, free vibrations at the natural frequency are essentially the process of exchanging energy between mass motions (kinetic energy) and strain variation (potential energy). The resonance occurs when the kinetic energy is maximized (i.e., $\frac{1}{2}m(\omega_n x)^2$) and the potential energy is minimized (i.e., $\frac{1}{2}kx^2$). Equating the two energy terms, one also reaches the natural frequency as:

$$\omega_n = \sqrt{\frac{k}{m}} = \sqrt{\frac{2 \times \text{strain energy}}{m \times \text{deflection}^2}} \quad (3.17)$$

With regard to modeling, potential energy may be minimized by an accurate modeling of soft links in the load path. This typically relates to the modeling of supports and structural discontinuities (e.g. joints, connections etc.). Kinetic energy is maximized by an accurate modeling of system mass, especially in situation with a large mass to stiffness ratio [42].

3.2 Undamped Free Vibrations of a SDOF System

Fig. 3.4 Collapse of top stories of Hotel Continental in Mexico City during the 8.1 magnitude Mexico City earthquake of 1985

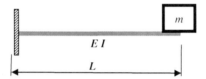

Fig. 3.5 A concentrated mass m attached to the end of a clamped (cantilever) beam

The SDOF shown in Fig. 3.2 can be realized in many types of physical modeling problems, such as a concentrated mass m attached to a clamped beam at one end shown in Fig. 3.5. By assuming that the mass of the beam can be omitted, the vibration of the beam-mass system is dominated by the inertia effects of the concentrated mass m. By definition, the stiffness of the system k, sometimes referred to as beam equivalent stiffness, is expressed as:

$$k = \frac{F}{\delta} \qquad (3.18)$$

In the equation above, F is the force applied to represent the inertia effects of the concentrated mass at the mass point. The transverse defection for the beam at the mass point is:

$$\delta = \frac{FL^3}{3EI} \qquad (3.19)$$

The natural frequency of the system is then:

$$\omega_n = \sqrt{\frac{k}{m}} = \sqrt{\frac{3EI}{L^3}} \qquad (3.20)$$

The equivalent stiffness of a beam-mass system with the most typical support conditions are summarized as below:

$$k = \begin{cases} \dfrac{48EI}{L^3} & \text{for simply supported beam at both ends} \\[2ex] \dfrac{3EI}{L^3} & \text{for beam clamped at one end} \\[2ex] \dfrac{192EI}{L^3} & \text{for beam clamped at both ends} \\[2ex] \dfrac{768EI}{7L^3} & \text{for beam clamped at one end and simply supported at the other end} \end{cases} \tag{3.21}$$

Note that, in the equation above, the stiffness is always referred to the position of the concentrated mass, where the natural vibration mode has its maximum amplitude. If one wants to measure the stiffness in another location along the beam, the applied force F must be shifted to that position.

The natural frequency for the system shown in Fig. 3.2 can also be calculated by using the static deflection approach:

$$\omega_n = \sqrt{\frac{k}{m}} = \sqrt{\frac{k}{W/g}} = \sqrt{\frac{kg}{W}} = \sqrt{\frac{g}{W/k}} = \sqrt{\frac{k}{\delta}} \tag{3.22}$$

where W is the weight of the mass m, and g is the acceleration of Earth's gravity.

In engineering practice, many types of problems can be simplified as an SDOF system. For example, for an initial estimation of the natural frequency and corresponding mode shape, a monopile or monotower structure can be simplified as a vertical beam vibrating in the horizontal direction, with a mass m at its top, shown in Fig. 3.6. The spring stiffness of the beam can be approximated as the beam's bending spring stiffness with appropriate support conditions. If one assumes a fixed support condition at the bottom of the monotower, the natural frequency (first eigenfrequency) of it can be calculated as:

$$f_n = \frac{1}{2\pi} \sqrt{\frac{3EI}{H^3(m + 0.23\mu H)}} \tag{3.23}$$

where μ is the unit mass of the tower per meter including both the structural mass and added mass due to the surrounding water, H is the equivalent tower height, and EI is the bending stiffness of the tower.

The derivation of the equation above is based on either an equivalent system analysis (Sect. 4.2.2) or Rayleigh energy method (Chap. 5).

3.3 Damped Free Vibrations of an SDOF

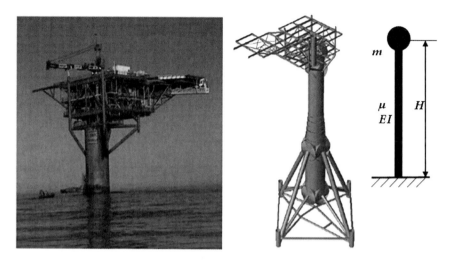

Fig. 3.6 Physical modeling realization from a realistic monotower wellhead platform (*left*) and its geometry modeling for finite element analysis (*middle*) to a simplified SDOF model (*right*) for hand calculation of eigenfrequencies (*Courtesy of Aker Solutions for the left and the middle figure*)

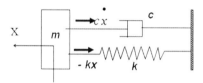

Fig. 3.7 An SDOF spring-mass-damper system

3.3 Damped Free Vibrations of an SDOF

The undamped free vibrations described in the previous section never occur in nature. More realistic free vibrations are modeled by adding a damper into the spring-mass system shown in Fig. 3.2, which results in an SDOF spring-mass-damper system as shown in Fig. 3.7. It is assumed that the damper is of viscous type with a damping coefficient c. This represents the conditions of damping due to the viscosity of oil in a dashpot fairly accurately, and it generates a damping force proportional to the velocity of the mass. The governing linear differential equation of motions for this system is then expressed as:

$$m\ddot{x}(t) + c\dot{x}(t) + kx(t) = 0 \tag{3.24}$$

The equation above can be solved by substituting harmonic motions in the exponent format $x(t) = e^{\lambda t}$ into the equation, which gives:

$$m\lambda^2 e^{\lambda t} + c\lambda e^{\lambda t} + k e^{\lambda t} = 0 \tag{3.25}$$

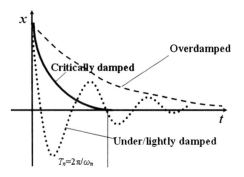

Fig. 3.8 Comparison of response decays at different damping levels

Because $e^{\lambda t} \neq 0$ (otherwise there would be no motion), dividing the equation above by $e^{\lambda t}$ gives:

$$m\lambda^2 + c\lambda + k = 0 \tag{3.26}$$

The roots of the equation are λ_1 and λ_2:

$$\lambda_{1,2} = -\frac{c}{2m} \pm \frac{\sqrt{c^2 - 4mk}}{2m} \tag{3.27}$$

Therefore, the general solution for the damped free vibrations is:

$$x(t) = Ce^{\lambda_1 t} + De^{\lambda_2 t} \tag{3.28}$$

where C and D are arbitrary constants and can be determined by the initial conditions of the mass, i.e. $x(t=0)$ and $\dot{x}(t=0)$.

The damping is very important in an oscillating system because it helps to limit the excursion of the system in a resonance situation. As a reference, we first define the critical damping c_c, which is the lowest damping value that gives no oscillation responses, i.e., the system does not vibrate at all and decays to the equilibrium position within the shortest time. This represents the dividing line between oscillatory and non-oscillatory motions:

$$c_c = 2\sqrt{km} = 2m\omega_n \tag{3.29}$$

The actual damping ratio can be specified as a percentage of critical damping:

$$\zeta = \frac{c}{c_c} \tag{3.30}$$

Therefore, Eq. (3.27) can be expressed in terms of damping ratio:

$$\lambda_{1,2} = \left(-\zeta \pm \sqrt{\zeta^2 - 1}\right)\omega_n \tag{3.31}$$

However, it should be noted that the expression above is valid only for $\zeta \leq 1$.

3.3 Damped Free Vibrations of an SDOF

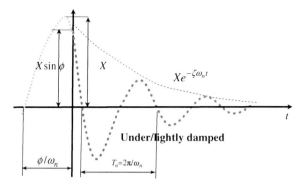

Fig. 3.9 Response decays for an underdamped/lightly damped system

Depending on the level of damping, five types of responses (Fig. 3.8) can be observed:

- Undamped ($\zeta = 0$): the responses exhibit harmonic vibrations without decay (see Sect. 3.2 and Fig. 3.3).
- Lightly damped/underdamped ($0 < \zeta < 1$): the damping is less than critical, the responses follow exponentially decaying ($e^{-\zeta \omega t}$) harmonic/sinusoidal oscillations ($\sin(\omega_d t + \phi)$) as shown in Fig. 3.9, λ is complex, the general solution of motions is:

$$\begin{aligned} x(t) &= e^{-\zeta \omega_n t} \left(C e^{-i\sqrt{1-\zeta^2} \omega_n t} + D e^{-i\sqrt{1-\zeta^2} \omega_n t} \right) \\ &= X e^{-\zeta \omega_n t} \sin\left(\sqrt{1-\zeta^2} \omega_n t + \phi \right) \\ &= X e^{-\zeta \omega t} \sin(\omega_d t + \phi) \end{aligned} \quad (3.32)$$

where $\omega_d = \omega \sqrt{1-\zeta^2}$ is the frequency of the damped oscillation. Initial displacement X and phase angle/phase lag ϕ ($\phi = \tan^{-1}\left(\frac{2\zeta(\omega/\omega_n)}{1-(\omega/\omega_n)^2}\right)$) depend on the initial condition as shown in Fig. 3.9.

Among all types of free vibration responses, lightly damped vibrations are the most typical responses. Fig. 3.10 shows a comparison of free vibration responses with various damping levels, with ζ ranging from 0 (undamped) to 50 %. The effectiveness of damping can be clearly observed, i.e., with only a small percent of damping, the response decay becomes significant.

- Critically damped ($\zeta = 1$): the damping equals the critical damping that will remove all vibration responses. One can obtain a double but equal root, and the general solution of motions is:

$$x(t) = e^{-\omega_n t}(C + Dt) \quad (3.33)$$

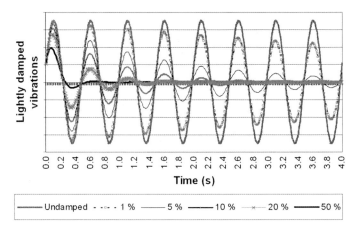

Fig. 3.10 Free vibration responses for undamped and lightly damped systems with a natural period of 0.5 s and phase angle $\phi = 0.3$

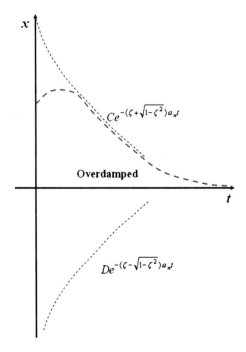

Fig. 3.11 Response decays for overdamped system

- Overdamped ($\zeta > 1$): the damping is larger than the critical, the motions exhibit smoothed exponential decay without any oscillatory vibration or harmonic components, which are shown in Fig. 3.11. Similar to the case of critical damping, the system does not oscillate and rests in the equilibrium position, but

3.3 Damped Free Vibrations of an SDOF

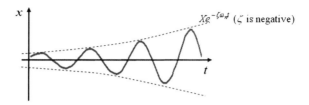

Fig. 3.12 Response increase for a system with negative damping

vibrates with a longer time (lower rate) than that of the critical damping case. λ is real, the general solution of motions (non-oscillatory) becomes:

$$x(t) = Ce^{-(\zeta+\sqrt{1-\zeta^2})\omega_n t} + De^{-(\zeta-\sqrt{1-\zeta^2})\omega_n t} \quad (3.34)$$

- Negatively damped ($\zeta < 0$): this is basically a special case of a lightly damped system. Instead, the responses in a linear system show an exponential increase as shown in Fig. 3.12, indicating that energy is added to the system. The negative damping is often related to the so-called self-excited vibrations (self-excitations), which are due to the sustaining alternating excitations that induce the instability of a system at its own natural or critical frequency. Different from resonance, the self-excited vibrations are in the class of nonlinear type and are essentially independent of the frequency of the external excitations. Typical examples of self-excited vibrations are the flutters of bridges (Fig. 1.12), masts and aircraft wing structures, vortex-induced vibrations (VIV) and friction-induced vibrations (vehicle braking or vehicle-bridge interactions). Self-excitations have caused abundant structural collapses. For example, many drilling risers hung from a drilling rig experiencing high speed ocean current have failed due to VIV. Cracks on tubular joints due to wind-induced VIV can also be found on many structures, as shown in Fig. 3.13.

It should be mentioned that, for the positively damped ($\zeta > 0$) cases above, even though the responses decay when the damping is modeled with viscous type, the responses never cease. Instead, they approach infinitely small amplitude. Therefore, for an actual structure, other types of damping (normally friction type) must exist to make the responses cease entirely. This will be discussed in Sect. 14.2.2.

For an SDOF system with more than one spring and damping system, one needs to calculate the resultant stiffness of the matrix, which requires the identification of the two basic types of combined stiffness of a spring-mass system, namely parallel and series systems, as shown in Fig. 3.14.

A resonance frequency is defined as the forcing frequency at which the largest response amplitude occurs. It is normally regarded as the equivalent term of natural frequency (first eigenfrequency). However, strictly speaking, the largest response for a lightly damped structure occurs at slightly different frequencies

Fig. 3.13 Cracks (marked with *circle*) found on a tubular joint due to wind-induced VIV on a high-rise flare boom in the North Sea (*Courtesy of Aker Solutions*)

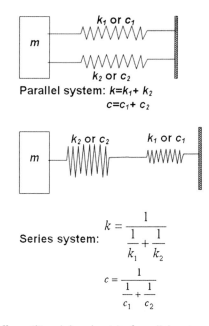

Fig. 3.14 Resultant stiffness (k) and damping (c) of parallel system and series system

when measured by different terms among displacement, velocity and acceleration as follows:

$$\text{Acceleration resonance frequency: } \omega_n \Big/ \sqrt{1 - 2\xi^2} \qquad (3.35)$$

$$\text{Velocity resonant frequency: } \omega_n \qquad (3.36)$$

$$\text{Displacement resonance frequency: } \omega_n \sqrt{1 - 2\xi^2} \qquad (3.37)$$

For typical engineering structures, since the damping ratio ξ is small, the differences can be neglected.

It should also be noticed that the natural frequency is defined slightly differently from that of any of the resonance frequencies above:

$$\omega_d = \omega_n \sqrt{1 - \xi^2} \qquad (3.38)$$

Chapter 4
Practical Eigenanalysis and Structural Health Monitoring

4.1 Eigenpairs, Global-, Local- and Rigid-Body Vibrations

"Eigen" is a Germanic term that means "property of." Therefore, the eigenfrequencies and eigenmode shapes together with damping are inherent properties of a specific structure or a vibrating system. Eigenfrequencies and mode shapes are determined by the structure/system's material properties (mass, stiffness and damping) and support conditions.

By neglecting the damping and assuming that the support conditions do not change, the equilibrium equation at the n degrees-of-freedom-system can be expressed as:

$$[M]\left\{\ddot{x}(t)\right\}_n + [K]\{x(t)\}_n = 0 \tag{4.1}$$

The solutions of the motions can then be written in a harmonic form:

$$\{x(t)\}_n = \{\phi\}_n \cos(\omega t) \tag{4.2}$$

where the amplitudes $\{\phi\}_n$ are independent of time and ω is the frequency of the vibration.

With the two equations above, one has:

$$[K - \omega^2 M]\{\phi\}_n = 0 \tag{4.3}$$

The equation above contains a set of n linear homogeneous equations in the unknowns $\phi_n^{(1)}$, $\phi_n^{(2)}$, ... $\phi_n^{(n)}$. The solution of ω^2 (square of eigenfrequency) and the associated $\{\phi\}_n$ (eigenvectors) in the equation above is known as "eigenproblem."

This equation always has the trivial solution when $\phi_n = 0$, which is not useful because it implies no motion. On the other hand, by vanishing the determinant of the coefficients, all these linear homogeneous equations should have a non-trivial solution (i.e. $\phi_n \neq 0$):

$$\det[K - \omega^2 M] = |K - \omega^2 M| = 0 \tag{4.4}$$

J. Jia, *Essentials of Applied Dynamic Analysis*, Risk Engineering,   55
DOI: 10.1007/978-3-642-37003-8_4, © Springer-Verlag Berlin Heidelberg 2014

The equation above is known as the characteristic equation or frequency equation. For a multi-degree of freedom system, the equation above gives a polynomial of degree n in the eigenfrequency ω^2. This polynomial equation has n roots ω_1^2, ω_2^2,...ω_n^2, which are called eigenvalues, square of eigenfrequencies, characteristic values, or normal values, and they are obviously real and positive or zero because the stiffness and mass matrices are symmetric and positively definite. Therefore, for N degrees-of-freedoms, there are N independent eigenvalues and the associated mode shapes.

For each eigenvalue ω^2 (often denoted by λ in the literature), a unique solution to Eq. (4.3) exists for $\{\phi\}_n$, which is known as eigenvectors or mode shapes. The eigenvalues and associated eigenvectors together are called eigenpairs.

For an elaborated presentation of eigenpairs, see Sects. 13.2 and 13.4.

Parameters affecting the eigenfrequencies are the stiffness of a system/structure and the mass distribution in it, support conditions, and damping. If these properties do not change, the eigenfrequencies and mode shapes will not change. It is not difficult to imagine that outer dimensions, such as height and length of the frame, significantly affect the stiffness. Also, increasing beam sections will result in a higher stiffness. A change in elastic modulus will give a linear increase of the stiffness. The influence of mass distribution on eigenfrequencies is due to the complexity of the structure. For example, when the mass of a structure is increased, the stiffness may also increase.

If the eigenvalues are arranged in an escalating order, the lowest eigenfrequency and the corresponding mode shape is called the fundamental eigenfrequency and fundamental mode shape.

For a redundant structure, with the order increase of vibration modes, the number of anti-nodes (locations with peak responses) in the corresponding mode shape is also increased as shown in Fig. 4.1.

Even though a number of eigenpairs is equal to the number of degrees-of-freedoms, depending on the application, normally only a number of eigenpairs (typically the lowest ones) are of engineering interest. This will be discussed in Sect. 13.4.

Solving eigenfrequencies normally requires an assumption or definition of corresponding mode shapes. Compared to eigenfrequencies, the exact solutions of mode shape are of secondary importance in many vibration problems. As will be illustrated later on, a reasonable approximation of mode shape will result in a calculated eigenfrequency close to the exact one.

In a general sense, the eigenmodes include not only the global and local vibration through deformation, but also up to six rigid body motion modes along the six degrees-of-freedoms, all of which are in a form of harmonic type. Rigid body motions indicate that the stiffness is not positively defined.

Rigid body vibrations normally occur when a structure is installed with flexible mounting; the structure can then move as an entire solid at the frequency of zero or close to zero (e.g. craft in flight). However, it is normally related to the vibration analysis of structures with intentionally designed flexibilities, such as aerospace or

4.1 Eigenpairs, Global-, Local- and Rigid-Body Vibrations 57

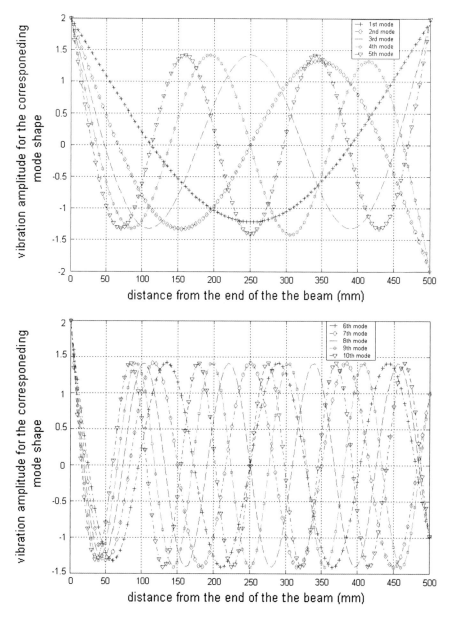

Fig. 4.1 Mode shapes for a beam simply supported at its two ends (0 and 500 mm in the horizontal axis indicate two ends of the beam)

airplane structures, or certain types of machinery. Therefore, rigid body vibration is of minor importance for most of the civil and mechanical engineering structures because they only cause small structural deformations and stresses. Some exceptions exist: firstly, the rigid body vibration may sometimes cause problems on structural connections if the structural acceleration and/or mass are high. Secondly, it may also compromise the integrity of slender and flexible structures such as pipelines.

Global vibrations occur when a structure vibrates as a whole while having various deformations on each part that are aligned together and strongly coupled. This is normally the most important type of vibration since it can cause amplification of global structural responses leading to catastrophic structural failure.

Local vibrations mean that a local part of the structure is vibrating without significant interference from vibrations from other parts. An example of this is the vibration of conductors, local frame or beam of an offshore structure, or vibrations of individual deck plates on a ship. This type of vibration sometimes causes local failure of a structure corresponding to the eigenfrequencies at the corresponding local vibration modes. Figure 4.2 illustrates an example of eigenmodes for both the global jacket vibrations and local topside frame vibrations.

4.2 Hand Calculation of Natural Frequency for Systems with Distributed Masses

It is noticed that a simple hand calculation for eigenfrequencies is based on an idealized model as rigid masses joined by massless stiffness items (i.e., spring or beam) and massless damping. However, this assumption can greatly compromise the accuracy of a hand calculation simply because an insufficient number of DOFs are modeled. An increase in DOFs can improve the calculation accuracy but also results in a distributed mass system.

Exact solutions of eigenpairs can be performed by the classical method, in which the equilibrium equations based on Newton's second law of motion are solved by differential equations (ordinary ones for discrete system and partial ones for distributed system). This provides exact solutions. However, this can only be applied to systems with relatively low configurations [146]. For continuous systems or systems with large degrees-of-freedoms, other methods such as Rayleigh energy method, equivalent system analysis, or finite element method etc. have to be adopted.

4.2.1 Classical Method for Exact Solutions

Consider the cantilever beam shown in Fig. 4.3. By assuming a small deflection of the beam, the beam's shear deformation can be neglected, resulting in a Euler–Bernoulli beam formulation. The differential equation of the deflection curve for beams with any types of end support conditions can be expressed as:

$$\frac{d^2}{dx^2}\left(EI\frac{d^2z}{dx^2}\right) = -\mu\frac{d^2z}{dt^2} \tag{4.5}$$

where μ is the density of the beam per unit length.

4.2 Hand Calculation of Natural Frequency for Systems with Distributed Masses

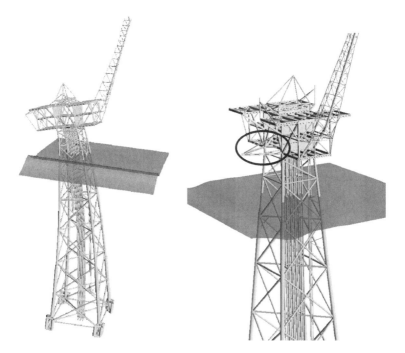

Fig. 4.2 The global (*left*) and local frame vibrations of a jacket platform at two distinct eigenfrequencies of 0.2 and 1.3 Hz, respectively (courtesy of Aker Solutions)

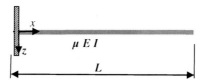

Fig. 4.3 A beam with uniformly distributed mass of μ

Dividing the equation above by μ, one can write the equilibrium equation for the lateral (along z in Fig. 4.3) vibrations of the beam as:

$$\frac{EI}{\mu}\frac{d^4 z}{dx^4} + \frac{d^2 z}{dt^2} = 0 \qquad (4.6)$$

By assuming an approximate vibration mode shape $X(x)$, the deflection of the beam at position x and time t is $z(x,t) = X(x)\sin(\omega_n t)$. We then have:

$$\frac{EI}{\mu}\frac{d^4(X(x))}{dx^4} - \omega_n^2 X(x) = 0 \qquad (4.7)$$

Rearranging the equation above gives:

$$\frac{d^4(X(x))}{dx^4} = \frac{\mu \omega_n^2}{EI} X(x) = \kappa^4 X(x) \tag{4.8}$$

To fulfill the required conditions of the equation above, $X(x)$ can be expressed in the form of trigonometric function:

$$X(x) = A_1 \sin(\kappa x) + A_2 \cos(\kappa x) + A_3 \sinh(\kappa x) + A_4 \cosh(\kappa x) \tag{4.9}$$

The constants A_1, A_2, A_3, and A_4 are determined from the boundary conditions of the beam. For a convenient solution of these constants, the equation above can be rewritten in a form with the zero constants for each of the typical boundary conditions [146]:

$$X(x) = A[\sin(\kappa x) + \cosh(\kappa x)] + B[\cos(\kappa x) - \cosh(\kappa x)] \\ + C[\sin(\kappa x) + \sinh(\kappa x)] + D[\sin(\kappa x) - \sinh(\kappa x)] \tag{4.10}$$

The boundary conditions are directly related to the different order of derivatives with respect to x: $X(x)$ is proportional to deflection, $\frac{d(X(x))}{dx}$ is proportional to the slope, $\frac{d^2(X(x))}{dx^2}$ is proportional to the moment, and $\frac{d^3(X(x))}{dx^3}$ is proportional to the shear force.

$$\frac{d(X(x))}{dx} = \kappa \left\{ \begin{array}{l} A[-\sin(\kappa x) + \sinh(\kappa x)] + B[-\sin(\kappa x) - \sinh(\kappa x)] \\ + C[\cos(\kappa x) + \cosh(\kappa x)] + D[\cos(\kappa x) - \cosh(\kappa x)] \end{array} \right\} \tag{4.11}$$

$$\frac{d^2(X(x))}{dx^2} = \kappa^2 \left\{ \begin{array}{l} A[-\cos(\kappa x) + \cosh(\kappa x)] + B[-\cos(\kappa x) - \cosh(\kappa x)] \\ + C[-\sin(\kappa x) + \sinh(\kappa x)] + D[-\sin(\kappa x) - \sinh(\kappa x)] \end{array} \right\} \tag{4.12}$$

$$\frac{d^3(X(x))}{dx^3} = \kappa^3 \left\{ \begin{array}{l} A[\sin(\kappa x) + \sinh(\kappa x)] + B[\sin(\kappa x) - \sinh(\kappa x)] \\ + C[-\cos(\kappa x) + \cosh(\kappa x)] + D[-\cos(\kappa x) - \cosh(\kappa x)] \end{array} \right\}$$

$$\tag{4.13}$$

It is noticed that, for typical boundary conditions, two of the constants among A, B, C and D are zero, leaving only the other two equations to be solved. For the cantilever beam shown in Fig. 4.3, we have:

$X(x = 0) = 0$, $\frac{d(X(x=0))}{dx} = 0$, $\frac{d^2(X(x=L))}{dx^2} = 0$, and $\frac{d^3(X(x=L))}{dx^3} = 0$. This gives:

$$A = 0 \tag{4.14}$$

$$C = 0 \tag{4.15}$$

$$\frac{d^2(X(x=L))}{dx^2} = 0 = B[-\cos(\kappa L) - \cosh(\kappa L)] + D[-\sin(\kappa L) - \sinh(\kappa L)] \tag{4.16}$$

$$\frac{d^3(X(x=L))}{dx^3} = 0 = B[\sin(\kappa L) - \sinh(\kappa L)] + D[-\cos(\kappa L) - \cosh(\kappa L)] \tag{4.17}$$

4.2 Hand Calculation of Natural Frequency for Systems with Distributed Masses 61

From Eq. (4.8), it is noted that the exact solutions for the eigenfrequencies are:

$$\omega_n = \sqrt{\kappa^4 \frac{EI}{\mu}} \tag{4.18}$$

The objective is to find the solutions for κ, which can be used to calculate both the eigenfrequencies and mode shapes ($X(x)$). We then have:

$$\frac{B}{D} = -\frac{\sin(\kappa L) - \sinh(\kappa L)}{\cos(\kappa L) + \cosh(\kappa L)} = \frac{\cos(\kappa L) + \cosh(\kappa L)}{\sin(\kappa L) - \sinh(\kappa L)} \tag{4.19}$$

The equation above is reduced to:

$$\cos(\kappa L) \cosh(\kappa L) = -1 \tag{4.20}$$

The value of κL can be calculated by checking handbooks of mathematics for solving hyperbolic and trigonometric functions. We here list the first three values corresponding to the first three eigenfrequencies for a clamped-free beam:

$$\kappa_1 L = 1.8751 \tag{4.21}$$

$$\kappa_2 L = 4.6941 \tag{4.22}$$

$$\kappa_3 L = 7.8548 \tag{4.23}$$

Table 4.1 summarizes the exact solutions for beams with uniformly distributed mass and various support conditions.

4.2.2 Equivalent System Analysis for Approximate Solutions

Previously, we discussed how, if a system can be idealized as an SDOF system, the natural frequency of the system can be conveniently calculated with simple hand calculations:

$$\omega_n = \sqrt{\frac{k}{m}} \tag{4.24}$$

However, an actual engineering structure or system often possesses many degrees-of-freedoms, and a very common scenario is that concentrated masses are connected by a series of stiffness members such as springs or beams. In order to use the formula above, one has to find its equivalent system counterpart with only an SDOF, i.e., equivalent stiffness k_{eq} and mass m_{eq}. This can be performed by finding the terms of stiffness and mass in the standard kinetic and potential energy formulation:

$$T = \frac{1}{2} m_{eq} \dot{x}(t)^2 \tag{4.25}$$

Table 4.1 Exact solution for eigenfrequencies of beams with uniformly distributed mass and various support conditions

$\omega_i = \sqrt{\kappa^4 \frac{EI}{\mu}}$	$\kappa_1 L$	$\kappa_2 L$	$\kappa_3 L$
	$i\pi$	$i\pi$	$i\pi$
	4.7300	7.8532	10.9956
	1.8751	4.6941	7.8548
	3.9266	7.0686	10.2102
	4.7300	7.8532	10.9956
	2.3650	5.4978	8.6394
	3.9266	7.0686	10.2102
	2.3650	5.4978	8.6394
	$(i\text{-}0.5)\pi$	$(i\text{-}0.5)\pi$	$(i\text{-}0.5)\pi$
	$i\pi$	$i\pi$	$i\pi$

4.2 Hand Calculation of Natural Frequency for Systems with Distributed Masses

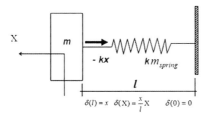

Fig. 4.4 An SDOF spring-mass system (the spring has a distributed mass of $\frac{m_{spring}}{l}$ per unit length)

$$V = \frac{1}{2} k_{eq}\, x(t)^2 \qquad (4.26)$$

In addition, the equivalent damping c_{eq} can also be evaluated through the work done by the viscous damping force between two arbitrary locations x_1 and x_2:

$$E = -\int_{x_1}^{x_2} c_{eq}\, \dot{x}(t)\, dx \qquad (4.27)$$

In the SDOF system analyzed in Sect. 3.2, the spring is assumed to be massless. Here we make a more realistic assumption that the spring has a mass of m_{spring}, as shown in Fig. 4.4. Let X be a coordinate along the spring's axis in its un-stretched position ($0 \le X \le l$). First we assume that the deflection (displacement) of the spring along it axial direction is:

$$\delta(X,t) = \frac{x}{l} X \qquad (4.28)$$

The kinetic energy of the spring is then:

$$T = \int dT = \frac{1}{2} \int_0^l \left(\frac{\partial \delta(X,t)}{\partial t}\right)^2 dm_{spring} + \frac{1}{2} m \dot{x}^2 = \frac{1}{2} \int_0^l \left(\frac{\dot{x}}{l} X\right)^2 \frac{m_{spring}}{l} dX + \frac{1}{2} m \dot{x}^2$$

$$= \frac{\dot{x}^2 m_{spring}}{2 l^3} \int_0^l X^2 dX + \frac{1}{2} m \dot{x}^2 = \frac{1}{2} \left(\frac{m_{spring}}{3} + m\right) \dot{x}^2$$

$$(4.29)$$

Therefore, if the mass of the spring $\frac{m_{spring}}{3}$ is placed at its end where the mass m is located, the kinetic energy is the same as that of the linear spring m_{spring} with linear vibrations of deflections [147].

The equivalent system analysis is essentially another form of Rayleigh energy analysis, as will be elaborated in Chap. 5, which is basically a means of calculating the eigenfrequency by equating the maximum kinetic and potential energy.

Table 4.2 summarizes the natural frequencies of spring and beams with uniformly distributed mass and a concentrated mass and with various support conditions.

Table 4.2 Natural frequencies of spring and beams with uniformly distributed mass and a concentrated mass and with various support conditions

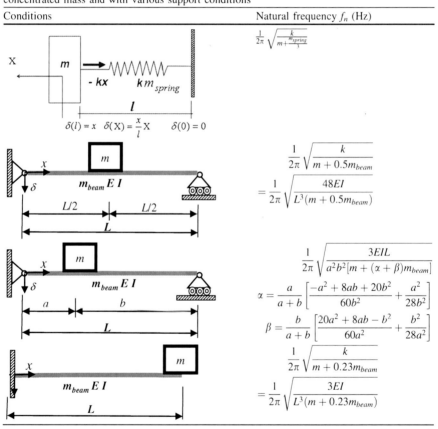

(continued)

4.2 Hand Calculation of Natural Frequency for Systems with Distributed Masses

Table 4.2 (continued)

Conditions	Natural frequency f_n (Hz)

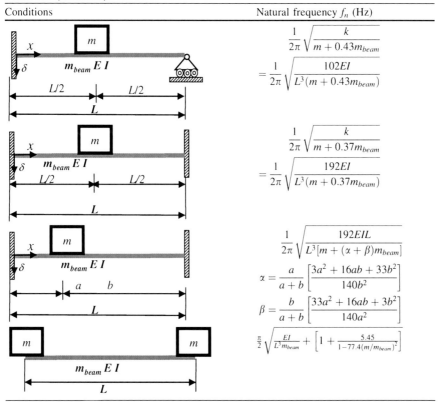

$$\frac{1}{2\pi}\sqrt{\frac{k}{m+0.43m_{beam}}}$$
$$=\frac{1}{2\pi}\sqrt{\frac{102EI}{L^3(m+0.43m_{beam})}}$$

$$\frac{1}{2\pi}\sqrt{\frac{k}{m+0.37m_{beam}}}$$
$$=\frac{1}{2\pi}\sqrt{\frac{192EI}{L^3(m+0.37m_{beam})}}$$

$$\frac{1}{2\pi}\sqrt{\frac{192EIL}{L^3[m+(\alpha+\beta)m_{beam}]}}$$
$$\alpha=\frac{a}{a+b}\left[\frac{3a^2+16ab+33b^2}{140b^2}\right]$$
$$\beta=\frac{b}{a+b}\left[\frac{33a^2+16ab+3b^2}{140a^2}\right]$$

$$\frac{\pi}{2}\sqrt{\frac{EI}{L^3m_{beam}}}+\left[1+\frac{5.45}{1-77.4(m/m_{beam})^2}\right]$$

Example: Derive the equivalent mass for a simply supported beam with a mass (m) attached to it as shown in Fig. 4.5. The density, the cross-section area, the Youngs' modulus and the moment of inertia (about the strong axis) of the beam are ρ, A, E, and I.

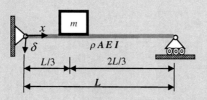

Fig. 4.5 A mass attached to a beam at a distance of 1/3 beam length from the *left end* of the beam

Solution: For the beam-mass system shown in Fig. 4.5, the vibration mode shape corresponding to its natural frequency can be assumed to be identical to the static deflection of the beam under the gravity load:

$$\delta(x) = \begin{cases} \frac{1}{81}(5L^2x - 9x^3) & \text{for } 0 \le x \le L/3 \\ \frac{1}{81}(5L^2x - 9x^3) + \frac{1}{6}(x - L/3)^3 & \text{for } L/3 < x \le L \end{cases}$$

The deflection at $x = L/3$ can be calculated as:

$$\delta(x = L/3) = \frac{4FL^3}{243EI}$$

where F is the force at the mass position to cause the deflection. Rewriting the equation above in terms of $\delta(x = L/3)$:

$$F = \frac{243EI\delta(x = L/3)}{4L^3}$$

By resembling the beam-mass vibration as the vibrations of the SDOF shown in Fig. 4.4, the kinetic energy of the mass-beam system is then:

$$T = \frac{1}{2}\int_0^L \rho A \left(\frac{\partial\delta}{\partial t}\right)^2 dx$$

$$= \frac{1}{2}\rho A \left(\frac{243EI}{4L^3}\right)^2 \left[\dot{\delta}(x = L/3)\right]^2 \cdot$$

$$\left\{ \int_0^{L/3} \left[\frac{1}{81}(5L^2x - 9x^3)\right]^2 dx + \int_{L/3}^{L} \left[\frac{1}{81}(5L^2x - 9x^3) + \frac{1}{6}(x - L/3)^3\right]^2 dx \right\}$$

$$= 0.586\rho AL$$

It is obvious from the equation above that the equivalent mass of the beam is 0.586 times the beam mass, provided that the vibration modes are controlled by a mass m located 1/3 of the way from one end of the beam.

4.2 Hand Calculation of Natural Frequency for Systems with Distributed Masses

Example: An offshore drilling jacket platform located in the North Sea has a topside weight of 21,600 t (m_{top}), the jacket weight together with the added mass is 8,700 t (m_{jack}), the distance between the jacket bottom and the center of gravity of the topside is 136 m. From the response measurement, the eigenfrequencies corresponding to two perpendicular principal directions (horizontally) are 0.36 Hz and 0.40 Hz. Calculate the stiffness of the jacket structure based on the assumption that the vibration modes are controlled by the topside mass.

Solution: Assuming the weight of the jacket is evenly distributed along its height, and the jacket bottom has a fixed support condition, by checking Table 4.2, one obtains:

$$f_n = \frac{\sqrt{\frac{k_{jack}}{m_{top} + 0.23 m_{jack}}}}{2\pi}$$

where k_{jack} is the bending stiffness of the jacket.

With the measured eigenfrequencies, the stiffnesses of the jacket along the two principal directions are: 1.21×10^5 and 1.49×10^5 kN/m (Fig. 4.6).

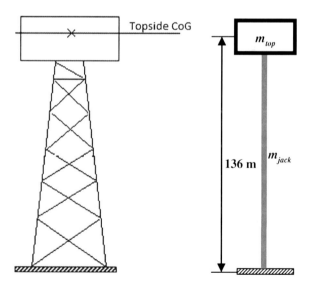

Fig. 4.6 An SDOF model (*right*) representing a topside-jacket system (*left*)

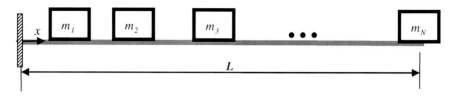

Fig. 4.7 N concentrated masses attached to a cantilever beam at discrete points along the length of the beam

4.2.3 Natural Frequency with Distributed Masses: Dunkerley Method for Approximate Solutions

For systems with distributed masses, a convenient method to estimate the natural frequencies is the Dunkerley method.

Consider a beam with N concentrated masses $m_1, m_2, m_3, \ldots m_N$ shown in Fig. 4.7. First remove all the masses except the first mass m_1, the natural frequency under this condition is denoted as f_{n1}. Similarly, remove all the masses except the second mass m_2, the natural frequency under this condition is denoted as f_{n2}. Repeat this operation for all the N masses. The natural frequency for the entire system can then be calculated using the Dunkerley method:

$$\frac{1}{f_n^2} \approx \frac{1}{f_{n1}^2} + \frac{1}{f_{n2}^2} \cdots + \frac{1}{f_{nN}^2} = \sum_{i=1}^{N} \frac{1}{f_{ni}^2} \quad i = 1, 2, N \qquad (4.30)$$

Provided that the assumption of mode shape is exact, Dunkerley's equation gives a lower-bound natural frequency to the exact one, and provides a good approximation if the mode shapes associated with different mass distributions are similar to each other and to the fundamental mode shape of the system [148]. The accuracy of results also depends on the beam's boundary conditions, number of masses and relative values of the masses. From an engineering point of view, it is accurate if the individual frequencies f_{ni} are not close to each other. Specifically, the natural (fundamental) frequency of the system needs to differ significantly from the higher-order eigenfrequencies.

Example: A machine with a mass M has rather large vibration amplitude. To solve this problem, a small mass m ($m = M/80$) is mounted on the top of the machine with a stiffness that is also 1/80 of the machine's stiffness (as shown in Fig. 4.8). This gives the natural frequency ($\frac{1}{2\pi}\sqrt{\frac{k}{m}}$) for the isolated small mass (m)-spring (k) system identical to that of the machine mass (M)-spring ($80\,k$) system. Therefore, when the machine reaches a resonance condition at its natural frequency, the resonance of the small mass also

4.2 Hand Calculation of Natural Frequency for Systems with Distributed Masses

occurs, which can efficiently absorb energy from the machine's vibration, thus decreasing the vibration amplitude of the machine. Use Dunkerley's equation to calculate the natural frequency of the complete system (m–k–M–80 k) and compare it with the exact solutions.

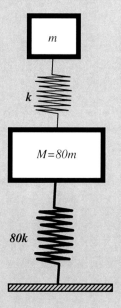

Fig. 4.8 Machine (M)-small mass (m) system

Solution: Using the Dunkerley method, the natural frequency for the system shown in Fig. 4.8 can be calculated based on a separated calculation of the natural frequency (f_{n1}) of the machine with its own mass distribution and the natural frequency (f_{n2}) of the concentrated mass m.

$$\frac{1}{f_n^2} \approx \frac{1}{f_{n1}^2} + \frac{1}{f_{n2}^2}$$

$$f_{n1} = \frac{1}{2\pi}\sqrt{\frac{80k}{80m}} = \frac{1}{2\pi}\sqrt{\frac{k}{m}}$$

$$f_{n2} = \frac{1}{2\pi}\sqrt{\frac{\frac{80k \cdot k}{80k+k}}{m}} = \frac{0.994}{2\pi}\sqrt{\frac{k}{m}}$$

$$\frac{1}{f_n^2} \approx \frac{1}{f_{n1}^2} + \frac{1}{f_{n2}^2} = \frac{1}{\frac{1}{4\pi^2}\frac{k}{m}} + \frac{1}{\frac{0.988}{4\pi^2}\frac{k}{m}}$$

$$= \frac{8.049\pi^2}{\frac{k}{m}}$$

Therefore, we have:

$$f_n = 0.112\sqrt{\frac{k}{m}}$$

The exact solution for the natural frequency is calculated as:

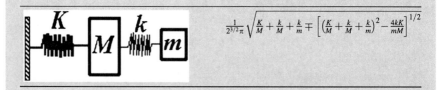

$$\tfrac{1}{2^{3/2}\pi}\sqrt{\tfrac{K}{M}+\tfrac{k}{M}+\tfrac{k}{m}\mp\left[\left(\tfrac{K}{M}+\tfrac{k}{M}+\tfrac{k}{m}\right)^2-\tfrac{4kK}{mM}\right]^{1/2}}$$

We obtain that the exact natural frequency for the current example:

$$f_n = 0.151\sqrt{\frac{k}{m}}$$

This result is 34 % higher than the one calculated from the Dunkerley method. Therefore, the application of the Dunkerley method is limited to systems with the individual eigenfrequencies f_{ni} far from each other, as discussed previously.

Example: Calculate the natural frequency of the beam-mass system shown in Fig. 4.9.

Solution: Using the Dunkerley method, the natural frequency for the system shown in Fig. 4.9 can be calculated based on a separated calculation of natural frequency (f_{n1}) of the beam with its own mass distribution and the natural frequency (f_{n2}) of the concentrated mass m attached to the beam without the beam mass as shown in Fig. 4.10.

$$\frac{1}{f_n^2} \approx \frac{1}{f_{n1}^2} + \frac{1}{f_{n2}^2}$$

4.2 Hand Calculation of Natural Frequency for Systems with Distributed Masses

From Table 4.1, we have:

$$f_{n1}^2 = \frac{1}{4\pi^2} \cdot \kappa^4 \frac{EI}{\mu} = \frac{1}{4\pi^2} \left(\frac{1.8751}{L}\right)^4 \frac{EI}{m_{beam}/L} = \frac{12.362EI}{4\pi^2 L^3 m_{beam}}$$

$$f_{n2}^2 = \frac{1}{4\pi^2} \cdot \frac{k_2}{m_{beam}} = \frac{1}{4\pi^2} \cdot \frac{3EI}{(L)^3 m_{beam}} = \frac{3EI}{4\pi^2 L^3 m}$$

$$\frac{1}{f_n^2} \approx \frac{1}{f_{n1}^2} + \frac{1}{f_{n2}^2} = \frac{4\pi^2 L^3 m_{beam}}{12.362EI} + \frac{4\pi^2 L^3 m}{3EI}$$

$$= \frac{4\pi^2 L^3 (0.080893 m_{beam} + 0.33333 m)}{EI}$$

Therefore, the calculated natural frequency is:

$$f_n \approx \sqrt{\frac{EI}{4\pi^2 L^3 (0.080893 m_{beam} + 0.33333 m)}} = \frac{1}{2\pi} \sqrt{\frac{3EI}{L^3 (m + 0.24 m_{beam})}}$$

This agrees well with the value shown in Table 4.2.

However, if one wrongly assumes that the beam in Fig. 4.9 can be idealized as a beam mass attached to a massless beam with half span length, as shown in the middle figure of Fig. 4.11, the calculations following this assumption are:

$$f_{n1}^2 = \frac{1}{4\pi^2} \cdot \frac{k_1}{m_{beam}} = \frac{1}{4\pi^2} \cdot \frac{3EI}{(L/2)^3 m_{beam}} = \frac{6EI}{\pi^2 L^3 m_{beam}}$$

$$\frac{1}{f_n^2} \approx \frac{1}{f_{n1}^2} + \frac{1}{f_{n2}^2} = \frac{\pi^2 L^3 m_{beam}}{6EI} + \frac{4\pi^2 L^3 m}{3EI} = \frac{\pi^2 L^3 m_{beam} + 8\pi^2 L^3 m}{6EI}$$

This leads to an incorrect calculation of the natural frequency of the system because the flexibility/stiffness of the other half beam span is ignored:

$$f_n \approx \sqrt{\frac{6EI}{\pi^2 L^3 m_{beam} + 8\pi^2 L^3 m}} = \frac{1}{2\pi} \sqrt{\frac{3EI}{L^3 (m + 0.125 m_{beam})}}$$

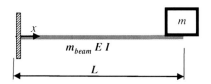

Fig. 4.9 A mass attached to a cantilever beam at the free end of the beam

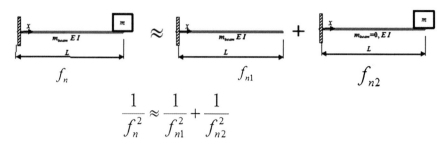

Fig. 4.10 Breakdown of the beam-mass system into two systems to calculate the natural frequency of the coupled system shown in Fig. 4.9 using the dunkerley method

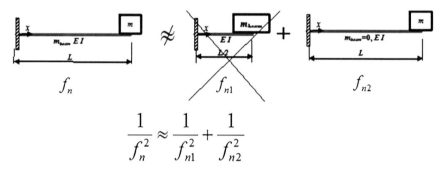

Fig. 4.11 Wrong assumption of the divided system used to calculate the natural frequency of the coupled system shown in Fig. 4.9

4.3 Using Symmetry and Anti-Symmetry in Eigenanalysis

With the presence of vibrations, the utilization of symmetry or anti-symmetry is in many cases restricted. This is because, even if the geometry is symmetrical or anti-symmetrical, the different order of vibration eigenmodes may show both

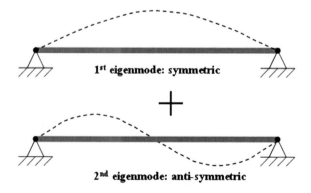

Fig. 4.12 A simply supported beam with a symmetrical fundamental eigenmode (*dotted line*) and an anti-symmetrical 2nd eigenmode (*dotted line*)

4.3 Using Symmetry and Anti-Symmetry in Eigenanalysis

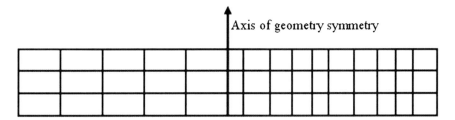

Fig. 4.13 A symmetric structure geometry modeled by an asymmetric mesh

symmetrical and anti-symmetrical responses, as shown in Fig. 4.12, making the combination of the responses neither symmetrical nor anti-symmetrical. If one enforces symmetry boundary conditions, this will eliminate all anti-symmetric modes, and vice versa. In certain cases, one is able to utilize the symmetry by combining the responses obtained from the same half model with both symmetric and anti-symmetric boundary conditions. However, this will require too high an overhead for combining two sets of results, and it is normally not used in engineering practice.

For performing dynamic responses using the finite element analysis, the calculated eigenfrequencies may be rather sensitive to the symmetry or anti-symmetry of the structure where it is applicable, i.e. slight mis-modeling of these

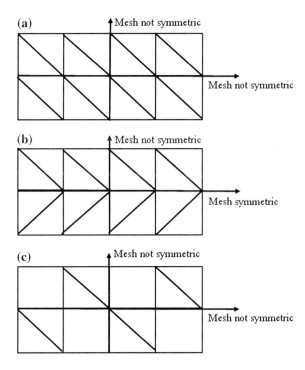

Fig. 4.14 A symmetric structure geometry modeled with asymmetric mesh

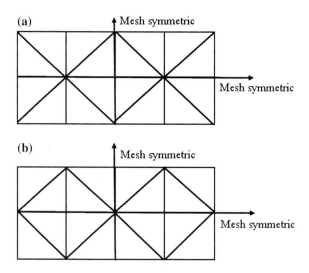

Fig. 4.15 A symmetric structure geometry modeled with symmetric mesh

characteristics will induce unrealistic vibration modes. Therefore, not only the structure geometry but also the meshing needs to be modeled symmetrically or anti-symmetrically whenever it is applicable.

Figure 4.13 shows an example of wrongly modeled mesh with an asymmetrical mesh (using shell element), which will lead to the right part of the structure being softer than the left part, i.e. some symmetrical or anti-symmetrical modes will disappear or appear at eigenmodes of wrong mode shapes and/or frequencies.

When one performs meshing with triangle elements, attention should also be paid to the symmetrical or anti-symmetrical characteristics. Figure 4.14 shows examples of meshing that is not symmetric, while their symmetric counterparts are illustrated in Fig. 4.15.

4.4 Vibration-Based Structural Health Monitoring

Traditionally, structural damages are detected through periodic visual inspection. Visual inspection can impose high cost. For example, for offshore structures, since they are surrounded by water, inspections such as subsea inspection involves certain risks for inspection divers and significant cost, and the application may require a lengthy approval process from the regularity authorities. Furthermore, even though periodical visual inspections are mandatory for important infrastructures, the reliability of this method is questionable and strongly depends on the inspection condition, and how experienced and dedicated the participating inspectors and engineers are. A survey by Moore and his coworkers [149] from the US Federal Highway Administration revealed that, at most, only 68 % of the condition ratings were correct and in-depth inspections could not find interior deficiencies given that visual examinations are very seldom carried out by inspectors. In some cases, termination of normal operation activities on structures may be required solely for the sake of allowing an inspection to be carried out, such as the possible production shutdown if one wants to inspect a tip of a flare tower for an offshore platform.

All the drawbacks above motivate the technology of structural health monitoring (SHM), which is the process of assessing the state of health (e.g., damage) of instrumented structures from measurements. It can be either a short-term (e.g. reparation) or a long-term (monitoring parameters continuously or periodically) process. The safety and reliability of structures can be improved by detecting damage before it reaches a critical state [150], thus minimizing the probability of catastrophic failures and allowing reduced efforts on inspection services and maintenance. The reduction of downtime and improvement in reliability enhances the productivity of the structure. The cost of SHM is relatively low, accounting for between only 2–5 % of total structures' cost for a period of 10 years [151]. For many infrastructures with high value equipment and contents, this percentage is even lower. Furthermore, SHM can also provide quick assessment after a major accident such as an earthquake, hurricane, explosion, or ship collision etc., which is essential for property owners, operators and relevant insurance companies to make immediate decisions. In addition, the immediacy and sensitivity of SHM also allows for the short-term verification of new designs, for which there is no or limited service experiences and which require more rigorous monitoring and inspection until adequate confidence is gained. This is because new designs often involve a high utilization of force, new structural details (possibly with high stress concentration), as well as large uncertainties in the responses [13]. SHM also serves to confirm the design parameters and perform quality assurance through the detection of damages that cause any changes in properties other than expected by original designs, and in the meantime can sometimes approve a better structural performance than expected by the design, which provides flexibility for the structure capacity.

Vibration-based SHM detect structural damage when noticeable changes in the measured eigenfrequencies and mode shapes occur. In addition, more detailed assessment can be reached by studying the peak responses (accelerations, velocity and displacement) and drift.

The physical diagnostic tool of an SHM system is the integration of various sensing devices and ancillary systems including sensors, data acquisition and processing systems, communication systems and health evaluation systems (including diagnostic algorithms and information management) for damage detection and modeling system [152, 153] as shown in Fig. 4.16.

Sensors are used to measure various mechanical and physical parameters. Typical sensors include the accelerometers, gyroscopes, inclinometer, strain gauges, fiber optic gauges, deflection transducers, curvature sensor, etc. For vibration-based SHM, accelerometers are the typically used sensors. Important specifications of accelerometers are measurable acceleration range and resolutions, size, noise level, stability over time and temperature, frequency response and

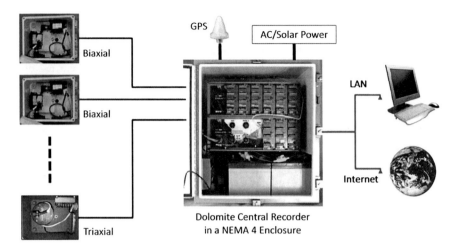

Fig. 4.16 SHM system including sensors (*left*, accelerometers in the current case), connection system, data acquisition (*middle*, the GPS is used for precise timing), communication system (LAN or Internet), data process system (computer in this case), and damage detection and modeling system (computer) (courtesy of Kinemetrics Inc.)

filters, and power consumption. As a general rule, more sensors with an appropriate data processing can secure a more reliable assessment.

The transfer of measurement data from sensors to the data acquisition system is important. Lead wires are usually used to transfer data. Note that long wires may have a negative influence due to the noise collected with the measured data; therefore, one needs to pay attention to the allowable length of the wire corresponding with the data acquisition system. As an alternative to wires, wireless communication between the sensors and data acquisition can be used in cases when a large number of sensors are used. In addition, for underwater monitoring, acoustic transmissions, which are subject to underwater noise but are relatively easy to install and operate, are also used.

The data acquisition system can accommodate many channels (typically with 12, 24, and 36 channels). It is also the most expensive single piece of equipment in an SHM task. In some cases for onsite measurement with a duration of less than a few days, the data acquisition system and process system are integrated together as one equipment package.

The communication system transmits the data from the acquisition system to the location where the data is processed (data process system). This can be a telephone lines, internet, transmitting devices, electric wires etc.

The processing of data is of great importance. It is basically a computer that processes the data from the sensor-acquisition system's input. The data from different sensor inputs will be related to each other. Typically, process tasks include the post-processing of inter-story drift, drift hysteresis loop, accelerations, velocity and displacement, calculation of transfer functions, etc. Ideally, the processing should be performed prior to the data storage.

Despite the appealing advantages of SHM over traditional visual inspection, significant uncertainties involved in the modal testing and its interpretation still pose major challenges to a reliable vibration-based SHM. This is more obvious for offshore structures, in which environmental conditions (temperature, humidity, and wind etc.) and prevailing excitations from waves and wind impart the uncertainties into ambient or forced vibration testing. The change of weight (mainly topside content weight) and marine growth varying with the season also add further uncertainties. In addition, measurement noise and bias errors arising from sensors, cables and the data acquisition system is an important source of uncertainty [154]. On top of this uncertainty, vibration-based SHM is sometimes unable to detect the local damages of structures simply because it measures the global vibration eigenfrequencies and mode shapes, and is thus insensitive to local stiffness changes. This is especially the case for rather redundant structures.

For more details about SHM, readers can refer to references [155] and [156].

Chapter 5
Solving Eigenproblem for Continuous Systems: Rayleigh Energy Method

Structural members or systems in the real world are usually of a type of continuous system with non-uniform mass or stiffness distribution. The closed form (exact) solutions in this case are not available. Therefore, approximate solutions have to be sought. Many methods are available for approximating the eigenvalues. All of these methods are based on a discretization of continuous systems, i.e., replacing the system by an equivalent discretized one. They are divided into two classes: the first one represents the solutions as a finite series consisting of space-dependent functions multiplied by time-dependent generalized coordinates, and applies to those systems for which the non-uniformity with regard to mass and stiffness is not significant, such as the classical method presented in Sect. 4.2.1. The second one simply represents the system with lumped mass (as will be elaborated in Sect. 13.6) at discrete points of the continuous system, such as the one presented in Sect. 4.2.2.

There are several approximation methods for estimating the natural frequency without solving the differential equation of the system, such as the Rayleigh energy method, the Dunkerley method, the Southwell method, the Galerkin method, the collocation method, the Holzer method, the Myklestad method, the integral formulation method, the Lumped-parameter method, and the Kantorovich method etc., [157]. Among all these approximate methods, the most popular is the Rayleigh energy method [158–160], simply due to its generality, convenience and efficiency. The method is based on the conservation of energy: for a system, when the kinetic energy is the zero, the potential energy reaches its maximum, and vice versa. By equating the maximum kinetic energy with the maximum potential energy of the system, the approximate estimation for the system's natural frequency can be calculated.

Consider the spring-mass system shown in Fig. 3.2. The deflection of the mass at time t can be written as:

$$x(t) = X \sin(\omega t) \tag{5.1}$$

The maximum potential energy in the system is:

$$V_{\max} = \frac{1}{2} kX^2 \tag{5.2}$$

J. Jia, *Essentials of Applied Dynamic Analysis*, Risk Engineering,
DOI: 10.1007/978-3-642-37003-8_5, © Springer-Verlag Berlin Heidelberg 2014

And the maximum kinetic energy is:

$$T_{\max} = \frac{1}{2}m\left(\frac{d^2x}{dt^2}\right)^2_{\max} = \frac{1}{2}m\omega_n^2 X^2 \tag{5.3}$$

Equating the two equations above gives:

$$\frac{1}{2}kX^2 = \frac{1}{2}m\omega_n^2 X^2 \tag{5.4}$$

Rewriting the equation above:

$$\frac{T_{\max}}{V_{\max}} = \frac{\frac{1}{2}m\omega_n^2 X^2}{\frac{1}{2}kX^2} = \frac{m\omega_n^2}{k} = 1 \tag{5.5}$$

We finally conclude that:

$$\omega_n^2 = \frac{k}{m} \tag{5.6}$$

This expression is the same as the one derived from the manipulation of the equation of equilibrium presented in Sect. 3.2.

In order to estimate the natural frequency of the structure, when using the Rayleigh energy method, a common procedure is to first assume a static deflection of the structure with concentrated load at the mass points. Based on this deflection, the kinetic and potential energy are calculated. By equating the two energy terms, the natural frequency can be calculated.

Consider the physical realization of a monotower with varied stiffness and mass as shown in Fig. 5.1. First, by assuming an approximate vibration mode shape $X(z)$, the deflection of the beam at position z and time t can be written as:

$$x(z,t) = X(z)\sin(\omega t) \tag{5.7}$$

The maximum kinetic energy T_{max} is:

$$\begin{aligned}
T_{\max} &= \frac{1}{2}\int_0^H \rho(z)A(z)\left(\frac{dx}{dt}\right)^2 dz + \frac{1}{2}\sum_{i=1}^N \left[m_i(\frac{dx}{dt})^2_{z=z_n}\right] \\
&= \frac{\omega^2}{2}\left[\int_0^H \rho(z)A(z)X^2(z)dz + \sum_{i=1}^N m_i X^2(z_n)\right]
\end{aligned} \tag{5.8}$$

The maximum potential energy V_{max} is:

$$V_{\max} = \frac{1}{2}\int_0^H E(z)I(z)\left(\frac{d^2X}{dz^2}\right)^2 dz \tag{5.9}$$

where $\rho(z)$ and $A(z)$ are the density and cross-section area in location z.

5 Solving Eigenproblem for Continuous Systems

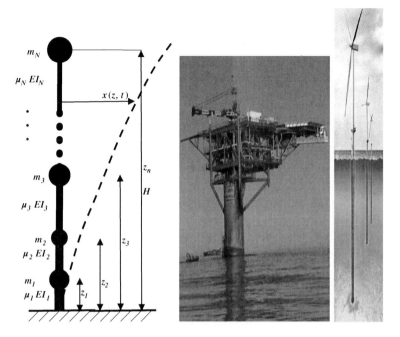

Fig. 5.1 The physical modeling (*left*) of a monotower and monopile structures for offshore wind turbines

To obtain the natural frequency, equate T_{max} and V_{max}, which gives:

$$\omega^2 = \frac{\int_0^H E(z)I(z)\left(\frac{d^2X}{dz^2}\right)^2 dz}{\int_0^H \rho(z)A(z)X^2(z)dz + \sum_{i=1}^N m_i X^2(z_n)} \quad (5.10)$$

By reviewing the equations above, it is noticed that, in the Rayleigh energy method, an important step in obtaining the natural frequency is to assume a reasonable vibration mode shape $X(z)$. The mode shape is normally chosen based on the displacement shape of free vibrations due to the inertial forces, and is proportional to the mass distribution and to the displacement amplitude. In principle, any shape that satisfies the geometric boundary conditions of the structure can be chosen.

The Rayleigh energy method can give the exact solutions provided an exact mode shape is given. However, note that if the exact mode shape is not known, its determination involves the solution of the vibration problem by the classical method. If the classical solution is available, the natural frequency is already included in it, and the adoption of the Rayleigh energy method is then not necessary [146]. Any shape other than the realistic vibration shape would require the action of additional external constraints to maintain the equilibrium, which would stiffen the system (adding strain energy) and subsequently increase the calculated frequency [161]. Therefore, in practice, this method does not provide an exact solution. For example, for a cantilever beam shown in Fig. 4.3, compared to a simple parabolic mode shape curve ($\delta(x) = cx^2$, where c is a constant), if one were to calculate the deflection curve on the basis of the dynamic load ($\omega^2 \mu c x^2$), the error between the calculated natural frequency and the exact one $\left(f_n = \sqrt{\frac{3EI}{L^3 \cdot 0.24 m_{beam}}}\right)$ decreases from 27.0 % to 0.2 % [18], which indicates that the assumption of a simple parabolic mode shape is far from reality. However, if the assumed mode shape curve is reasonable, the calculated eigenvalue is not very sensitive to the approximation of the mode shape and would be accurate enough from an engineering point of view. This is mainly because the vibration is stationary when perturbed around any of the actual system eigenvectors [180].

For a cantilever beam dominated by bending, as shown in Fig. 5.1, one may assume a unit vibration mode shape (deflection shape) of one of the following two:

$$X(z) = \left(1 - \cos\left(\frac{\pi z}{2H}\right)\right) \tag{5.11}$$

$$X(z) = (3Hz^2 - z^3) \tag{5.12}$$

Both the equations above satisfy the boundary conditions of the free tip and fixed bottom end, at least geometrically. We also have:

$$M(z = H) = EI \frac{d^2 X(z = L)}{dz^2} = 0 \tag{5.13}$$

$$X(z = 0) = \frac{dX(z = 0)}{dz} = 0 \tag{5.14}$$

In addition, it is noted that the mode shape assumption gives a non-zero shear force at the tip, which is important as the concentrated mass m_N at the tip induces the inertia force:

$$V_{shear}(z = H) = EI \frac{d^3 X(z = L)}{dz^3} \tag{5.15}$$

The first assumption of the mode shape expressed in Eq. (5.11) gives the natural frequency:

$$\omega_n^2 = \frac{\frac{\pi^4}{16H^4} \int_0^H E(z)I(z) \cos^2\left(\frac{\pi z}{2H}\right) dz}{\int_0^H \rho(z)A(z)\left[1 - \cos\left(\frac{\pi z}{2H}\right)\right]^2 dz + \sum_{i=1}^N m_i \left[1 - \cos\left(\frac{\pi z_i}{2H}\right)\right]^2} \tag{5.16}$$

The second assumption of the mode shape Eq. (5.12) gives the natural frequency:

$$\omega_n^2 = \frac{36 \int_0^H E(z)I(z)(H - z)^2 dz}{\int_0^H \rho(z)A(z)\left[z^4(3H - z)^2\right] dz + \sum_{i=1}^N \left[m_i(3Hz_i^2 - z_i^3)^2\right]} \tag{5.17}$$

The results from the two equations above should be approximately the same.

If the tip of the beam is also restrained (such as is the case for the legs of a jackup, the top parts of which are connected to the deck hull with certain fixtures (Fig. 5.2)), a judicious choice of vibration mode can be:

$$X(z) = \left(1 - \cos\left(\frac{\pi z}{H}\right)\right) \tag{5.18}$$

This satisfies the boundary conditions with fixed bottom and non-zero tip restraints.

If both ends of the beam are fixed, the vibration mode shape can be assumed as:

$$X(z) = \frac{1}{2}\left(1 - \cos\left(\frac{2\pi z}{H}\right)\right) \tag{5.19}$$

Compared to the Dunkerley method (Sect. 4.2.3), which is limited to positive definite systems with lumped masses (Sect. 13.6) and yields lower-bound solutions, the Rayleigh energy method applies equally well to both discrete and continuous systems [181]. Also, as mentioned before, it normally gives an upper-bound solution, i.e., the natural frequency obtained from the Rayleigh energy method leads to an overestimate of the exact value.

Example: Siri platform is a three-leg jackup structure located in the North Sea, which is shown in Fig. 5.2. The legs are made of steel (Young's modulus E is 2.1×10^{11} Pa and density is $\rho = 7,850$ kg/m^3) and have a cross-section of $\phi 3500 \times 80$ mm ($2r \times t$). The effective height/length of the leg H is 86 m and the vertical distance d between the water level and the bottom of the legs is 62 m. All three legs are submerged. Calculate the natural period of the jackup structure with regard to global bending. The total mass (operational weight) of the deck hull m_{deck} is 11,600 tons. The density of sea water $\rho_{seawater}$ is 1,025 kg/m^3.

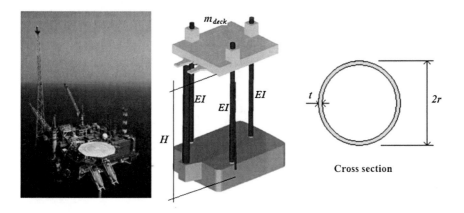

Fig. 5.2 Siri jackup structure with three legs (courtesy of Dong Energy)

Solution: using the Rayleigh energy method, by equating the maximum kinetic energy to the maximum potential energy of the system, the natural period can be calculated.

The three legs are only restrained at the bottom and connected at the deck level. Therefore, the resultant bending stiffness of the three legs together can be calculated as the sum of stiffness contribution from each individual leg. First assume a reasonable mode shape as $X(z) = (1 - \cos(\frac{\pi z}{H}))$, which satisfies the actual boundary conditions at both ends of the legs, as mentioned before. The maximum kinetic energy T_{max} is:

$$T_{max} = \frac{1}{2} \int_0^d 3\rho(z)_{submerg} A(z) \left(\frac{dx}{dt}\right)^2 dz + \frac{1}{2} \int_d^H 3\rho(z)A(z) \left(\frac{dx}{dt}\right)^2 dz$$
$$+ \frac{1}{2} m_d \left(\frac{dx}{dt}\right)^2_{Z=H}$$
$$= \frac{1}{2} \int_0^d 3\overline{m}_{submerg} \left(\frac{dx}{dt}\right)^2 dz + \frac{1}{2} \int_d^H 3\overline{m}_{leg} \left(\frac{dx}{dt}\right)^2 dz$$
$$+ \frac{1}{2} m_{deck} \left(\frac{dx}{dt}\right)^2_{Z=H}$$

The maximum potential energy V_{max} is:

$$V_{max} = \frac{1}{2} \int_0^H 3EI \left(\frac{d^2X}{dz^2}\right)^2 dz - \frac{1}{2} m_{deck} g \int_0^H \left(\frac{d^2X}{dz^2}\right)^2 dz$$

where $\overline{m}_{leg}$ and $\overline{m}_{submerg}$ are the mass per unit length for legs that are not submerged and legs that are submerged (flooded part of legs), respectively.

$$\overline{m}_{leg} = \rho \cdot 2\pi r t = 7850\text{kg/m}^3 \times 0.88\text{m}^2 = 6905 \text{ kg/m}$$

For flooded members the added mass is modeled as twice the mass of the water with the volume of the structure members, i.e., the added mass coefficient $C_M = 2.0$:

$$\overline{m}_{submerg} = C_M \pi r^2 \rho_{seawater} + \overline{m}_{leg} = 2 \times 9862 \text{ kg/m} + 6905 \text{ kg/m}$$
$$= 26628 \text{ kg/m}$$

By equating T_{max} and V_{max}, the natural angular frequency is calculated as:

$$\omega_n = \sqrt{\frac{\pi^2 \left(\frac{3\pi^2 EI}{H^2} - m_{deck} g\right)/(8H)}{m_{deck} + 3 \left(\overline{m}_{submerg} \lambda + \overline{m}_{leg} \chi\right)}}$$

5 Solving Eigenproblem for Continuous Systems

where λ and χ are two parameters related to the length for submerged and un-submerged parts of legs:

$$\lambda = 0.375d - \frac{H}{2\pi}\sin\left(\frac{\pi d}{H}\right) + \frac{H}{16\pi}\sin\left(\frac{2\pi d}{H}\right)$$

$$\chi = 0.375(H - d) + \frac{H}{2\pi}\sin\left(\frac{\pi d}{H}\right) - \frac{H}{16\pi}\sin\left(\frac{2\pi d}{H}\right)$$

The calculated angular natural frequency is therefore:

$$\omega_n = 1.02 \text{ rad/s}$$

The calculated natural frequency is:

$$f_n = 0.16 \text{ Hz}$$

The natural period is then:

$$T_n = \frac{2\pi}{\omega_n} = 6.14 \text{ s}$$

The calculated natural frequency is 5 % higher than the measured one (0.154 Hz) from a modal testing, which is accurate enough from an engineering point of view. Possible reasons for the errors are that the legs' bases in reality have some degrees of flexibility, while in the current example they are assumed to be rigid. Also, the connection between the topside and three legs are not perfectly rigid, as is assumed. Furthermore, the weight from marine growth and hydrodynamic damping are not accounted for. In addition, in reality the deviation between the center of gravity of the topside and the stiffness center may induce slight torsional vibration in the actual measurement, which cannot be captured based on the current assumption. In addition, the Rayleigh energy method is an upper-bound approach due to the approximation of the mode shape. Finally, the actual condition during the modal testing may differ to some extent from the assumed condition in the current hand calculation example.

It is noticed from the natural frequency equation above that if the weight of the deck $(m_{deck}g)$ equals the Euler buckling load of the three legs, $\frac{3\pi^2 EI}{H^2}$, the natural frequency becomes zero. This phenomenon is called stiffness softening, which will be introduced in Sect. 6.3. Fortunately, the current deck weight is only around 10 % of the Euler buckling load.

At this point, readers have already learned four approaches to finding the natural frequency of a vibrating system:

1. Using Newton's second law (Sect. 3.2): either write the equilibrium equation of motions and reduce it to a standard form or from the static deflection δ.
2. Using Hamilton's principle (Lagrange's equation, Sect. 2.3) to establish the equilibrium equation of motions.
3. Using the conservation of energy (Sect. 4.2.2 and in this chapter): either realize the equivalent mass and stiffness, or use $V_{max} = T_{max}$, or write $\frac{dV}{dt} + \frac{dT}{dt} = 0$.
4. For a continuous system, the classical method (Sect. 4.2.1) can also be used by assuming a reasonable mode shape and boundary conditions, this enables the differential equations to be solved for obtaining the eigenfrequencies.

Chapter 6
Vibration and Buckling Under Axial Loading

6.1 Vibration Versus Buckling

When a beam deforms transversely, its cross-sections suffer both translations and rotations in such a manner that if axial loads are acting, they input energy to the beam's potential energy. The presence of axial loads on a structure tends to increase (in case of tension) or decrease (in case of compression) the stiffness of a structure. This effect is called stress stiffening/softening. Consequently, the change of the stiffness will further change the eigenfrequencies of the structure. In case of compression load, if the load magnitude is large enough, dynamic buckling can occur.

The eigenfrequencies and mode shapes of a structure under axial loads can be calculated in the same manner as for one without axial load effects.

$$m\ddot{x} + c\dot{x} + (k - k_N)x = 0 \qquad (6.1)$$

where $k_N = \lambda k$ is the nonlinear strain (geometric or initial stress) stiffness, λ is the eigenvalue for the linearized buckling solutions.

By neglecting the damping term, the equation for calculating the eigenfrequency under combined elastic stiffness k and nonlinear strain (geometric or initial stress) stiffness k_N is:

$$\left|(k - k_N) - \omega^2 m\right| = 0 \qquad (6.2)$$

We first study the relationship between vibration and buckling. The formulation of linearized buckling and vibration generalized eigenproblem can be expressed as:

$$k_N \varphi = \lambda k \varphi \qquad (6.3)$$

$$k\phi = \omega^2 m\phi \qquad (6.4)$$

where φ is the eigenvector for the linearized buckling solutions; λ is the eigenvalue for both the buckling eigenproblem and ϕ is the eigenvector for the vibration eigenproblem solutions.

J. Jia, *Essentials of Applied Dynamic Analysis*, Risk Engineering,
DOI: 10.1007/978-3-642-37003-8_6, © Springer-Verlag Berlin Heidelberg 2014

It is known that the buckling eigenproblem is only related to the stiffness of the structure, while the dynamic eigenproblem has a relation with both the stiffness and mass distribution of structures. Both the dynamic eigenanalysis and buckling analysis are eigenorientated: the buckling takes place when, as a result of subtracting stress stiffness induced by axial load from elastic stiffness $(k - k_N)$, the resultant structure stiffness drops to zero; and the vibration modes occur when, as a result of subtracting inertia stiffness from the initial elastic stiffness $(k - \omega^2 m)$, the natural frequency of a structure decreases with the increase of the compressive stress in the structure; when the frequency reaches zero (the inertial force term $m\ddot{x}$ vanishes), the corresponding load magnitude equals the buckling load [186]. In practice, normally, only the first buckling load and shape are of real significance, because the system would have failed when the load exceeds the first buckling load [161]. However, vibration eigenfrequencies above the first one can also be important.

6.2 Vibration and Buckling Under Harmonic Axial Loads

Under harmonic axial loading, a range of different buckling loads can be defined. Consider the spring-mass-damper system shown in Fig. 6.1 that is subjected to harmonic external forces with an amplitude of F_0 and an angular frequency of Ω. The governing linear differential equation of motions for this system can be written as:

$$m\ddot{x}(t) + c\dot{x}(t) + (k - k_N(t))x(t) = F_0 \sin(\Omega t) \tag{6.5}$$

The steady-state responses can be expressed as:

$$x(t) = X_0 \sin(\Omega t) \tag{6.6}$$

Combining the two equations above and, by neglecting the damping and dividing both sides of the equation above by m, we have:

$$(k - k_N)X_0 - \Omega^2 m X_0 = F_0 \tag{6.7}$$

Inserting $k_N = \lambda k$ into the equation above and rearrange it, one obtains:

$$(k - \Omega^2 m)X_0 - \lambda k X_0 = F_0 \tag{6.8}$$

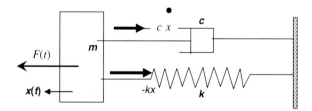

Fig. 6.1 An SDOF spring-mass-damper system under an external force $F(t)$

6.2 Vibration and Buckling Under Harmonic Axial Loads

The first item $(k - \Omega^2 m)$ in the equation above is called the dynamic stiffness.

If the excitation amplitude is approaching zero, the buckling can occur under the condition expressed in the eigenvalue equation:

$$\left|(k - \omega^2 m) - \lambda k\right| = 0 \tag{6.9}$$

An infinite number of combinations of vibration frequency ω and buckling load λ satisfy the equation above [161].

In summary, under a buckling load λ, one can calculate the vibration eigen-frequencies using Eq. (6.2). For any given frequency of vibration ω, the corresponding buckling load can be obtained using Eq. (6.9).

6.3 Eigenvalues Under the Influence of Axial Loads

Consider a simply supported beam under a compressive force F at its tip as shown in Fig. 6.2. Assuming that the beam, with a mass per unit length of μ, is vibrating transversely (along X direction) only due to its own mass inertia loads without contribution from other mass/inertia sources, and the beam only experiences small deformation, the differential equation of the beam's deflection curve can be expressed as:

$$\frac{\partial^2}{\partial z^2}\left(EI \frac{\partial^2 x}{\partial z^2}\right) - F \frac{\partial^2 x}{\partial z^2} = -\mu \frac{\partial^2 x}{\partial t^2} \tag{6.10}$$

It is noticed that the discrete models introduced in Chaps. 2, 3 and Sect. 4.2.2 only result in an ordinary differential equation. However, the beam modeled in Fig. 6.2 is a continuous system, and both the kinematical relation terms (left hand side) and the inertia term (right hand side) need to appear in the equilibrium equation simultaneously, leading to a partial differential equation, which is more difficult to solve.

We divide the above equation by μ:

$$\frac{EI}{\mu} \frac{\partial^4 x}{\partial z^4} - \frac{F}{\mu} \frac{\partial^2 x}{\partial z^2} + \frac{\partial^2 x}{\partial t^2} = 0 \tag{6.11}$$

By assuming an approximate vibration mode shape $X(z)$, the deflection of the beam at position z and time t is $x(z, t) = X(z) \sin(\omega t)$. We then have:

$$-\frac{EI}{\mu} \frac{d^4(X(z))}{dz^4} + \frac{F}{\mu} \frac{d^2(X(z))}{dz^2} - \omega^2 X(z) = 0 \tag{6.12}$$

It is noted that the equation above is essentially a representation of the modal pattern. Multiplying both sides of the equation with $-\mu/EI$ gives:

$$\frac{d^4(X(z))}{dz^4} - \frac{F}{EI} \frac{d^2(X(z))}{dz^2} + \frac{\mu\omega^2}{EI} X(z) = 0 \tag{6.13}$$

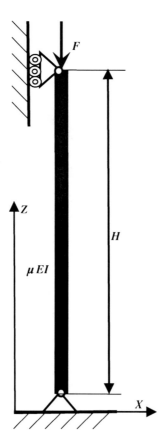

Fig. 6.2 A simply supported beam loaded with axial force F

The equation above can be rewritten in a standard differential equation form:

$$\frac{d^4(X(z))}{dz^4} - \alpha^2 \frac{d^2(X(z))}{dz^2} + \beta^4 X(z) = 0 \tag{6.14}$$

where $\alpha^2 = \frac{F}{EI}$ and $\beta^4 = \frac{\mu \omega^2}{EI}$.

The solution is:

$$X(z) = \psi_1 \sinh(r_1 z) + \psi_2 \cosh(r_1 z) + \psi_3 \sin(r_2 z) + \psi_4 \cos(r_2 z) \tag{6.15}$$

where $r_1 = \sqrt{\frac{\alpha^2}{2} + \sqrt{\frac{\alpha^2}{4} + \beta^4}}$ and $r_2 = \sqrt{-\frac{\alpha^2}{2} + \sqrt{\frac{\alpha^2}{4} + \beta^4}}$

For the simply supported beam, the support conditions at both ends have a zero translations and bending moments, i.e., $X(z = 0) = 0, X(z = H) = 0$, $\frac{d^2(X(z=0))}{dz^2} = 0$, and $\frac{d^2(X(z=H))}{dz^2} = 0$.

6.3 Eigenvalues Under the Influence of Axial Loads

Table 6.1 Euler buckling (critical) load for beams with various support conditions

Support conditions	Buckling (critical) load P_e
Clamp-free (cantilever beam)	$\dfrac{0.25\pi^2 EI}{H^2}$
Simply supported at both ends	$\dfrac{\pi^2 EI}{H^2}$
Clamp-simply supported (propped cantilever)	$\dfrac{2.048\pi^2 EI}{H^2}$
Clamp-clamp	$\dfrac{4\pi^2 EI}{H^2}$

The support conditions above require that $\sin(r_2 H) = 0$, which can be satisfied if:

$$r_2 H = i\pi \tag{6.16}$$

where i is any positive integer.

Therefore, we have:

$$H\sqrt{-\frac{\alpha^2}{2} + \sqrt{\frac{\alpha^2}{4} + \frac{\mu\omega^2}{EI}}} = i\pi \tag{6.17}$$

The solution of the equation above is:

$$\omega_i = \sqrt{\frac{EI}{\mu}\frac{i^2\pi^2}{H^2}}\sqrt{1 + \frac{FH^2}{i^2\pi^2 EI}} \tag{6.18}$$

Note that, as the Euler buckling (critical) load for a simply supported beam is $P_e = \frac{\pi^2 EI}{H^2}$, the equation above can be rewritten as:

$$\omega_i = \omega_{i,noF} \cdot \sqrt{1 + \frac{F}{n^2 \cdot P_e}} \tag{6.19}$$

The above equation is called Galef's formula [187], which illustrates the relationship between the eigenfrequency ω_i under axial force F ($F < 0$ for compressive force and $F > 0$ for tension force) and the eigenfrequency of the uncompressed/untensioned beam $\omega_{i,noF}$. Essentially, this reflects a change of initial stiffness under axial load, i.e., the compressive axial force decreases while the tensile one increases the eigenfrequency of the beam.

The Euler buckling (critical) load P_e for beams with various support conditions is listed in Table 6.1.

Figure 6.3 shows the eigenfrequency ratio due to the presence of the axial compressive force. It is noticed that, due to the moderating influence of the factor $\frac{1}{i^2}$, the effects of the axial force are much more pronounced for the natural frequency (first eigenfrequency) than the higher-order eigenfrequencies. The effects

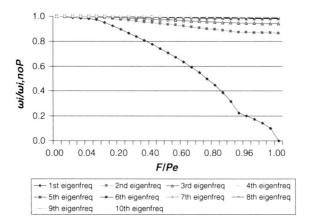

Fig. 6.3 The eigenfrequency ratio due to variation of the axial compressive force P, Pe: Euler critical buckling load, $\omega_i/\omega_{i\mathrm{noP}}$: the ratio of the eigenfrequency between that with and without the presence of the axial compressive force

are also more significant for slender members (such as offshore risers) due to its relatively low P_e.

It should also be noted that the support conditions of a beam limit the applicability of Galef's formula: for the fundamental eigenfrequency ($i = 1$), Galef's formula is exact for pinned–pinned, sliding-pinned and sliding–sliding beams, approximate for sliding-free, clamped-free (the case for the current analysis), clamped-pinned, clamped–clamped and clamped-sliding beams, but not valid for pinned-free and free–free beams. However, for the third and higher modes of vibrations, Galef's formula can be applied for all types of support conditions [188].

The effects of axial load can also be studied by using the Rayleigh energy method, in which the strain energy term due to the presence of effective axial force is added to the potential energy. This is already illustrated in the example in Chap. 5.

By calculating with or without gravity effects, Table 6.2 shows the comparison of the first ten eigenperiods of an offshore jacket structure (Fig. 6.4). Some representative mode shapes are also shown in Fig. 6.4. The axial load's effects can be observed in almost all eigenmodes. Especially in the lower order of the eigenmodes, the eigenperiods for the case with gravity effects are higher than their counterpart without gravity effects. The higher the order of the eigenmode, the smaller is the eigenperiod difference between the two cases. One exceptional case is that for the ninth eigenmode, for which the eigenperiod with gravity effect is even lower than that without gravity effect. Another exceptional case is that a local vertical vibration at the bottom horizontal frames occurs at a lower eigenperiod (1.014 s) for the case with gravity effect than that without gravity effects (1.152 s). This is because, for a structure with large degrees of freedom, even though the gravity can induce global compression, it may also introduce tension at some local members. This is especially applied to the higher order of eigenmodes, where

6.3 Eigenvalues Under the Influence of Axial Loads

Table 6.2 The first ten eigenperiods of a jacket-topside-flare tower structure (hydrodynamic added masses are included) for the cases with gravity and without gravity effects, in calm sea conditions

Mode number	Eigenperiod (s)		
	With gravity (axial force)	Without gravity	Diff. (%)
1	4.173	4.120	1.3
2	4.115	4.065	1.2
3	2.453	2.442	0.4
4	1.203	1.194	0.7
5	1.193	1.187	0.5
	NA	1.152	NA
6	1.090	1.086	0.4
7	1.014	NA	NA
8	0.990	0.987	0.3
9	0.914	0.939	−2.7
10	0.902	0.901	0.1

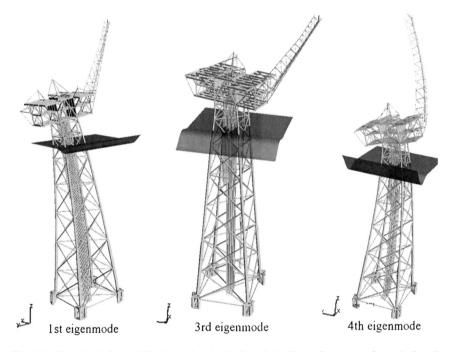

Fig. 6.4 The global flexural (*first*), torsional (*third*) and significant flare tower flexural (*fourth*) vibration eigenmodes

complex local vibration modes rather than global vibration modes are exhibited. The slight tuning of the local member stiffness can cause the eigenfrequencies of local vibration mode to increase (due to tension), decrease (due to compression) or even cross each other.

Based on Galef's work, Bokaian [189] investigated the applicability of Galef's formula under a set of support conditions and also extended his study to tension loads [190]. Using Galef's formula, Shaker [191] first investigated relevant practical problems in aerospace structures: a vibrating beam with arbitrary boundary conditions, a cantilever beam with tip mass under constant axial loads, and a cantilever beam with tip mass under axial loads applied on the tip directed to the root. In addition, he extended the analysis to the tension loads as well. Virgin [180, 192, 193] carried out a series of research work to investigate the dynamic behavior of axially loaded structures. He experimentally illustrated that a beam with an upward orientation experiences de-stiffening effects and a beam with a downward orientation is stiffened by the weight of the beam through the development of tension stress in the beam. By including the effects of large displacement equilibrium paths through a nonlinear moment–curvature relationship, Virgin and Plaut [193] further presented a formulation to obtain the eigenpairs of vertical cantilevers. Several pieces of research discuss the use of measured vibration frequencies obtained from the non-destructive modal testing to determine the approximate buckling loads [194–196]. All these researches conclude that changes in the measured vibration frequencies during increasing loading can be used to predict the buckling of a structure. With the aim of developing robust low-dimensional models, Mazzilli and his co-workers [197] used a nonlinear normal mode method to develop a rigorous derivation of nonlinear equations, which governs the dynamics of an axially loaded beam. They also applied the equations for a study of dynamic characteristics of offshore risers. Readers interested in this topic may read the relevant references cited above.

Chapter 7
Eigenfrequencies of Non-uniform Beams, Shallow and Deep Foundations

7.1 Non-uniform Beams

In engineering practice, non-uniform beams are sometimes constructed in the form of a stepped beam, for which the mass (m_i) and stiffness (E_iI_i) within each constant cross-section segment can be assumed to be constant. A concentrated mass M is located at the tip of the beam. With the mode shape assumed by Eq. (5.11), the natural frequency can be calculated as:

$$\omega_n^2 = \frac{\pi^4}{16H^4} \frac{\sum_{i=1}^{N} E_iI_il_i \cos^2\left(\frac{\pi z_i}{2H}\right)}{\sum_{i=1}^{N} \left\{ m_il_i\left[1 - \cos\left(\frac{i}{2H}\right)\right]^2 \right\} + M} \tag{7.1}$$

where l_i is the length for segment i.

Furthermore, for a realistic monotower structure as shown in Fig. 5.1, it has foundation stiffness at its base, as shown in Fig. 7.1.

The natural frequency of the rigid beam with only foundation's rotation stiffness k_r and tip mass M presented is:

$$\omega_r^2 = \frac{k_r}{\left(\int_0^H \rho(z)A(z)z^2dz + M \right)} \tag{7.2}$$

The natural frequency of the rigid beam (bending stiffness is infinitely high) with only foundation's lateral stiffness k_h and tip mass M presented is:

$$\omega_h^2 = \frac{k_l}{\left(\int_0^H \rho(z)A(z)dz + M \right)} \tag{7.3}$$

The stiffness of the system by including the flexibility of the beam and foundation can be taken as the individual beam stiffness (k_b), the foundation's lateral stiffness (k_h) and rotation stiffness (k_r) working in series. The natural frequency is then [182]:

J. Jia, *Essentials of Applied Dynamic Analysis*, Risk Engineering,
DOI: 10.1007/978-3-642-37003-8_7, © Springer-Verlag Berlin Heidelberg 2014

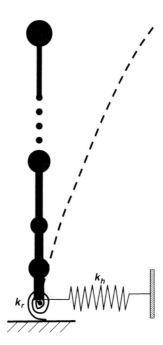

Fig. 7.1 The physical modeling of a non-uniform beam with lateral (k_h) and rotation (k_r) stiffness at its bottom to resemble the foundation stiffness

$$\omega_{total} = \frac{\text{resultant stiffness of the entire system}}{\text{sum of the equivalent stiffness for springs in series}}$$

$$= \frac{3\sum_{i=1}^{N} E_i I_i l_i \cos^2(\frac{\pi z_i}{2H})/H}{\left(\sum_{i=1}^{N}\left\{m_i l_i \left[1 - \cos(\frac{\pi z_i}{2H})\right]^2\right\} + M\right)H^3 \cdot \left[\frac{H^3}{\frac{\pi^4}{16}\sum_{i=1}^{N} E_i I_i l_i \cos^2(\frac{\pi z_i}{2H})/H} + \frac{H^2}{k_r} + \frac{1}{k_h}\right]} \quad (7.4)$$

$$= \frac{3(EI)_{eq}}{(M + m_{eq}H)H^3\left[\frac{48}{\pi^4} + C_{foundation}\right]}$$

where $(EI)_{eq} = \sum_{i=1}^{N} E_i I_i l_i \cos^2(\frac{\pi z_i}{2H})/H$

$$m_{eq} = \sum_{i=1}^{N}\left\{m_i l_i \left[1 - \cos(\frac{\pi z_i}{2H})\right]^2\right\}/H$$

$$C_{foundation} = \frac{3(EI)_{eq}}{K_{eq}H}$$

$$K_{eq} = \frac{k_r k_h H^2}{k_r + k_h H^2}$$

7.1 Non-uniform Beams

The term $C_{foundation}$ reflects the flexibility of the foundation, which varies between 0, for a very stiff foundation, and 0.5 for a reasonably flexible foundation.

7.2 Shallow and Deep Foundations

Foundations are critical components for ensuring stability, transferring the load from the upper structure down to base soils or rocks. There are mainly two types of foundations, shallow and deep foundation. A shallow foundation is used when the earth directly beneath a structure has sufficient capacity to sustain the load transferred from the upper structure. Typical examples of shallow foundations are footings (spread and combined), soil retaining structures (retaining walls, sheet piles, excavations and reinforced earth) and the foundation of a gravity-based structure shown in Fig. 7.2. A deep foundation is used when the soil (such as clay) near the ground surface does not have sufficient capacity to bear the weight. Typical examples of deep foundation are piles and shafts.

For a shallow foundation such as that shown in Fig. 7.2, the physical modeling illustrated in Fig. 7.1 can be applied. Here the essential task is to calculate lateral (k_h) and rotational (k_r) stiffness. By separating the lateral (sliding) and rotational motions as two independent components, Nataraja and Kirk [183] present the lateral and rotational stiffness as:

$$k_h = \frac{8G_s}{2-\upsilon}\left(1 - 0.05\omega\sqrt{\frac{\rho_s}{G_s}}\right)R \qquad (7.5)$$

$$k_r = \frac{8G_s}{3(1-\upsilon)}\left(1 - 0.215\omega R\sqrt{\frac{\rho_s}{G_s}}\right)R^3 \qquad (7.6)$$

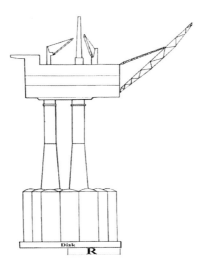

Fig. 7.2 A GBS structure with a shallow foundation directly resting on the soil. (*Courtesy of Aker Solutions*)

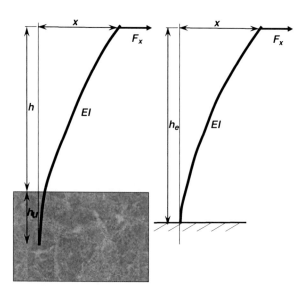

Fig. 7.3 Modeling of pile-soil stiffness from foundation-soil system (*left*) to its equivalent counterpart with a rigid foundation

where G_s, v and ρ_s are the shear modulus, Poisson's ratio and density of the soil, respectively.

The relevant damping at the corresponding degrees-of-freedom is:

$$c_h = \frac{8}{2-v}\sqrt{\rho_s G_s}\left(0.67 + 0.02\omega R \sqrt{\frac{\rho_s}{G_s}}\right) R^2 \tag{7.7}$$

$$c_r = \frac{0.375}{1-v}\omega \rho_s R^5 \tag{7.8}$$

where ω is the angular frequency of the disk as shown in Fig. 7.2.

For pile foundations, the bending stiffness can simply be expressed as:

$$k = \frac{3EI}{h_e^3} \tag{7.9}$$

in which h_e is called the equivalent length (height). As shown in Fig. 7.3, it is a hypothetical pile fully fixed at the base, which gives the same horizontal deflection x under the same horizontal force F_x as the pile is inserted into flexible soil.

h_e can be approximated as [184]:

$$h_e = h_u \left[0.4 + 1.353\left(\frac{h}{h_u}\right) + 1.875\left(\frac{h}{h_u}\right)^2 + \left(\frac{h}{h_u}\right)^3\right]^{\frac{1}{3}} \tag{7.10}$$

where $h_u = \left(\frac{102.9EI}{\eta_0}\right)^{\frac{1}{5}}$, η_0 is the horizontal subgrade reaction constant. This ranges from 141 tons/m^3 for relatively loose to 1,201 tons/m^3 for dense sand [185].

Chapter 8
Deterministic and Stochastic Motions

> *Essentially, all models are wrong, but some are more useful than others.*
>
> Statistician George Edward Pelham Box 1987.

8.1 Category of Motions

Any motions or signals can be classified as either deterministic or random as shown in Fig. 8.1. The deterministic motions are those that can be exactly predicted at any time instant, such as the rotation of a propeller shaft.

In contrast, random motions are those whose instantaneous value cannot be predicted at any time instant or reproduced, while their essential features of process can be described using probabilistic concepts (Sect. 8.3 and Chap. 10). To understand these processes "one should conceptually think in terms of all time history records that could have occurred" [43]. Examples of random/stochastic processes are ocean wave height, wind speed variation for a year, or temperature variation in the future. The complexity of these natural phenomena are such that they can be modeled as random processes [24].

The choice of selecting a deterministic or random approach depends on the type of motions or loads, and this also influences the type of structural analysis methods chosen. Treating deterministic motions and structural responses is normally performed by a straightforward structural dynamic analysis in the time domain. While describing random motions and analyzing the structural responses under random motions requires a utilization of probability methods with certain type of random functions described statistically, this is a pillar for stochastic dynamic analysis.

In conclusion, rather than being an alternative approach to represent motions or loads, deterministic and stochastic modeling are two sides of the same coin, of which the former is used to represent the known motions; the latter one is to describe the missing/unknown information of motions. They are both useful for the prediction of structural responses.

J. Jia, *Essentials of Applied Dynamic Analysis*, Risk Engineering,
DOI: 10.1007/978-3-642-37003-8_8, © Springer-Verlag Berlin Heidelberg 2014

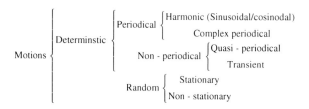

Fig. 8.1 Classification of dynamic motions and signals [44]

8.2 Deterministic Motions

As illustrated in Fig. 8.1, deterministic motions can be categorized as either periodical or non-periodical. The former are defined as motions that repeat themselves at regular time intervals, such as harmonic (sinusoal or cosisual) motions, triangular motions or complex periodical motions of an arbitrary shape. As illustrated in Fig. 8.2, all periodical motions repeat themselves with a period of T_p, and have discrete components at one (harmonic) or more (complex periodical) defined frequencies. Examples of periodical motions are the free vibrations of a bridge or a building structure.

Non-periodical motions, as their name implies, do not repeat themselves. This type of vibration occurs when a system has two or more incommensurable natural frequencies, or contains nonlinearities, or has time-variant parameters and/or unsteadiness of the excitations. Among non-periodical motions, quasi-periodical

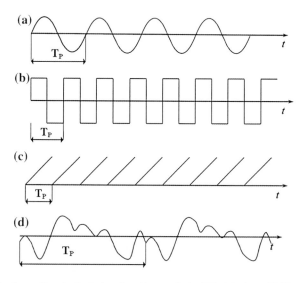

Fig. 8.2 Illustrations of periodical signals with a period of T_p. (**a**) sinusoidal/harmonic motion, (**b**) rectangular motion, (**c**) triangular motion, (**d**) complex periodic motion

8.2 Deterministic Motions

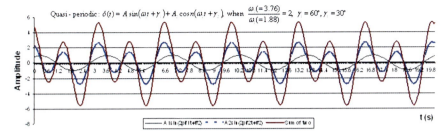

Fig. 8.3 The sum of sines and cosines motions never repeat themselves exactly

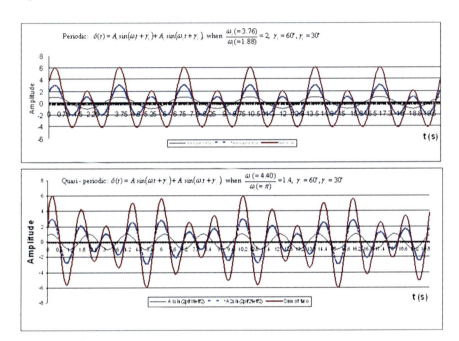

Fig. 8.4 The sum of two sine functions

motions (also called almost periodical motions) look like periodical motions but are not strictly speaking periodical if observed closely. For example, the sum of sines and cosines motions never repeat itself exactly as shown in Fig. 8.3, even if it looks like periodical motions. The sum of two or more sines is periodical only if the ratios of all pairs of frequencies are found to be rational numbers (integers) as shown in the upper (rational number) and lower (non-rational number) figure of Fig. 8.4. Quasi-periodical motions also have discrete components in the frequency domain, and are normally considered periodical motions with an infinitely long period. An example of such type of motions is the acoustic signal created by tapping a slightly asymmetric wine glass [44].

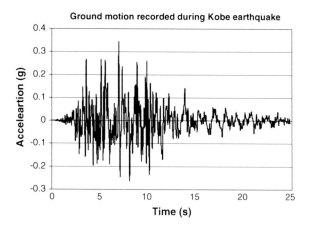

Fig. 8.5 The ground accelerations recorded at a station during Kobe earthquake, Japan, 17th of January, 1995

Transient motions have the characteristics of limited duration i.e., the motion $\delta(t) = 0$ for $t \rightarrow \pm\infty$ [44]. Examples are earthquake ground motions of a duration generally less than one minute (shown in Fig. 8.5); impact loading of ship colliding with an offshore platform structure; impact loading of dropped objects on a deck, etc.

8.3 Random/Stochastic Process

The random/stochastic process is essentially an extension from the random variables to the sequences of random variables. Mathematically, the quantity $X(t)$ can be defined as a random/stochastic process if $X(t)$ is a random variable for each value of stochastic/random process.

Analysis of random data is enhanced when the statistical properties of a random process remain constant for every realization of a certain event. Strictly speaking, the probability distribution of all physical motions/loadings/responses changes with time, i.e., their mean and variance are not constant. However, in the engineering world, one often encounters a wide class of problems where the type and intensity of the motions/loadings/responses vary rather slowly in comparison to the actual random fluctuations [45], i.e., the probability distribution does not change over the time intervals of interest [46]. Such random phenomena are suitably modeled by the stationary stochastic/random process.

8.3 Random/Stochastic Process

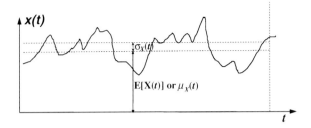

Fig. 8.6 Mean value (E[X(t)] or $\mu_X(t)$) and standard deviationStandard deviation ($\sigma_X(t)$) of a randomRandom process

Examples of stationary loadings and motions are sea wave (not wave impact loading) and wind loadings within a short term (in fact, wave and wind turbulence loads are non-stationary in the long term because their mean values are not constant over time), platform vibrations subjected to wave loadings, engine vibrations or random ground excitations on running vehicle tires due to imperfections of the road surface or railway tracks etc. It should also be mentioned that earthquake ground motions are in general a non-stationary process of short duration. Therefore, the mathematical treatment applicable for stationary processes may not be applied directly to the earthquake ground motions.

Before mathematically illustrating a stationary process, we first define two most important statistical properties: mean $\mu_X(t)$ and standard deviation $\sigma_X(t)$ (a measure of the dispersion or spread about the mean) of a random process, which are illustrated in Fig. 8.6 and expressed as:

$$\mu_X(t) = E[X(t)] = \lim_{T \to \infty} \frac{1}{T} \int_0^T X(t)dt \qquad (8.1)$$

$$\sigma_x(t) = \sqrt{\overline{[X(t) - \mu_X(t)]^2}} = \sqrt{\lim_{T \to \infty} \frac{1}{T} \int_0^T [X(t) - \mu_X(t)]^2 dt} \qquad (8.2)$$

where E(.) stands for the mean (expected) value.

The completely/strictly stationary process is warranted if the following four conditions are met:

1. The mean of the process is constant over time, i.e., $\mu_X(t) = \mu_X$
2. The variance of the process is constant over time, i.e., $\sigma_X^2(t) = \sigma_X^2$
3. The probability of the process is a function of time difference $(t_2 - t_1)$ and does not depend on individual times t_2 and t_1, i.e., $p(x_1, t_1; x_2, t_2) = p(x_1, t_1 + T; x_2, t_2 + T)$

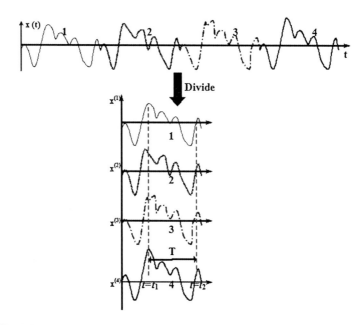

Fig. 8.7 Divide the time history into units of equal duration

4. The joint probability of the process at any time is identical to the joint probability of the same vaiable displaced with an arbitrary amount of T: i.e., $p(x_1, t_1; x_2, t_2; \ldots; x_n, t_n) = p(x_1, t_1 + T; x_2, t_2 + T; \ldots; x_n, t_n + T)$.

If the process only satisfies the conditions 1, 2, and 3 above, it is called weakly/simply stationary, i.e., both $E(X(t))$ and $E(X(t) X(t + T))$ are independent of time t and only dependent on T. Here E stands for the mean (expected) value. $E(X(t) X(t + T))$ is often referred to as autocorrelation function or autocorrelation, denoted $R_X(T)$.

The stationary process can also be elaborated by studying Fig. 8.7, which shows a long time history $X(t)$, if one cuts it into N pieces of samples ($X^{(1)}(t)$, $X^{(2)}(t)$,..., $X^{(N)}(t)$) with equal duration, and the duration is of a sufficient while also finite length to capture the essential characteristics of the motions/loadings/responses. Here, two dashed lines across the ensemble at arbitrary times t_1 and t_2 are used. By the collection of samples, each of which can be regarded as a result from a separate experiment, one can then create a random process $X(t)$ as:

$$X(t) = X^{(1)}(t) + X^{(2)}(t) + \cdots + X^{(N)}(t) \quad (8.3)$$

8.3 Random/Stochastic Process

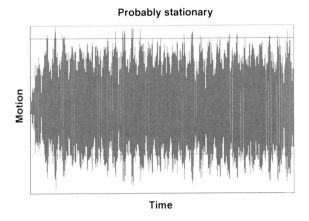

Fig. 8.8 An example of a "probably" stationary process

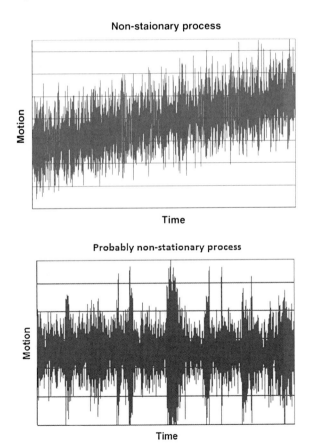

Fig. 8.9 Non-stationary process with varying mean (*upper*) and variance (*lower*). Note that the process in the lower figure could probably be stationary as well

The stationary process can be identified if the following two conditions are fulfilled:

1. The value of $\frac{1}{N}\sum_{i=1}^{N} X^i(t_1)$ should be approximately the same for any arbitrary time t_1, provided N is a rather large number.

2. At a preselected constant time interval T, $\frac{1}{N}\sum_{i=1}^{N} \left[X^i(t_1) \cdot X^i(t_2)\right]$ should be approximately the same for any value of t_1 and t_2, provided N is a rather large number.

The stationary process can be represented exactly by a sum of infinite number of sinusoidal functions with random phase angles. However, using a finite number of the functions for the summing can also give sufficient accuracy.

Figure 8.8 shows an artificial history of "probably" stationary motions. Figure 8.9 shows two examples of artificial non-stationary motions. By visualizing this figure, it is quite obvious that the motions in the upper figure show an increase of mean value, while the lower one shows that the variance of motions is neither stationary nor periodically stationary.

If the average calculated from each sample is the same as those of any other samples and is equal to the average of the ensemble $x(t)$, the random process is then ergodic. This characteristic allows for the use of only one single time series (one realization) to estimate all statistical parameters and characterize the random process. An ergodic process must be stationary; however, a stationary process is not necessarily ergodic. Many practical random processes that are nonergodic have both ergodic and nonergodic aspects. Nonergodic variation could easily dominate a problem if it is sensitive to small parameter variations or if the variations are large and not well controlled [46].

It should be noted that in reality, a strict identification of a stationary process is rather difficult. Therefore, judgment based on experiences is often needed so that a likely stationary process can be identified. Moreover, the identification of ergodicity is even more challenging and is often neglected in engineering practice, even if this is an important characteristic of a process in statistical investigation.

To transform most time series that are of interest in the field of engineering to their frequency domain counterpart, Fourier transformation can be used. It is based on the assumption that random signals can be represented by the sum of a number of sinusoids or wavelets, each with a specific amplitude, frequency and phase angle. This will be elaborated in Chap. 9. However, not every stationary process can be realized by a Fourier transformation. An example of this are processes with infinite variance.

Transient, complex periodical and non-periodical motions, which are of deterministic types, and random vibrations (Sect. 11.4), which are of a stochastic type, all do not have a single component in the frequency domain. For analysis dealing with those types of motions, Fourier analysis can be performed. It is based on the manipulation of loads and responses in the frequency domain with different amplitudes and corresponding frequencies. This will be introduced in Sects. 11.2, 11.3.3, and 11.4.

One can also treat the transient motions/excitations as the superposition of impulses of rather short duration, which will be discussed in Sect. 11.3.1.

Chapter 9
Time Domain to Frequency Domain: Spectrum Analysis

The concept of spectrum can be attributed to Isaac Newton, who, with the aid of a prism, discovered that sunlight can be decomposed into a spectrum of colors from red to violet in about 1700. This indicates that any light comprises numerous components of light of various colors (wave lengths). The earliest function most closely resembling the spectral density function was developed by Arthur Schuster, who investigated the presence of periodicities in meteorological [47], magnetic [48], and optical [49] phenomena. The spectrum provides a measure of the light's intensity varied with respect to its wavelength. This concept has been generalized to represent many physical phenomena by decomposing them into their individual components.

9.1 Fourier Spectrum

Invented by Baron Jean Baptiste-Joseph Fourier in 1807, but the subject of great skepticism from his contemporaries at that time, the Fourier transform has now become a major analysis method in the frequency domain across a wide range of engineering applications. It states that any periodical function $\delta(t)$ in the time domain, not necessarily harmonic, has an equivalent counterpart in the frequency domain, which can be represented by a convergent series of independent harmonic functions as a Fourier series:

$$\delta(t) = c_0 + \sum_{i=1}^{N} c_i \sin(\omega_i t + \gamma_i) \tag{9.1}$$

where:

c_0 is the average value of $\delta(t)$, $c_0 = \frac{1}{T_0} \int_0^{T_0} \delta(t)\, dt$

T_0 is the duration of the motions

c_i is the amplitude of the nth harmonic of Fourier series, $c_i = \sqrt{a_i^2 + b_i^2}$

J. Jia, *Essentials of Applied Dynamic Analysis*, Risk Engineering,
DOI: 10.1007/978-3-642-37003-8_9, © Springer-Verlag Berlin Heidelberg 2014

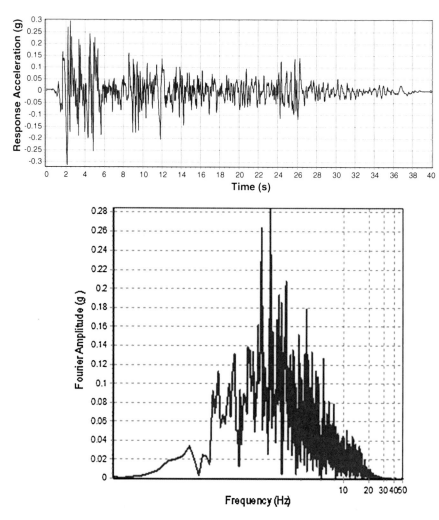

Fig. 9.1 Fourier amplitude spectrum (*lower*) of the strong ground motions (*upper*) recorded at Imperial Valley, California

$a_i = \frac{2}{T_0} \int_0^{T_0} \delta(t) \cos(\omega_i t) \, dt$ is the amplitude of cosinusoidal excitations

$b_i = \frac{2}{T_0} \int_0^{T_0} \delta(t) \sin(\omega_i t) \, dt$ is the amplitude of sinusoidal excitations

ω_i is the nth frequency of component, with the lowest one being $\omega_0 = \frac{2\pi}{T_0}$

γ_i is the phase angle, $\gamma_i = \tan^{-1}\left(\frac{a_i}{b_i}\right)$, which defines the stagger related to time origin, and controls the times at which the peaks of harmonic motions/loadings/responses occur, and influences the variation of $\delta(t)$ with time.

9.1 Fourier Spectrum

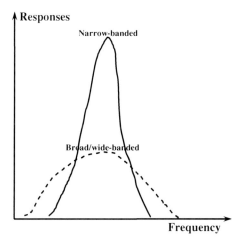

Fig. 9.2 Fourier amplitude spectrum for a narrow-banded and a broad-banded random process with both of their origins of ordinate at the mean value of each time series

Physically, the equation above is a representation of irregular records by the sum of an N sine waves of amplitudes c_i and frequency ω_i.

The Fourier series gives a complete description of motions since the motions can be recovered by the inverse Fourier transform.

For most motion records, such as sea wave elevations, by judicious choice of the datum level of the measurements, its average value (c_0) can be assumed to be zero. Equation (9.1) can then be reduced to:

$$\delta(t) = \sum_{i=1}^{N} c_i \sin(\omega_i t + \gamma_i) \tag{9.2}$$

Total energy is proportional to the average of the squares $\delta(t)$, which is the sum of the energy contents for each individual component at each frequency ω_i (Parseval theorem).

A plot of c_i versus ω_i from Eq. (9.2) is called a Fourier amplitude spectrum (normally referred to as a Fourier spectrum). To further explain the application of this concept, let's take the analysis of earthquake ground motions as an example. For a given earthquake and site, in order to obtain a complete picture of the strength of seismic ground motions (upper figure in Fig. 9.1) in each individual frequency or period, the Fourier spectrum is introduced as the Fourier amplitude (c_i) that varies with frequency or period, as shown in Fig. 9.1. From Fourier analysis of abundant strong ground motion time histories, it is found that the major motion contents are below the period of 30 s.

As will be presented in Chap. 10 and Fig. 10.4, narrow-banded time series have a dominant frequency, and it is typically the result of resonance or near-resonance responses, while a broad/wide-banded motions or excitations have a noticeable variety of frequencies. Since the mean values of the time series are constant and do not contribute to the Fourier amplitude in the frequency domain, the narrow- and

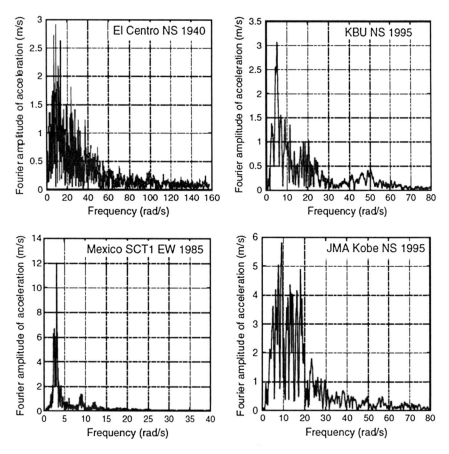

Fig. 9.3 Fourier amplitude spectrum of strong ground motions recorded at four different earthquake events [73]

broad-banded time series can be illustrated in the frequency domain using the Fourier amplitude spectrum as shown in Fig. 9.2.

Figure 9.3 shows the Fourier amplitude spectrum of the ground acceleration histories (Fig. 9.4) for four earthquake events of El Centro NS (Imperial Valley 1940), Kobe University NS (Hyogoken-Nanbu 1995), SCT1 EW (Mexico Michoacan 1985), JMA Kobe NS (Hyogoken-Nanbu 1995). It is clearly shown that, within the period range longer than 0.25 s, where the majority of seismic motion energy is concentrated, spectrum peaks appear at different frequencies. The Fourier amplitude spectrum for Mexico SCT1 EW record is comparatively narrow banded, with the majority of energy concentrated at a period of around 2.4 s. For the JMA Kobe NS record, the energy content is spread at a wide range of frequencies.

The ups and downs in a Fourier spectrum, for example for an earthquake ground motion record, can be smoothed and plotted in a logarithmic scale. The smoothed spectrum shown in Fig. 9.5 has a standard shape with the largest

9.1 Fourier Spectrum

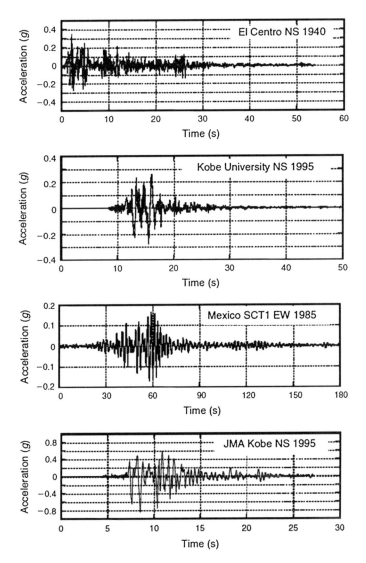

Fig. 9.4 Time history of strong ground accelerations recorded during the four different earthquake events [73]

acceleration over an immediate range of frequency. This immediate range of frequency is defined by its upper boundary frequency (cutoff frequency) f_{cutoff}, and the lower boundary frequency (corner frequency) f_{corner} as shown in Fig. 9.5.

In earthquake engineering, if a Fourier amplitude spectrum is used to represent ground motions, f_{corner} is the one above, in which earthquake radiation spectra are inversely proportional to the cube root of the seismic moment [74, 75] (which is the indication of the energy release due to the rupture of the faults during an earthquake,

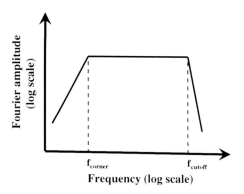

Fig. 9.5 Smoothed Fourier amplitude spectrum with corner and cutoff frequency logarithmic space [125]

see reference [23] for its definition). The ground motions at corner frequency are higher for large earthquakes than that for small ones. Below f_{corner}, the spectra are proportional to the seismic moment [23]. However, the characteristics of cutoff frequency are unfortunately much less clearly understood. Hanks [76] and Papageorgiou and Aki [77] indicate that the cutoff frequency relates to the near-site effects and source effects and can be regarded as constant for a given geographic region [78].

A plot of γ_i versus ω_i from Eq. (9.2) is called a Fourier phase spectrum. Different from the Fourier amplitude spectrum, the Fourier phase spectra from actual earthquake records do not have any standard shape [125].

9.2 Power Spectrum Density

In most cases, engineers are only interested in the absolute value of the Fourier amplitude instead of whether it is part of the sine or cosine series. This is because the absolute value provides the total amount of information contained at a given frequency. Since the square of the absolute value is considered to be the power of the signal, instead of the Fourier spectrum the motions can then be expressed in terms of power $P(\omega_i)$, defined as:

$$P(\omega_i) = \frac{1}{2} c_i^2 \qquad (9.3)$$

where c_i is the amplitude of the ith harmonic of the Fourier series.

Imagining that $\zeta(t)$ is voltage, the power dissipated across a 1 ohm resistor is then $[\zeta(t)]^2$, and the total power dissipated across the resistor is $\int_0^{T_0} [\zeta(t)]^2 dt$. By assuming that the total power of motions calculated from the sum of each individual frequency component (Parseval's theorem) equals that of the time domain, one reaches:

9.2 Power Spectrum Density

$$\sum_{n=1}^{+\infty} P(\omega_i) = \int_0^{T_0} [\zeta(t)]^2 dt = \frac{1}{\pi} \int_0^{\omega_N} c_i^2 \, d\omega = \frac{1}{\pi} \sum_{i=1}^{N} c_i^2 \Delta\omega_i \tag{9.4}$$

where $\omega_N = \dfrac{\pi}{\text{sample time interval } \Delta t \text{ over the time history}}$ is the highest frequency in the Fourier series, or Nyquist frequency, i.e., the frequency range beyond which the motion content cannot be accurately represented. In such a condition, a distorted Fourier spectrum called aliasing will be introduced. $\Delta\omega_i$ is half of the spacing between two adjacent harmonics ω_{i+1} and ω_{i-1}.

For an efficient (optimal) signal sampling, in order to extract valid frequency information, one must bear in mind that the sampling of the motion/loading/response signals must occur at a certain rate: (1) for a time record with the duration of T seconds, the lowest frequency component measurable is $\Delta\omega_{min} = \frac{2\pi}{T}$ or $\Delta f_{min} = \frac{1}{T}$. (2) The maximum observable frequency is inversely proportional to the time step, i.e., $\omega_{obs} = \dfrac{2\pi}{\text{sample time interval } \Delta t \text{ over the time history}}$, and the sampling rate must be at least twice the desired frequency (ω_{max} or f_{max}) to be measured, i.e., $\omega_{obs} > 2\omega_{max} = 2\omega_N$, where ω_N is the Nyquist frequency. With the two properties (1) and (2) above, the sampling parameters can be expressed as:

$$\omega_{max} = \omega_N = \frac{\pi}{\Delta t} \text{ or } f_{max} = \frac{1}{2\Delta t} \tag{9.5}$$

$$\Delta t = \frac{\pi}{\omega_{max}} = \frac{\pi}{\omega_N} \text{ or } \Delta t = \frac{1}{2f_{max}} \tag{9.6}$$

The description above is often referred to as the Shannon or Nyquist sampling theorem.

By dividing the total power in the equation above with the duration T_0, one gets the average power intensity λ_0:

$$\lambda_0 = \frac{1}{T_0} \int_0^{T_0} [\zeta(t)]^2 dt = \frac{1}{\pi T_0} \int_0^{\omega_N} c_i^2 \, d\omega = \frac{1}{\pi T_0} \sum_{i=1}^{N} c_i^2 \Delta\omega_i \tag{9.7}$$

By observing this equation, it is also noticed that the average power intensity λ_0 is equal to the mean squared motion record (σ_ζ^2).

The power spectral density $S(\omega)$ is therefore defined such that the following equation can be fulfilled:

$$\lambda_0 = \int_0^{\omega_N} S(\omega) \, d\omega \tag{9.8}$$

It is obvious that:

$$S(\omega_i) = \frac{1}{\pi T_0} c_i^2 \tag{9.9}$$

The expression above also shows the relationship between the power spectral density $S(\omega)$ and the Fourier amplitude c_i.

The benefits of using power spectral density lie in the fact that it can characterize many different motion records and identify their similarities, and can be used for further computation to obtain the responses. This is especially the case for characterizing stationary Gaussian (Sect. 10.2) type motions. For example, for sea wave elevation or wind velocity, even though a one-to-one wave elevation or wind velocity does not generally exist, all records that result in identical spectral density do have the same statistical properties, i.e., the details of records that vary greatly may have identical spectral density.

The calculation of power spectrum can be used to estimate the statistical properties of many records, such as wave elevations, wind velocities, ground surface roughness, seismic ground motions, etc., and these can then be further used to compute stochastic responses using random vibration techniques (Sect.11.4). It also has the merit of executing the computation much faster than the Fast Fourier Transformation (FFT), because the computation is performed in place without allocating memory to accommodate complex results. However, since phase information is lost and cannot be reconstructed from the power spectrum's output sequence, power spectrum cannot be utilized if phase information is desired.

When using the power spectrum, various terminologies that are slightly different from each other exist in different fields or different purposes of applications. For example, the ordinate of the wave spectral density can be based on an amplitude spectrum $\left(\left(\frac{\zeta}{2}\right)^2\right)$, an amplitude half-spectrum $\left(\left(\frac{\zeta}{2}\right)^2/2\right)$, a height spectrum ($\zeta^2$), or a height double spectrum ($2\,\zeta^2$) etc., [85]. The abscissa can be chosen as angular frequency, cyclic frequency or period.

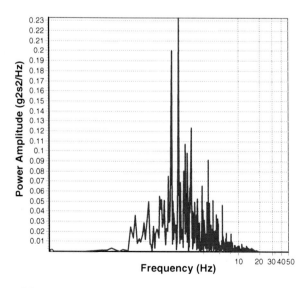

Fig. 9.6 Power of the strong ground motion (Fig. 9.1) recorded at Imperial Valley; g represents the acceleration of Earth's gravity

9.2 Power Spectrum Density

Figure 9.6 shows an example of the power spectrum density of strong ground motions (Fig. 9.1) recorded at Imperial Valley.

In some applications, the power spectral density $S(\omega)$ is normalized by its area (λ_0), which gives:

$$S_\zeta(\omega) = \frac{1}{\lambda_0} S(\omega) \qquad (9.10)$$

Example:
To characterize the global vibrations of a high-rise building and identify the local vibrations and acoustic performance on certain floors in this building, measurements of acceleration are planned. The lowest frequency Δf_{min} component measurable and the maximum desired frequency (f_{max}) are designed as 0.2 and 100 Hz, respectively. Determine a suitable time duration, the number of sampling points and the sample time interval for the measurement.

Solution: One can determine the time duration for measurement using the lowest frequency or the highest period measurable as: $T = \frac{1}{\Delta f_{min}} = 5$ s.

The number of sampling points is: $N = \frac{f_{obs}}{\Delta f_{min}} = \frac{2 f_{max}}{\Delta f_{min}} = \frac{2 \times 100 \text{ Hz}}{0.2 \text{ Hz}} = 1000$,

and the sample time interval over the time history is: $\Delta t = \frac{T}{N} = \frac{5 \text{ s}}{1000} = 0.005$ s (or $\Delta t = \frac{1}{2 f_{max}} = \frac{1}{2 \times 100} = 0.005$ s).

Readers may bear in mind that only processes with finite variance can be represented by spectra. Fortunately, even if theoretically the variance of a process can be infinite, almost every process in the engineering world has a finite variance.

Although using the Fourier transformation is the most common way of generating a power spectrum, other techniques such as the maximum entropy method can also be used [78].

Chapter 10
Statistics of Motions and Loads

The study of motion statistics is useful in many branches of dynamic analysis, giving a quantitative measure of motions in a concise manner and therefore proving efficient for mathematical treatment. Furthermore, it also paves the way for structural integrity (e.g., fatigue assessment) and reliability assessment.

This chapter will go through the basic concepts of statistics, which are essential for readers to further engage in the application of random vibrations and fatigue assessment, which will be elaborated in later chapters.

The three most important probability density distributions, the Gaussian, Weibull and Poisson distributions, are presented. The first one is elaborated since it directly describes a large group of dynamic phenomena in a statistical manner. As modern engineers work within the framework of load-resistance and performance-based design philosophy, a proper understanding of load level from the probability point of view is important for integrity and further risk assessment. For example, one may need to extrapolate from a few years of measured wind speed data in order to estimate extreme wind speed for, say, 50 years. Therefore, the latter two distributions are also introduced because they serve to characterize the probability of load level.

It should be mentioned that, rather than being based on a theoretical background, all three probability distributions are based on experiments or real populations from various fields of study.

When reading this chapter, readers need to understand that a random process is a sequence of random variables defined over a period of time. It is not unusual for new learners to mix up these two concepts, which are presented in Sect. 8.3

10.1 Narrow- and Wide-Banded Process

As an introduction, we begin with the study of narrow- and wide/broad-banded motions, as already briefly mentioned in Sect. 9. As illustrated in Figs. 10.1 and 10.2, narrow-banded motions have a dominant frequency and are characterized by a close to sinusoidal and smoothed time history, and the motions' process is relatively smooth and regular. Responses with narrow-banded characteristics are typically the results of a resonance or near-resonance response, which have the following properties [50]:

J. Jia, *Essentials of Applied Dynamic Analysis*, Risk Engineering,
DOI: 10.1007/978-3-642-37003-8_10, © Springer-Verlag Berlin Heidelberg 2014

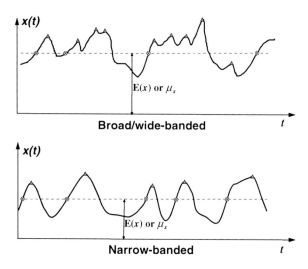

Fig. 10.1 Two random motion time series showing wide- (*upper*) and narrow-banded (*lower*) characteristics, respectively; ▲ indicates a local maximum (positive or negative peak), ⊗ indicates an upcrossing

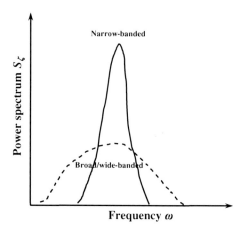

Fig. 10.2 Power spectrum for a *narrow-banded* and a *broad-banded* random process

- Only positive local and global maxima
- Only one maximum for each positive zero crossing (number of upcrossing equals the number of maxima).

Given that the mean positive zero-crossing periods between the narrow- and broad-banded processes are the same, the number of maxima for the narrow-banded time series within a specific time interval is less than that of the broad-banded one.

10.1 Narrow- and Wide-Banded Process

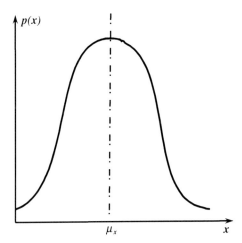

Fig. 10.3 Gaussian/normal distribution

In contrast, a broad/wide-banded process has noticeable varieties of frequencies and exhibits a jagged and irregular time history, as shown in the upper figure of Fig. 10.1.

10.2 Gaussian Distribution

Many physical problems such as wind velocity fluctuations can be approximated as the sum of a sufficiently large number of independent random variables under fairly general condition (each with finite mean and standard deviations). According to the Central Limit Theorem [51], they can be assumed to follow Gaussian/normal distribution, after Carl Gauss (1777–1855), who applied it to the calculation of measurement accuracy. The Gaussian process is characterized by the probability density function (Fig. 10.3) as:

$$p(x) = \frac{1}{\sigma_x \sqrt{2\pi}} e^{\left[-\frac{[x(t)-\mu_x]^2}{2\sigma_x^2}\right]} \qquad (10.1)$$

Gaussian distribution is symmetrical about its mean and has its deviations from the mean.

One important feature of Gaussian distribution is that it can be completely defined by its mean and standard deviation, and the responses of a linear system to this form of excitations still follow the Gaussian distribution.

For describing a non-normal (non-Gaussian) probability distribution of a random process, two additional statistical properties, namely skewness (k_3), which is the average of $[x(t) - u_x]^3$ normalized by σ_x^3; and kurtosis (k_4), also called peakedness, which is the average of $[x(t) - u_x]^4$ normalized by σ_x^4, are also often used:

$$k_3 = E\left[\left(\frac{x(t) - u_x}{\sigma_x}\right)^3\right] = \lim_{T \to \infty} \frac{1}{T} \int_0^T \left[\left(\frac{x(t) - u_x}{\sigma_x^3}\right)^3\right] dt \qquad (10.2)$$

$$k_4 = E\left[\left(\frac{x(t) - u_x}{\sigma_x}\right)^4\right] = \lim_{T \to \infty} \frac{1}{T} \int_0^T \left[\left(\frac{x(t) - u_x}{\sigma_x^4}\right)^4\right] dt \qquad (10.3)$$

where E stands for the mean (average) value.

From the equations above, it is clear that the skewness characterizes the degree of asymmetry of a distribution around its mean. And kurtosis characterizes the relative peakness or flatness of a distribution compared with the normal distribution. Compared to a Gaussian distribution with skewness equal to 0.0, positive skewness indicates a distribution with an asymmetric tail extending toward more positive values. Negative skewness indicates a distribution with an asymmetric tail extending toward more negative values. A kurtosis of more than 3.0 indicates a relatively peaked distribution. A kurtosis of less than 3.0 indicates a relatively flat distribution. The non-Gaussian distribution of responses can be identified by a non-zero skewness and a kurtosis of the responses unequal to 3.0. A larger deviation of the kurtosis from the value of 3.0 indicates more significant deviation of the responses from that of the Gaussian distribution.

Here, the question arises as to how to judge whether a process is significantly non-Gaussian from an engineering point of view. There are no widely accepted criteria. A possible rule of thumb is to find a range with plus and minus the double standard error of skewness and kurtosis from the perfect Gaussian process skewness (0.0) and kurtosis (3.0). If the skewness and kurtosis of the time series are outside this range, the process may be regarded as significantly non-Gaussian.

There are certain types of loading acting on a structure that are of non-Gaussian characteristics. A typical loading is the wave loading on offshore structures, which is primarily due to the nonlinear relations induced from drag forces in Morison's equation [52, 53, 201], the variation of water surface causing the intermittency of wave loading, and the variation of buoyancy forces on members in the splash zone [53–57]. In addition, even if the loading follows the Gaussian distribution, the nonlinearities in structural response (as will be presented in Sect. 15.2) may also modify the probability distribution with a trend toward a non-Gaussian status.

For mathematical convenience, one can move the origin of ordinate (vertical) in Fig. 10.1 to the mean value of each time series, i.e., $u_x = 0$. The resulting time series are shown in Fig. 10.4.

For a general stationary Gaussian process (not necessarily narrow-banded) with zero mean, the amplitude of the process is distributed with Rice distribution [58], named after Stephen O Rice (1907–1986):

10.2 Gaussian Distribution

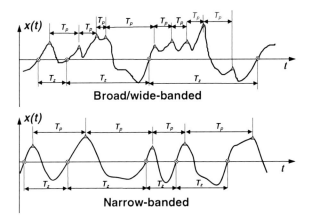

Fig. 10.4 Two random motion time series with both of their ordinate origins at the mean value of each time series; ⊗ indicates a local maximum (positive and negative peak), ▲ indicates a zero upcrossing, T_p is the period of peaks, and T_z is the period of zero crossing

$$p(x) = \frac{1}{\sigma_x \sqrt{2\pi}} \left\{ \psi e^{\left[-\frac{x^2(t)}{2\sigma_x^2 \psi^2}\right]} + \sqrt{1-\psi^2} \frac{x(t)}{\sigma_x} e^{\left[-\frac{x^2(t)}{2\sigma_x^2}\right]} \frac{x(t)\sqrt{1-\psi^2}}{\psi \sigma_x} \int_{-\infty}^{t} e^{-\frac{t^2}{2}} dt \right\}$$

(10.4)

where ψ is the bandwidth parameter or energy density spectrum [66, 67] given by:

$$\psi = \sqrt{1 - \frac{\lambda_2^2}{\lambda_0 \cdot \lambda_4}} \text{ for } 0.0 \leq \psi \leq 1.0 \quad (10.5)$$

A smaller ψ indicates a close to narrow-banded process and a larger one is close to the one that shows a broad-banded process.

We define the spectral moments that can be considered as moments of area of the energy spectrum about the vertical axis as below:

$$\lambda_i = \int_0^\infty \omega^i S_\zeta(\omega) d\omega \quad (10.6)$$

From the equation above, it is noticed that the zero moment is the area under the spectral curve and is also equal to the variance of the process:

$$\lambda_0 = \sigma_x^2 \quad (10.7)$$

The mean period of motions can then be calculated as:

$$\overline{T} = \frac{2\pi \lambda_0}{\lambda_1} \quad (10.8)$$

The mean period of motion peaks/crests (peak period) shown in Fig. 10.4 can be calculated as [58]:

$$\overline{T}_p = 2.0\pi\sqrt{\frac{\lambda_2}{\lambda_4}} \tag{10.9}$$

The average of zero crossing period (mean crest period) defined in Fig. 10.4 can be calculated as [88]:

$$\overline{T}_z = 2.0\pi\sqrt{\frac{\lambda_0}{\lambda_2}} \tag{10.10}$$

The bandwidth can also be indicated using an irregularity factor κ in terms of $\overline{T}_p$ and $\overline{T}_z$:

$$\kappa = \frac{\overline{T}_p}{\overline{T}_z} \tag{10.11}$$

The irregularity factor varies between 0.0 and 1.0. When it approaches 1.0, the relevant motions exhibit a regular sine form in time and are narrow-banded. When the factor approaches 0.0, the motions show a form close to white noise (the power spectral density is constant over the entire frequency range, see Sect. 11.4.2) that is broad-banded.

In dynamic analysis, usually, the average of the bandwidth parameter of energy density spectrum [67] ψ as aforementioned is used, and its relation with the irregularity factor is:

$$\psi = \sqrt{1 - \frac{\lambda_2^2}{\lambda_0 \cdot \lambda_4}} = \sqrt{1 - \left(\frac{\overline{T}_p}{\overline{T}_z}\right)^2} = \sqrt{1 - \kappa^2} \text{ for } 0.0 \le \psi \le 1.0 \tag{10.12}$$

As presented, $\psi = 1.0$ indicates a broad-banded process. Motions at various frequencies are presented. In such a condition, the probability density of peaks for this limiting case is therefore Gaussian:

$$p(x) = \frac{1}{\sigma_x\sqrt{2\pi}}\left\{e^{\left[-\frac{x^2(t)}{2\sigma_x^2}\right]}\right\} \tag{10.13}$$

In many applications, loads can be expressed as a spectrum $S(\omega)$, which is the sum of two individual spectra $S_1(\omega)$ and $S_2(\omega)$:

$$S(\omega) = S_1(\omega) + S_2(\omega) \tag{10.14}$$

The spectrum $S(\omega)$ has two peaks, and is thus called a two-peak spectrum.

For example, the two-peak spectrum is often used to combine waves due to local wind sea with spectrum $S_1(\omega)$ and waves due to swell generated at a distance with the spectrum $S_2(\omega)$.

10.2 Gaussian Distribution

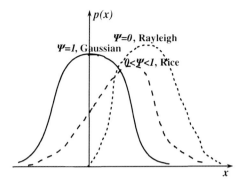

Fig. 10.5 Comparison of Gaussian/normal, Rice and Rayleigh distribution

Therefore, regardless of the order for the spectrum moment, the spectral moment λ for the combined spectrum $S(\omega)$ is also additive [106], i.e., the combined spectrum equals the sum of two spectrum moments λ_1 and λ_2:

$$\lambda = \lambda_1 + \lambda_2 \tag{10.15}$$

It should be noticed that, in case that a process cannot be modeled by Gaussian distribution (e.g., loads or responses exhibit significant nonlinear characteristics), other methods have to be adopted to describe the statistical information about the process. Among them, Monte Carlo simulation method [107] is the most direct approach.

10.3 Short-Term Distribution for Continuous Random Process: Rayleigh Distribution

As presented in Sect. 10.2, $\psi = 0.0$ indicates a narrow-banded Gaussian process. The probability density of peaks follows Rayleigh distribution (named after Lord Rayleigh (1842–1919), to describe the distribution of intensity of sound emissions from an infinite number of sources). That is, the motions concentrate around a frequency, and the probability density function is then:

$$p(x) = \frac{x(t)}{\sigma_x^2} e^{\left[-\frac{x^2(t)}{2\sigma_x^2}\right]} = \frac{x(t)}{\lambda_0} e^{\left[-\frac{x^2(t)}{2\lambda_0}\right]} \tag{10.16}$$

At the range $0.0 < \psi < 1.0$, the equation above becomes:

$$p(x) \rightarrow \sqrt{1 - \psi^2} \frac{x(t)}{\sigma_x^2} e^{\left[-\frac{x^2(t)}{2\sigma_x^2}\right]} = \sqrt{1 - \psi^2} \frac{x(t)}{\lambda_0} e^{\left[-\frac{x^2(t)}{2\lambda_0}\right]} \tag{10.17}$$

Figure 10.5 illustrates the difference between Gaussian, Rice and Rayleigh distribution. It is clearly shown that the Rayleigh distribution gives an upper limit for the distribution of maxima. The maximum values of a narrow-banded Gaussian process are essentially Rayleigh distributed. For example, if the wave elevations are an approximately narrow-banded Gaussian process (which is the case for short-term waves with a duration of up to 10 h), the individual wave height (amplitude) and crest height then follow the Rayleigh distribution [59]. This also means that if Rayleigh distribution is used for estimating the probability at a certain level, it always results in a conservative evaluation (with regard to values of a probability of exceedence at a certain level) compared to that of the Rice distribution (with a probability distribution closer to exact conditions).

The probability lying between $-\infty$ and x is defined as the cumulative probability distribution function, or distribution function:

$$P(x) = \int_{-\infty}^{x} p(x)dx \tag{10.18}$$

It is obvious that the probability covering between $-\infty$ and $+\infty$, i.e., the integral over the full range, is 1.0:

$$P(x) = \int_{-\infty}^{+\infty} p(x)dx = 1.0 \tag{10.19}$$

In addition, the accumulative probability distribution also possesses the following properties:

$$P(x + \Delta x) - P(x) = p(x)dx \tag{10.20}$$

$$P(x_2) - P(x_1) = \int_{x_1}^{x_2} p(x)dx \tag{10.21}$$

With the definition above, one can express the mean or expected value (first moment of x) as:

$$E(x) = \int_{-\infty}^{+\infty} xp(x)dx \tag{10.22}$$

The rth central moment about the mean is then calculated as:

$$E[(x - u_x)^r] = \int_{-\infty}^{+\infty} (x - u_x)^r p(x)dx \tag{10.23}$$

where u_x is the mean value of x defined in Sect. 8.3.

Therefore, it is obvious that the second, third and fourth moment about the mean are the variance, skewness and kurtosis of the random variable.

10.3 Short-Term Distribution for Continuous Random Process

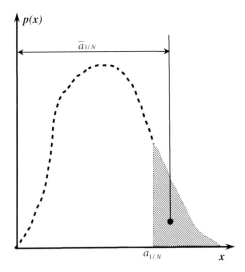

Fig. 10.6 Comparison between the average ($\bar{a}_{1/N}$) and the most probable ($a_{1/N}$) amplitude peak of the 1/N highest peaks

It is of particular interest for engineering to calculate the maximum value that is exceeded by a specific probability. The corresponding Rayleigh cumulative probability that is exceeded with the probability of 1/N is:

$$P(x \leq a_{1/N}) = \int_{-\infty}^{x} p(x)dx = 1 - e^{\left[-\frac{a_{1/N}^2}{2\sigma_x^2}\right]} = 1 - e^{\left[-\frac{a_{1/N}^2}{2\lambda_0}\right]} \quad (10.24)$$

where $a_{1/N}$ is the value of maximum value of x.

For calculating the most probable extreme/largest amplitude of N successive peaks:

$$1 - P(x \leq a_{1/N}) = 1/N \quad (10.25)$$

One finally obtains the value of one exceeded, on average, once in N cycles is that with an exceedence probability of 1/N:

$$a_{1/N} = \sqrt{2\sigma_x^2 \ln N} = \sqrt{2\lambda_0 \ln N} \quad (10.26)$$

The average amplitude peak of the highest 1/N proportion of peaks is also commonly used in statistical analysis, which is actually the center of the shaded area in Fig. 10.6:

$$\bar{a}_{1/N} = \frac{\int_{a_{1/N}}^{+\infty} xp(x)dx}{\int_{a_{1/N}}^{+\infty} p(x)dx} = N \int_{a_{1/N}}^{+\infty} xp(x)dx \quad (10.27)$$

Cartwright and Longuet-Higgins [58] show that $\bar{a}_{1/N}$ is calculated as:

$$\bar{a}_{1/N} = \sigma_x \sqrt{2 \ln N} + N \sigma_x \sqrt{\frac{\pi}{2}} \left[1 - \text{erf}\left(\sqrt{\ln N}\right)\right] \tag{10.28}$$

where $erf(x) = \frac{2}{\sqrt{\pi}} \int\limits_{0}^{x} e^{-k^2} dk$ is the error function.

Figure 10.6 illustrates the definition of the average ($\bar{a}_{1/N}$) and the most probable ($a_{1/N}$) amplitude peak of the $1/N$ highest peaks.

With the equation above, the average value of all peaks ($N = 1$) is calculated as:

$$\bar{a}_1 = \sigma_x \sqrt{\frac{\pi}{2}} = 1.25\sigma_x = 1.25\sqrt{\lambda_0} \tag{10.29}$$

The average value of 1/3 highest peaks ($N = 3$), also named the significant value, is:

$$\bar{a}_{1/3} = \sigma_x \sqrt{2 \ln 3} + 3\sigma_x \sqrt{\frac{\pi}{2}} \left[1 - \text{erf}\left(\sqrt{\ln 3}\right)\right] = 2.0\sigma_x = 2.0\sqrt{\lambda_0} \tag{10.30}$$

And the average of 1/10 peaks ($N = 10$) is:

$$\bar{a}_{1/10} = 2.55\sigma_x = 2.55\sqrt{\lambda_0} \tag{10.31}$$

From the three equations above, the relationship among the average of $1/N$ highest peaks is:

$$\bar{a}_{1/3} = 1.60\bar{a}_1 = 0.78\bar{a}_{1/10} \tag{10.32}$$

Example: In ocean engineering, the wave height is typically defined as twice the wave amplitude (peaks). Calculate the average, the probability and cumulative density function of $1/N$ highest wave height $\overline{H}_{1/N}$.

Solution:

$$\overline{H}_{1/N} = \frac{\int_{H_{1/N}}^{+\infty} H p(H) dH}{\int_{H_{1/N}}^{+\infty} p(H) dH} = N \int_{H_{1/N}}^{+\infty} H p(H) dH$$

From the equation above, it is obvious that the average of $1/N$ highest wave height $\overline{H}_{1/N}$ is obtained by multiplying the average of $1/N$ highest amplitude peak $a_{1/N}$ with a factor of 2.0, this gives:

$$\overline{H}_1 = 2.5\sigma_x = 2.5\sqrt{\lambda_0}$$

$$\overline{H}_{1/3} = 4.0\sigma_x = 4.0\sqrt{\lambda_0}$$

$$\overline{H}_{1/10} = 5.1\sigma_x = 5.1\sqrt{\lambda_0}$$

The probability density function in terms of wave height H is:

$$p(x) = \frac{H}{4\lambda_0} e^{\left[-\frac{H^2}{8\lambda_0}\right]}$$

The cumulative probability distribution function in terms of wave height H is then:

$$P(x) = 1 - e^{\left[-\frac{H^2}{8\lambda_0^2}\right]}$$

10.4 Long-Term Distribution for Continuous Random Process: Weibull distribution

While the Rayleigh distribution may be suitable for modeling individual waves in the short term (2–10 h), Weibull distribution [61], named after Walloddi Weibull (1887–1979), is normally used to model the life distributions (long term) of mechanical units or extreme environmental loads [62]:

$$p(x) = \beta\delta(x-a)^{\beta-1}e^{-\delta(x-a)^{\beta}} \qquad (10.33)$$

when $x \geq a$, $\delta > 0$ and $\beta > 0$.

The parameter a is the lower boundary of the values that the random variable X can take.

The distribution gives a straight line if plotted on a Weibull probability paper.

The Weibull cumulative probability distribution function (with a lower bound, below which the probability is zero) is then:

$$P(x) = \int_{-\infty}^{x} p(x)dx = 1 - e^{-\delta(x-a)^{\beta}} \qquad (10.34)$$

The Weibull distribution is rather useful for expressing the long-term distribution for a physical process that results in a limit on the possible value of x. Examples of using it is to calculate the extreme value distribution of wave height (for both significant $H_{1/3}$ and individual wave height) that a marine structure may experience in its lifetime (long term) or the extreme wind speed at a location within 50 years. In such cases, the long-term probability distribution of wave height or wind speed can be modeled in terms of three parameter Weibull distribution:

$$F(u) = 1 - e^{-\left(\frac{u-\alpha}{\gamma}\right)^{\beta}} \qquad (10.35)$$

where u is for example the wind speed or significant wave height (typically 1 h mean for wind speed and 3 h for wave height); α is a location parameter; β is a shape parameter; and γ is a scale parameter.

The extreme values u_R, corresponding to a return period of R, are obtained by inverting the equation above for a cumulative probability $F = 1 - \frac{\tau}{pR}$:

$$u_R = \alpha + \gamma \left[-\ln\left(\frac{\tau}{pR}\right) \right]^{\frac{1}{\beta}} \qquad (10.36)$$

where τ is the duration of the event (typically 1 h for mean wind speed and 3 h for significant wave height); and, p is a sector or monthly probability, i.e., 1/12 for monthly omni(all)-directional distributions).

A long-term observation of sea wave elevation may be performed by recording the significant wave height $H_{1/3}$ and the corresponding zero crossing period T_z for 20 min each third hour. The $H_{1/3}$ and T_z are estimated for each observation period either by directly using the short term wave statistics (zero crossing analysis) [60] defined in Fig. 10.4 or by calculating the wave spectrum and applying the method illustrated in the last example.

Other long-term probability distributions exist for various applications, such as the Lognormal distribution, Fisher-Tippett I or Gumbel distribution. For a complete summary of the topic, readers may read references [62, 81, 166].

It should be noted that there is no strict rule stating that the Weibull distribution only fits long-term events or the Rayleigh distribution can only be applied to short-term events. Their applicability is always problem dependent.

Example: In order to calculate the directional 1 h mean wind speed at 10 m above the ground (or sea surface) at a specific site, the relevant directional Weibull parameters in terms of location, shape and scale are obtained from the relevant report, as given in Table 10.1. Calculate the cumulative probability for 1 h mean wind speed from 5 to 29 m/s.

Solution: The long-term cumulative probability distribution of wind speed can be assumed to follow Weibull distribution:

$$F(u) = 1 - e^{-\left(\frac{u-\alpha}{\gamma}\right)^{\beta}}$$

By taking u as 1 h mean wind speed from 5 to 29 m/s, the corresponding cumulative probabilities calculated for each mean wind speed are illustrated in the white area in Table 10.1. From the sum of probability from all

10.4 Long-Term Distribution for Continuous Random Process

directions at each wind speed, it is shown that 91 % of wind occurs at a wind speed below 16 m/s.

For a further screening, the scatter diagram shown in Table 10.1 may be categorized with fewer wind blocks, and each block has an individual probability of occurrence instead of the cumulative probability. This is shown in Table 10.2.

Table 10.1 Directional cumulative probability based on the given Weibull parameters

Direction	N	NE	E	SE	S	SW	W	NW	Sum	Omni
location parameter	0.545	0.775	0.000	0.000	0.682	0.000	0.000	1.440		0.000
shape parameter	2.074	1.707	2.182	2.606	2.220	2.536	2.310	1.878		2.100
scale parameter	9.125	7.472	10.757	11.437	10.337	12.042	11.935	8.907		10.364
2 m/s	0.0220	0.0446	0.0251	0.0106	0.0103	0.0105	0.0160	0.0055	0.018	0.0311
3 m/s	0.0636	0.1188	0.0598	0.0301	0.0355	0.0290	0.0403	0.0372	0.052	0.0713
4 m/s	0.1249	0.2120	0.1091	0.0627	0.0771	0.0593	0.0769	0.0917	0.102	0.1267
5 m/s	0.2023	0.3147	0.1713	0.1093	0.1341	0.1020	0.1254	0.1636	0.165	0.1946
6 m/s	0.2911	0.4190	0.2440	0.1699	0.2044	0.1571	0.1847	0.2475	0.240	0.2719
7 m/s	0.3860	0.5192	0.3240	0.2429	0.2848	0.2233	0.2529	0.3382	0.321	0.3551
8 m/s	0.4819	0.6110	0.4079	0.3256	0.3716	0.2985	0.3276	0.4305	0.407	0.4404
9 m/s	0.5742	0.6921	0.4922	0.4146	0.4606	0.3799	0.4061	0.5205	0.493	0.5246
10 m/s	0.6592	0.7614	0.5738	0.5058	0.5481	0.4643	0.4855	0.6047	0.575	0.6045
11 m/s	0.7345	0.8188	0.6500	0.5948	0.6306	0.5484	0.5632	0.6809	0.653	0.6780
12 m/s	0.7986	0.8651	0.7190	0.6781	0.7056	0.6289	0.6367	0.7476	0.722	0.7434
13 m/s	0.8514	0.9015	0.7795	0.7525	0.7714	0.7031	0.7043	0.8044	0.783	0.8000
14 m/s	0.8933	0.9294	0.8309	0.8162	0.8271	0.7690	0.7644	0.8514	0.835	0.8475
15 m/s	0.9255	0.9503	0.8733	0.8683	0.8727	0.8254	0.8165	0.8894	0.878	0.8862
16 m/s	0.9493	0.9656	0.9073	0.9092	0.9088	0.8720	0.8603	0.9193	0.911	0.9170
17 m/s	0.9665	0.9766	0.9338	0.9397	0.9364	0.9091	0.8961	0.9422	0.938	0.9408
18 m/s	0.9785	0.9844	0.9538	0.9616	0.9569	0.9374	0.9245	0.9594	0.957	0.9587

(continued)

10 Statistics of Motions and Loads

Table 10.1 (continued)

19 m/s	0.9866	0.9898	0.9686	0.9766	0.9716	0.9584	0.9465	0.9721	0.971	0.9719
20 m/s	0.9918	0.9934	0.9791	0.9863	0.9818	0.9732	0.9629	0.9811	0.981	0.9813
21 m/s	0.9952	0.9958	0.9865	0.9923	0.9887	0.9834	0.9750	0.9875	0.988	0.9878
22 m/s	0.9972	0.9974	0.9915	0.9959	0.9932	0.9901	0.9835	0.9919	0.993	0.9922
23 m/s	0.9985	0.9984	0.9948	0.9979	0.9960	0.9943	0.9894	0.9948	0.996	0.9952
24 m/s	0.9992	0.9990	0.9969	0.9990	0.9977	0.9968	0.9934	0.9967	0.997	0.9971
25 m/s	0.9996	0.9994	0.9982	0.9995	0.9987	0.9983	0.9960	0.9980	0.998	0.9983
26 m/s	0.9998	0.9997	0.9990	0.9998	0.9993	0.9991	0.9976	0.9988	0.999	0.9990
27 m/s	0.9999	0.9998	0.9994	0.9999	0.9997	0.9996	0.9986	0.9993	1.000	0.9994
28 m/s	0.9999	0.9999	0.9997	1.0000	0.9998	0.9998	0.9992	0.9996	1.000	0.9997
29 m/s	1.0000	0.9999	0.9998	1.0000	0.9999	0.9999	0.9996	0.9998	1.000	0.9998
30 m/s	1.0000	1.0000	0.9999	1.0000	1.0000	1.0000	0.9998	0.9999	1.000	0.9999
31 m/s	1.0000	1.0000	1.0000	1.0000	1.0000	1.0000	0.9999	0.9999	1.0000	1.0000
32 m/s	1.0000	1.0000	1.0000	1.0000	1.0000	1.0000	0.9999	0.9999	1.0000	1.0000

Table 10.2 Directional probability for each individual wind block based on the cumulative probability calculated in Table 10.1

Direction		N	NE	E	SE	S	SW	W	NW	Sum	Omni
$F(u) = 1 - e^{-\left(\frac{u}{c}\right)^{\beta}}$	5 m/s	0.2023	0.3147	0.1713	0.1093	0.1341	0.1020	0.1254	0.1636	0.1654	0.1946
	7 m/s	0.1837	0.2045	0.1527	0.1335	0.1507	0.1212	0.1275	0.1745	0.1560	0.1605
	9 m/s	0.1882	0.1730	0.1682	0.1718	0.1758	0.1566	0.1532	0.1823	0.1711	0.1695
	10 m/s	0.0850	0.0693	0.0816	0.0911	0.0875	0.0844	0.0794	0.0842	0.0828	0.0800
	11 m/s	0.0753	0.0574	0.0763	0.0891	0.0826	0.0841	0.0777	0.0762	0.0773	0.0735
	12 m/s	0.0642	0.0463	0.0690	0.0832	0.0750	0.0805	0.0736	0.0667	0.0698	0.0654
	14 m/s	0.0419	0.0279	0.0514	0.0637	0.0557	0.0659	0.0602	0.0470	0.0517	0.0475
	17 m/s	0.0411	0.0264	0.0605	0.0714	0.0637	0.0836	0.0796	0.0528	0.0599	0.0546
	19 m/s	0.0200	0.0131	0.0348	0.0368	0.0352	0.0493	0.0504	0.0299	0.0337	0.0311
	29 m/s	0.0134	0.0102	0.0313	0.0234	0.0283	0.0415	0.0531	0.0277	0.0286	0.0280
	Sum	1.0000	0.9999	0.9998	1.0000	0.9999	0.9999	0.9996	0.9998	0.9999	0.9998

10.5 Number of Occurrence Within a Fixed Time or Space Interval: Poisson Distribution

Different from Gaussian distribution, which is for modeling a continuous random process, Poisson distribution, named after Simeon-Denis Poisson (1781–1840), is a discrete probability distribution that expresses the probability of a given number of events occurring within a fixed interval of time and/or space if these events occur with a known average rate and are independent of the time since the last event. It can also be used for estimating the number of events in specified intervals, such as distance, area or volume. Poisson distribution is expressed as:

$$P(N = n) = \frac{u^n e^{-u}}{n!} \tag{10.37}$$

where N and u are the number of occurrence for a particular event and the average number of occurrences of all events within the time interval, respectively.

The distribution above has the following four properties:

- Independency: the number of occurrences (variable) in one time interval is independent from that in any other time interval
- Non-simultaneously occurring: the probability that two or more events occur simultaneously (or within a rather short time interval) is zero (negligible)
- Random occurrence: the events occur randomly in time or space
- Uniform distribution: the mean number of events within a rather short time interval is proportional to the length of the time interval.

Examples of events that follow a Poisson distribution are the number of flaws in a given length of material, the number of phone calls arriving at a call center per hour, the number of visitors to a shopping mall within a week, and the number of major and great earthquakes (larger than Richter magnitude of 7.0) occurring within 1,000 years.

Essentially, Poisson probability distribution predicts the degree of spread around a known average rate of occurrence. It is shown that the inter-event time in a Poisson process is exponentially distributed. With the average rate of occurrence λ of the considered event and time period of interest t, the equation above can be rewritten as:

$$P(N = n) = \frac{(\lambda t)^n e^{-\lambda t}}{n!} \tag{10.38}$$

The probability of at least one event exceedance ($N \geq 1$) in a period of t can then be calculated as:

$$P[N \geq 1] = 1 - e^{-\lambda t} \tag{10.39}$$

If one set period of $t = 1$ year, the annual probability of exceedance can be calculated as:

$$q = 1 - e^{-\lambda \cdot 1} = 1 - e^{-\frac{1}{R}} \tag{10.40}$$

where R is called return period, or recurrence interval, which is defined as the average time between the design conditions being exceeded, and is widely used as an estimate of the likelihood of an event.

Both return period and annual probability of exceedance are exchangeable and are important parameters for reliability and risk analysis. For example, for design of all types of structures, it is essential to estimate the worst load condition during the life time of the structure. This load condition is usually calibrated with an annual probability of exceedance or return period. This is normally carried out by fitting the measured load data (e.g. wave height, wind speed, or earthquake ground accelerations) to a decent probability distribution function, and by extending the probability distribution to the entire range of values. This process is illustrated in Figs. 10.7 and 10.8.

For example, based on the equation above, one may calculate that the annual probability of exceedance for return period of 1, 10, 100 and 600 years are 0.63, 10^{-1} and 10^{-2}, 1.665×10^{-3}, respectively. In many design codes, the load levels are given with corresponding annual probability of exceedance. Readers may transfer this into the return period, which provides more sensible information to the general public.

It should be noticed that the Poisson process is a memory-less model, in that it is independent of elapsed time. In addition, it is also independent of size or location of any previous events. Therefore, this model is only applicable for the statistical characterization of a region with a large area [63].

10.6 Joint Probability Distribution

The idea of probability density and cumulative probability distribution function for a single variable in the previous sections can be extended to any number of random variables, leading to the study of joint probability distribution. For two variables (X, Y), the probability density function $p(x, y)$ can be represented as a surface above a horizontal plane, and the cumulative probability $(P(x, y))$ of x lying in the range between x and $x + dx$ as well as y lying between y and $y + dy$ is $p(x, y)dxdy$. Some of the parameters that are of interest in engineering are the means (u_x and u_y), the variance (σ_x^2 and σ_y^2) and the covariance (σ_{xy}):

10.6 Joint Probability Distribution

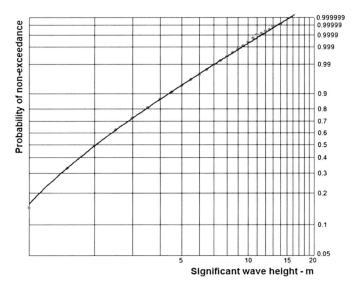

Fig. 10.7 Observed (*dots*) and fitted (*line*) cumulative probability distribution of significant wave heights for a typical site in the North Sea

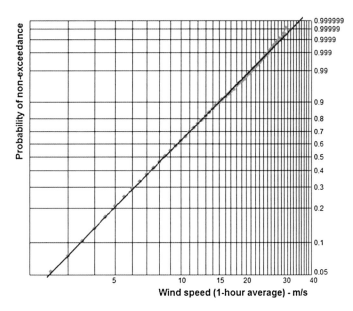

Fig. 10.8 Observed (*dots*) and fitted (*line*) cumulative probability distribution of 1 h mean wind speed for a typical site in the North Sea

$$u_x = E[x] = \int\limits_{-\infty}^{+\infty} \int\limits_{-\infty}^{+\infty} xp(x,y)dxdy \qquad (10.41)$$

Example: Based on a Poisson process, calculate the average rate of annual occurrence λ and return period of at least one event that has 10 % probability of being exceeded in the next 50 years.

Solution: With $P[N \geq 1] = 10\%$, $t = 50$, we then have: $10\% = 1 - e^{-\lambda \cdot 50}$, the average rate of annual occurrence λ is 0.0021, and the return period $R = 1/\lambda = 475$ years. It is noted that, in many seismic design codes, this is referred to as the "standard" return period.

$$u_y = E[y] = \int\limits_{-\infty}^{+\infty} \int\limits_{-\infty}^{+\infty} yp(x,y)dxdy \qquad (10.42)$$

$$\sigma_x^2 = E\left[(x - u_x)^2\right] = \int\limits_{-\infty}^{+\infty} \int\limits_{-\infty}^{+\infty} (x - u_x)^2 p(x,y)dxdy \qquad (10.43)$$

$$\sigma_y^2 = E\left[(y - u_x)^2\right] = \int\limits_{-\infty}^{+\infty} \int\limits_{-\infty}^{+\infty} (y - u_y)^2 p(x,y)dxdy \qquad (10.44)$$

$$\sigma_{xy} = E\left[(x - u_x)(y - u_y)\right] = \int\limits_{-\infty}^{+\infty} \int\limits_{-\infty}^{+\infty} (x - u_x)(y - u_y)p(x,y)dxdy \qquad (10.45)$$

Here, we introduce the definition of normalized covariance, which is used to identify the correlations between two variables:

$$\rho_{xy} = \frac{\sigma_{xy}}{\sigma_x \sigma_y} \qquad (10.46)$$

$\rho_{xy} = \pm 1.0$ indicates a full correlation (linear dependence) between the variables x and y; while if $\rho_{xy} = 0.0$, the variables x and y are uncorrelated. In engineering practice, if the $|\rho_{xy}| \geq 0.9$, the two variables x and y are regarded as highly correlated.

10.7 Long-Term Prediction

Let's take calculating the wave-induced structural responses as an example. A short-term sea state is for most practical purposes reasonably characterized by the significant wave height H_s, and the corresponding spectral wave period T_s. However, when the JONSWAP spectrum (Sect. 12.1.2) is used, the spectral wave period Ts is often replaced by spectral peak period T_p.

In order to provide a better fit to the data in the lower tail of the distribution, the long-term variation of the wave climate can be described by replacing the three-parameter Weibull distribution (Sect. 10.4) with a combination of a Lognormal and a Weibull distribution (LoNoWe):

$$f_{H_S}(h_s) = \frac{1}{\sqrt{2\pi}\alpha h_s} e^{-\frac{(\ln h_s - \theta)^2}{2\alpha^2}} \quad \text{for } h_s \le \eta \tag{10.47}$$

$$f_{H_S}(h_s) = \frac{\beta}{\rho} \left(\frac{h_s}{\rho}\right)^{\beta-1} e^{-\left(\frac{h_s}{\rho}\right)^{\beta}} \quad \text{for } h_s > \eta \tag{10.48}$$

where η is called the transition parameter that separates the Lognormal model for small values of H_s from the Weibull model for the larger values [24]. It is geographical location dependent.

The two equations above give the marginal probability distribution for the significant wave height H_s. However, the corresponding spectral peak period T_p needs to be determined using the conditional probability for given values of H_s:

$$f_{T_p|H_S}(t_p|h_s) = \frac{1}{\sqrt{2\pi}\sigma \cdot t_p} e^{-\frac{(\ln t_p - \mu)^2}{2\sigma^2}} \tag{10.49}$$

where both μ and σ depend on the wave height:

$$\mu = a_1 + a_2 h_s^{a_3} \tag{10.50}$$

$$\sigma^2 = b_1 + b_2 \cdot e^{-b_3 \cdot h_s} \tag{10.51}$$

where a_1, a_2, a_3, b_1, b_2 and b_3 are constants.

The joint probability of wave height H_s and spectral peak period T_p are then obtained by multiplying the marginal probability density function of significant wave height (Eqs. (10.47) or (10.48)) with the conditional probability density function of the spectral peak period (Eq. (10.49)):

$$f_{H_S T_p}(h_s, t_p) = f_{H_S}(h_s) f_{T_p|H_S}(t_p|h_s) \tag{10.52}$$

The LoNoWe distribution is fitted to the data such that the extreme values corresponding to a required annual probability of exceedance (e.g., 10^{-2} for a return period of 100 years) is equal to the corresponding values obtained when fitting a three-parameter Weibull distribution to the data (e.g., Table 10.1).

The environmental parameters may differ from one location to another. For example, at the Oseberg field in the North Sea, $\beta = 1.363$ and $\alpha = 0.569$, while at the Statfjord field less than one hundred kilometers north-west of the Oseberg field, $\beta = 2.691$ and $\alpha = 0.657$.

10.8 Environmental Contour Line Method

For a very complex response problem, a full long-term analysis elaborated in Sect. 10.7 will typically be out of reach. As an alternative, one can estimate the q-annual probability response using the environmental contour line method.

For example, in the application of ocean engineering, this method is based on an appropriate formulation of the design storm concept with combinations of significant wave height and spectral peak period, which are located along a contour line in the H_s and T_P plane [71]. Figure 10.9 shows a contour line plot for an area located in the North Sea.

This method is described in detail in Ref. [65]. It has been claimed to be a rational basis for deciding the short-term design storms [64]. The major steps of the method are [71, 72]:

- The q-annual probability contour lines are established, which provide all combinations of H_s and T_p corresponding to an annual probability of being "exceeded" by q, shown in Fig. 10.9.
- For a given response problem, one has to find the most unfavorable sea state along the q-probability contour line.
- For the worst sea state along the contour lines, the distribution function for the 3-h duration maximum response is established.
- Finally, the q-probability value of the selected response quantity is estimated by the value of the 3-hour extreme value distribution that is exceeded by probability 1-α. For $q = 10^{-2}$ (return period of 100 years), NORSOK standard N-003 [71] recommends $\alpha = 0.85$–0.9.

Obviously, by using the contour line method, only a few most unfavorable sea states need to be identified and used for calculating structural responses. However, readers need to realize that this method is an approximate one. An appropriate selection of values for parameter α may be challenging. Some researchers show that the best estimate would be reached for an α-value lower than 0.85 and others show that the "correct" value of α is larger than 0.9 [72]. The "true" value of α can only be found if it is calibrated to the result of a long-term analysis described in Sect. 10.7.

10.8 Environmental Contour Line Method

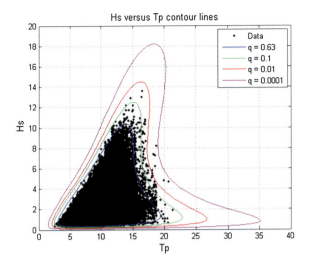

Fig. 10.9 q—probability contour lines of H_S—T_p for q = 0.63 (1 year of return period), 10^{-1} (10 year of return period), 10^{-2} (100 year of return period) and 10^{-4} (10,000 year of return period) for omni-directional waves at a field in the North Sea. The duration of sea state is 3 h

For more thorough knowledge of statistics, readers may read references [24, 44–46].

Chapter 11
Forced Vibrations

By applying an external force $F(t)$ on the system shown in Fig. 3.7, an SDOF spring-mass-damper system under forced excitation is constructed as illustrated in Fig. 11.1. The equation of motions for the system is expressed as:

$$m\ddot{x}(t) + c\dot{x}(t) + kx(t) = F(t) \tag{11.1}$$

Depending on the types of loading that are harmonic, periodical, transient or random, special mathematical treatment can be used to find the solutions of the equation above. This will be illustrated in the subsequent sections.

11.1 Forced Vibrations Under Harmonic Excitations

11.1.1 Responses to Harmonic Force

By exerting an external harmonic force $(F(t) = F_0 \sin(\Omega t))$ with an amplitude of F_0 and an angular frequency of Ω shown in Fig. 11.1, or displacement excitations in a harmonic form on the spring-mass-damper system, an SDOF spring-mass-damper system under forced harmonic excitation is constructed. The governing linear differential equation of motions for this system in case of harmonic force excitations can then be written as:

$$m\ddot{x}(t) + c\dot{x}(t) + kx(t) = F_0 \sin(\Omega t) \tag{11.2}$$

Dividing both sides of the equation above by m, this equation is rewritten as:

$$\ddot{x}(t) + \frac{c}{m}\dot{x}(t) + \omega_n^2 x(t) = (F_0/m) \sin(\Omega t) \tag{11.3}$$

By realizing that $c = 2\omega_n m\zeta$, the equation above finally gives:

$$\ddot{x}(t) + 2\omega_n \zeta \dot{x}(t) + \omega_n^2 x(t) = (F_0/m) \sin(\Omega t) \tag{11.4}$$

J. Jia, *Essentials of Applied Dynamic Analysis*, Risk Engineering,
DOI: 10.1007/978-3-642-37003-8_11, © Springer-Verlag Berlin Heidelberg 2014

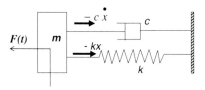

Fig. 11.1 An SDOF spring-mass-damper system under an external force $F(t)$

As the equation above is a second-order non-homogeneous equation, the general solution for it is the sum of the two parts: the complementary solution $x_c(t)$ to the homogeneous (free vibrations) equation and the particular solution $x_p(t)$ to the non-homogeneous equation:

$$x(t) = x_c(t) + x_p(t) \tag{11.5}$$

The complementary solution exhibits transient vibrations at the system's natural frequency and only depends on the initial condition and the system's natural frequency, i.e., it represents free vibrations as discussed in Sect. 3.3 and does not contain any enforced responses:

$$x_c(t) = X e^{-\zeta \omega_n t} \sin\left(\sqrt{1-\zeta^2}\omega_n t + \phi\right) \tag{11.6}$$

It is noticed that this aspect of the vibration dies out due to the presence of damping, leaving only the particular solution exhibiting steady-state harmonic oscillation at excitation frequency Ω. This particular solution is also called the steady-state solution that depends on the excitation amplitude F_0, the excitation frequency Ω as well as the natural frequency of the system, and it persists motions for ever:

$$x_p(t) = E \sin(\Omega t) + F \cos(\Omega t) \tag{11.7}$$

By substituting the equation above and its first and second derivatives into Eq. (11.4), one obtains the coefficients E and F as:

$$E = \frac{F_0}{k} \frac{1 - (\Omega/\omega_n)^2}{\left[1 - (\Omega/\omega_n)^2\right]^2 + [2\zeta(\Omega/\omega_n)]^2} \tag{11.8}$$

$$F = \frac{F_0}{k} \frac{-2\zeta\Omega/\omega_n}{\left[1 - (\Omega/\omega_n)^2\right]^2 + [2\zeta(\Omega/\omega_n)]^2} \tag{11.9}$$

By inserting the expression for coefficient E and F into Eq. (11.7) and rearranging it, one can rewrite the steady-state solution as:

$$x_p(t) = \frac{F_0}{km} \frac{\sin(\Omega t - \varphi)}{\sqrt{\left[1 - (\Omega/\omega_n)^2\right]^2 + [2\zeta(\Omega/\omega_n)]^2}} \tag{11.10}$$

11.1 Forced Vibrations Under Harmonic Excitations

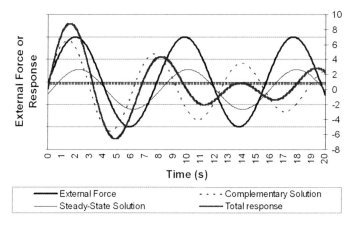

Fig. 11.2 Transient and steady-state responses due to external harmonic force excitations applied on a system with $\omega_n = 1.0$, $\Omega = 0.8$, $\zeta = 0.05$, and $\phi = 0.1$

where φ is the phase between the external input force and the response output, with the most noticeable feature being a shift (particularly for underdamped systems) at resonance. It can be calculated as:

$$\varphi = \tan^{-1}\left(\frac{2\zeta(\Omega/\omega_n)}{1-(\Omega/\omega_n)^2}\right) \quad (11.11)$$

It is clearly shown that the steady-state solutions are mainly associated with the excitation force and the natural frequency. Figure 11.2 shows an example of the dynamic responses due to the contribution from both transient and steady-state responses, with a Ω/ω_n ratio of 0.8, a damping value of 0.05 ($\varphi = 0.21$), and $\phi = 0.1$. Phases between the two types of responses can be observed.

Under harmonic excitations, the magnitude and phase of the displacement responses strongly depend on the frequency of the excitations, resulting in three types of steady-state responses, namely quasi-static, resonance, and inertia dominant responses, which are illustrated in Figs. 11.3 and 11.4.

The dynamic responses with respect to the ratio between the forcing and natural frequency can also be explained by the equilibrium of the spring-mass-damping system: the inertia and elastic forces counteract each other, among which the former ones act away from the neutral position and the latter ones act toward it. The damping forces act in the opposite direction of motions and therefore change their directions.

When the frequencies of excitations are well below the natural frequencies of the structure, i.e., $\Omega/\omega_n < < 1$, both the inertia and damping term are small, and the responses are controlled by the stiffness. Therefore, the excitations have a load effect equivalent to a static load at the same location with the same amplitude and direction, i.e., the displacement of the mass follows the time varying force almost instantaneously (the responses lag behind the excitation force by a phase angle $\varphi < 90°$ as shown in Figs. 11.3a and 11.5). The dynamic responses are therefore rather insignificant and can normally be neglected. A quasi-static analysis approach is routinely

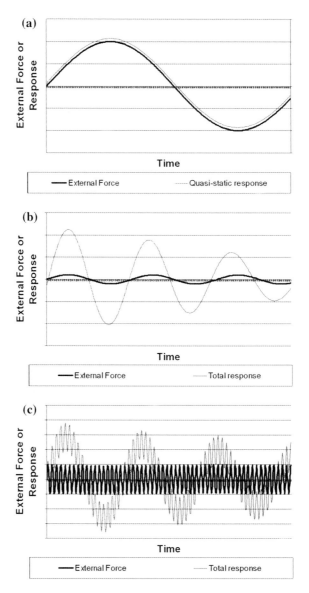

Fig. 11.3 Damped responses due to harmonic excitations with the characteristics of (**a**) quasi-static ($\Omega/\omega_n << 1$); **b** resonance (Ω/ω_n close to 1); and **c** inertia dominant responses ($\Omega/\omega_n >> 1$), for a system with $\omega_n = 1.0$ and $\zeta = 0.03$

adopted. However, this may not always be adequate, particularly when frequencies approach resonance condition or when there are transients with duration close to half the natural periods of structures [68]. The left figure in Fig. 1.29 illustrates a combination of complementary and steady-state solutions. The former shows fluctuations similar to background noises with the natural period of the structure (2.5 s), the

11.1 Forced Vibrations Under Harmonic Excitations

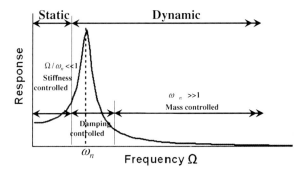

Fig. 11.4 Responses at different ranges of forcing frequency [81]

latter shows stiffness-controlled responses with the period of wave loading. From an energy point of view, this reflects the condition in which the maximum kinetic energy is lower than the maximum potential energy.

When the excitation frequencies are close to the natural frequency of the system, i.e., ($\Omega/\omega_n \approx 1$), the inertia term becomes larger. In the absence of excitation and damping, the spring and inertia forces cancel (balance) each other. Theoretically, this results in infinite responses in cases when $\Omega/\omega_n = 1$ (provided damping is absent). The external forces are only overcome (controlled) by the damping forces. In this situation, energy is being added to the system in a most efficient way. This also implies that the level of damping must be near the resonance frequency. The total dynamic responses can be calculated by the sum of both transient and steady-state oscillations, and both types of oscillations are important. Resonance then occurs by producing responses that are much larger than those from quasi-static responses as shown in Fig. 11.3b, and there is a dramatic change of phase angle, i.e., by neglecting the damping, the displacement is 90° out of phase with the force, while the velocity is in phase with the excitation forces. In a typical situation in which the damping is well below 1.0, the responses are much larger than their quasi-static counterparts. From an energy point of view, when the frequency of excitations is equal to the natural frequency, the maximum kinetic energy is equal to the maximum potential energy.

When the excitation frequencies are well above the natural frequency of the system ($\Omega/\omega_n >> 1$), the external forces are expected to be almost entirely overcome by the large inertia force, the excitations are so frequent that the mass cannot immediately follow the excitations as shown in Fig. 11.3c, i.e., the responses have a phase angle $\varphi > 90°$ to the excitation force as shown in Fig. 11.5. Even though the total dynamic responses can still be calculated by the sum of both transient and steady-state oscillations, the transient vibrations are normally more significant than the steady-state oscillations, i.e., the inertia of the system dominates the responses. The responses of the mass are therefore small and almost out of phase (phase angle approaches 180°) with the excitation forces [69]. From an energy point of view, this reflects the condition in which the maximum kinetic energy is larger than the maximum potential energy. When $\Omega/\omega_n > \sqrt{2}$, the response amplitude is less than the corresponding static deflection (F_0/k).

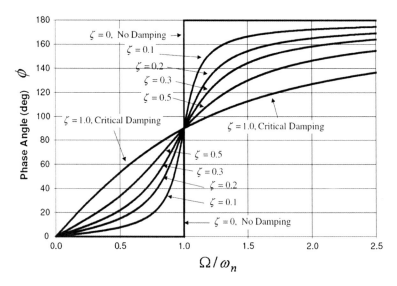

Fig. 11.5 Plot of phase angle between the response output and load input, with varied ω_n ratio and damping

For a convenient and quantitative discussion, the steady-state solution may be reduced to non-dimensional form as a ratio between the displacement amplitude of steady-state solution X_p and the maximum quasi-static deflection $X_0 = F_0/k$, under the excitation force F_0 (not to be confused with the initial condition X in Sect. 3.3.), which is also called the magnification factor or non-dimensional gain function:

$$\left|\frac{X_p}{X_0}\right| = \frac{1}{\sqrt{\left[1 - (\Omega/\omega_n)^2\right]^2 + [2\zeta(\Omega/\omega_n)]^2}} \tag{11.12}$$

And the phase angle between the external forces and displacement responses is:

$$\tan\varphi = \frac{2\zeta(\Omega/\omega_n)}{1 - (\Omega/\omega_n)^2} \tag{11.13}$$

With the two equations above, one can investigate the sensitivity of response characteristics only depending on the Ω/ω_n ratio and the damping level ζ. For example, the magnification factor is plotted as shown in Fig. 11.6. Obviously, with the increase of damping, the maximum amplitude decreases and the peak of magnification appears at a frequency increasingly below the natural frequency ω_n. Again, the damping is only effective at or close to the resonance condition.

At the excitation frequency $\omega_n\sqrt{1 - 2\zeta^2}$, the peak of the magnification factor appears, which is called the amplification factor or dynamic amplification factor (DAF):

$$Q = \frac{1}{2\zeta} \tag{11.14}$$

11.1 Forced Vibrations Under Harmonic Excitations

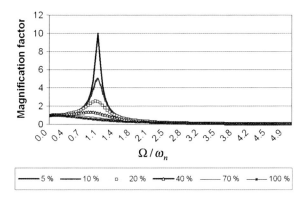

Fig. 11.6 Plot of magnification factor as a function of Ω/ω_n and damping due to the external applied forces

In structural analysis, if appropriate, quasi-static analysis is always preferred to dynamic analysis. When dynamic effects can be assumed to be approximately uniform throughout the structural systems and small, one static analysis or a series of static analyses, with a small correction to account for dynamic effects, is sometimes adequate. The action effects due to static and dynamic (inertia) effects can be added together. For example, in ISO 19902 [70], it is stated that, for fixed offshore structures, the correction for dynamic effects may be determined using one or the other of the following methods:

By performing one static analysis in which the actions are enhanced by a set of equivalent quasi-static inertial actions representing the dynamic responses, this method is normally applied to structural parts that normally do not experience any quasi-static action effects, so when multiplied by a DAF (the other method described below), the quasi-static action effects will still be negligible, whereas the properly calculated dynamic action effects resulting from inertial actions can be significant.

By performing a series of static analyses over an appropriate range of frequencies of excitations, where, at each frequency, the corresponding actions are applied and the calculated action effects are multiplied by the DAF of an SDOF system at that frequency. For example, for structures subjected to wave loading (except for quasi-static wave loading effects (Fig. 1.24b), the equivalent quasi-static action representing the dynamic inertia part of wave loading (Fig. 1.24c) can be obtained by a factor (DAF-1) multiplied with the static wave loading.

The DAF applied based on a static analysis should normally not be lower than 1.0 or higher than 1.5. For responses with larger DAFs, a direct dynamic analysis should be performed.

In cases in which the excitation period is far from a structure's natural period, the DAF can be omitted. For example, in the NORSOK standard N-003[71], it is specified that, for fully constrained (bottom supported) platforms, dynamic effects need not to be accounted for when the natural period is below 3.0 and 2.0 s for determining steady wave action effects for assessment toward ultimate limit state and fatigue limit state, respectively.

Another way of investigating responses due to the harmonic force excitations is to study the transfer function, also called frequency response function $H(\omega)$, which is defined as the complex displacement responses due to the complex input force of unit magnitude ($F_0 = 1$):

$$H(\omega) = \frac{1}{(k - m\omega^2) + ic\omega} = \frac{(k - m\omega^2) - ic\omega}{(k - m\omega^2)^2 + c^2\omega^2} \tag{11.15}$$

The transfer function in dynamic analysis is analogous to the stiffness inverse in a static analysis, i.e., when $\omega = 0$, $H(0) = \frac{1}{k}$, which is actually a simple statement of Hooke's law of elasticity. This indicates that the responses for static analysis at zero frequency and dynamic analysis can be treated separately, which is an essential procedure for random vibrations as discussed in Sect. 11.4.1.

Obviously, transfer functions in terms of the velocity and acceleration responses can be derived by multiplying the equation above by $i\omega$ and $-\omega^2$.

It should be noted that the definition of transfer functions differs depending on their application. For example, in ocean engineering, the transfer function is often defined as the hydrodynamic force or responses to unit wave height.

Sometimes, the gain function is also used to investigate the response characteristics, which is defined as the modulus of transfer function, and it is actually the amplitude of displacement for $F_0 = 1$:

$$|H(\omega)| = \frac{X_p}{F_0}(F_0 = 1) = \sqrt{H(\omega)H^*(\omega)} = \sqrt{(\mathrm{Re}H)^2 + (\mathrm{Im}H)^2}$$

$$= \frac{1}{\sqrt{(k - m\omega^2)^2 + c^2\omega^2}} \tag{11.16}$$

where H^* is the complex conjugate of H.

Example: An offshore wind turbine system comprises five physical components: rotor, transmission, generator, support structure, and control system. Each of these influences the dynamic behavior of the complete turbine system. It is noticed that the two most significant excitations are due to the rotation of blades and wave-induced forces. Assume that the turbine system typically has three blades, and the tip speed of each blade is around 90 m/s during operation. From the metocean report for the relevant offshore site, it is known that most of the waves occur at wave periods ranging from 5.5 to 8.5 s. Give a preliminary specification of the natural period of the support structure supporting a wind turbine with blade diameters of 60, 120, and 170 m, respectively.

11.1 Forced Vibrations Under Harmonic Excitations

Solution: With regard to blade excitations, the support structure's natural period should not coincide with the first and second excitation period of the blades. The first excitation period is calculated as (π × blade diameter)/(blade tip speed). This gives the first excitation period (T_{1p}) of 2.1, 4.2 and 5.9 s for the blades with diameters of 60, 120, and 170 m, respectively. The second excitation period (T_{3p}) is the blade's passing period, which is the first excitation period divided by the number of blades, i.e., 3 for the current example. Therefore, the second excitation period is 0.7, 1.4 and 2.0 s for the blades with diameters of 60, 120, and 170 m, respectively. This gives three possible natural period ranges for designing the support structure: a natural period larger than the first blade excitation period (soft–soft design), the one between the first and the second blade excitation period (soft-stiff design), and the one below the second blade excitation period (stiff–stiff design). This design methodology is illustrated in Fig. 11.7.

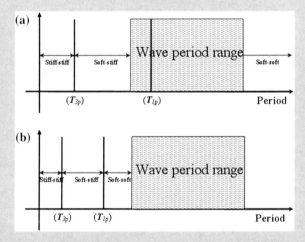

Fig. 11.7 Period interval for stiff–stiff, soft-stiff and soft–soft design of offshore wind turbine support structures. **a** Large blade diameter. **b** Normal blade diameter

Traditionally, the soft–soft design is preferred because this usually leads to an economical design due to the need for less material and construction cost. For the current example, this means that the natural period should be above 2.1, 4.2 and 5.9 s for the blades with diameters of 60, 120, and 170 m, respectively. It is noticed that the periods of 4.2 and 5.9 s are rather close to the wave excitation period. Therefore, the support structures with 120 and 170 m blade diameters need to be designed with a natural period above the wave period, say in the range above 12 s, which is essentially a compliant structure.

For soft-stiff design, this means that the support structure's natural period should be in the range of [0.7, 2.1 s] for 60 m blade diameter, [1.4, 4.2 s] for

120 m blade diameter, and [2.0, 6.0 s] for 170 m blade diameter. Since the period of 6.0 is also in the range of the wave excitation period, the natural period range for 170 m blade diameter has to be adjusted to, for example, [2.0, 5.5 s] as shown in the upper figure of in Fig. 11.7.

For stiff–stiff design, the structures' natural periods need to be below 0.7, 1.4 and 2.0 s for the blades with diameters of 60, 120, and 170 m, respectively. This usually leads to an uneconomical design due to the cost increase with regard to material and construction.

Note that, in the wind energy industry, the general trend is that the scale of turbines is becoming larger and larger. This would result in an increase of the blade's diameter, e.g., a 170 m diameter for a 7 MW wind turbine. The first and second excitation period are also significantly increased as shown in this example. This motivates the engineer to shift from a soft–soft to a soft-stiff or even to a stiff–stiff design. In addition, the variable tip speed of the turbine also becomes a design alternative, which adds additional restrictions on the natural period range of the structures.

As an innovative idea, it is also possible to convert the existing/abandoned offshore rigs into support structures for offshore wind turbine systems, which can avoid/delay enormous decommissioning costs for oil companies as well as avoid cost and pollution for constructing new support structures for offshore wind turbines. Most of the existing fixed offshore platform structures have natural periods below 3.5 s, after removing part of the topside, the natural periods will further decrease. This period range is relevant to the soft-stiff or even soft–soft design. For developing this idea, one also needs to account for the cost with respect to maintenance and power grid integration.

Example: An offshore jacket structure has a natural period of 3.2 s. When it is subjected to a 13 m design wave with the wave period of 10.7 s, a base shear of 39 MN along the direction of vibration corresponding to the natural period of the jacket is obtained. Calculate the corresponding equivalent quasi-static action on the jacket due to the inertia effects. The weights for the jacket and topside are 8,000 and 15,000 tons, respectively. The damping (structural damping + hydrodynamic damping + foundation damping, as will be elaborated in Sect. 14.5) can be assumed to be 2 %.

Solution: By assuming that the eigenperiods of the first and second eigenmodes are close, and the dynamic responses are dominated by those two eigenmodes, the DAF can be calculated based on the first eigenperiod of the jacket structure:

11.1 Forced Vibrations Under Harmonic Excitations

$$\left|\frac{X_p}{X_0}\right| = \frac{1}{\sqrt{\left[1 - (\Omega/\omega_n)^2\right]^2 + [2\zeta(\Omega/\omega_n)]^2}}$$

$$= \frac{1}{\sqrt{\left[1 - (T_n/T_w)^2\right]^2 + [2\zeta(T_n/T_w)]^2}}$$

$$= \frac{1}{\sqrt{\left[1 - (3.2/10.7)^2\right]^2 + [2 \times 0.02 \times (3.2/10.7)]^2}}$$

$$= 1.098$$

The equivalent quasi-static action on the jacket and topside is 39 MN $\times$ (1.098-1) = 3.822 MN.

It is noted that the equivalent quasi-static action should be applied to the entire structure (Fig. 1.24c) instead of on the jacket where the wave loading is applied, so it is applied as an acceleration field on the entire structure along the direction of the wave loading. The acceleration field is tuned such that the amplification of base shear is fairly correct for the corresponding design wave:

$$(3.822\text{MN} \times 10^6)/(23{,}000\text{tons} \times 10^3) = 0.167 \text{ m/s}^2.$$

More realistically, the acceleration can be applied as a uniform rotational acceleration $\ddot{\theta}$ [rad/s^2] (Fig. 11.8) with the rotation center located at the horizontal center of the jacket bottom. Based on the distance between the rotation center and the center of gravity (CoG) of topside (H_{ts}) and jacket (H_{jk}) respectively, the relative horizontal acceleration between the jacket and the topside is H_{jk}/H_{ts}, respectively. The acceleration on the CoG of the jacket can then be calculated as:

$$a_{jk} = \frac{3.822 \times 10^6 \text{ N}}{\left[8000 \times 10^3\right] + \left[\left(\frac{H_{ts}}{H_{jk}}\right) \times 15000 \times 10^3\right]}$$

The rotation acceleration on both jacket and topside is calculated as

$$\ddot{\theta} = a_{jk}/H_{jk}.$$

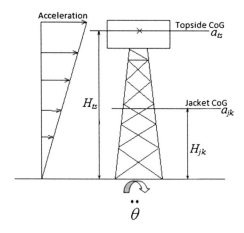

Fig. 11.8 The equivalent quasi-static action may be approximated using angular acceleration $\ddot{\theta}$ at the bottom of the jacket

It is noted that the DAF can be different not only due to the variation of excitation period, but also due to the direction of excitation and different parts of a structure under investigation. For example, if a structure is subjected to excitations along the direction of an eigenmode other than the first one, the first eigenmode may make little contribution to the DAF. Another example is that even though a DAF calculated based on the wave period and eigenperiod can reasonably be applied to the members within a narrow face or horizontal plane shown in Fig. 11.9, it cannot be applied for jacket braces on a broad face that is perpendicular to the wave excitations. Furthermore, different components of forces on a member and a joint may have different DAFs. Therefore, only the relevant DAF should be used, which may need sophisticated judgment on the part of the engineer.

11.1.2 Responses to Harmonic Base Excitations

Instead of external forces, consider the system shown in Fig. 11.1 excited by prescribed harmonic motions (absolute motions) $y(t) = Y\sin(\Omega t)$ at its base as shown in Fig. 11.10. This model represents a large number of practical engineering problems, such as a structure subjected to earthquake loading, running vehicles subjected to ground excitations due to road roughness etc. With a slight modification of the solutions presented previously for the applied forces, the responses under the base excitations can be obtained. The force on the spring is proportional to the relative displacement $z(t) = x(t) - y(t)$ between the base and the mass. The equation of motions can then be written as:

$$m\ddot{z}(t) + c\dot{z}(t) + kz(t) = -m\ddot{y}(t) \qquad (11.17)$$

11.1 Forced Vibrations Under Harmonic Excitations 153

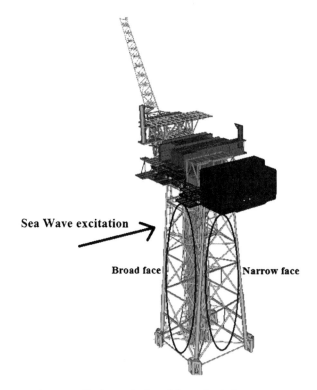

Fig. 11.9 Sea wave is perpendicular to the broad face of a jacket structure (courtesy of Aker Solutions)

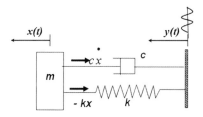

Fig. 11.10 An SDOF spring-mass-damper system under base excitations $y(t) = Y \sin(\Omega t)$

Assuming that the responses are also harmonic, the steady-state solution of the equation above is:

$$z(t) = Z \sin(\omega t - \varphi) \tag{11.18}$$

Dividing the equation of motions with m leads to:

$$\ddot{z}(t) + 2\zeta \omega \dot{z}(t) + \omega^2 z(t) = \Omega^2 Y \sin(\Omega t) \tag{11.19}$$

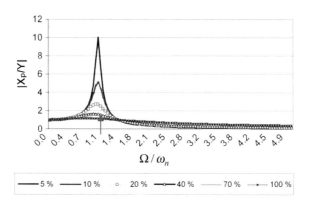

Fig. 11.11 Plot of non-dimensional gain function varying with Ω/ω_n ratio and damping, due to base excitations

where φ is the phase between the base excitation input and the response output:

$$\varphi = \tan^{-1}\left(\frac{2\zeta(\Omega/\omega_n)^3}{1-(\Omega/\omega_n)^2+2\zeta(\Omega/\omega_n)^2}\right) \quad (11.20)$$

The ratio between the steady-state amplitude (absolute motions) and the excitation amplitude is:

$$\left|\frac{X_p}{Y}\right| = \sqrt{\frac{1+[2\zeta(\Omega/\omega_n)]^2}{\left[1-(\Omega/\omega_n)^2\right]^2+[2\zeta(\Omega/\omega_n)]^2}} \quad (11.21)$$

The value above is a non-dimensional gain function, which is illustrated in Fig. 11.11. It is noticed that, regardless of damping level, the function always reaches unity when Ω/ω_n equals to $\sqrt{2}$. This characteristic is important for the design of vibration absorber or isolator for various types of devices: by choosing a decent stiffness k so that $\Omega/\omega_n > \sqrt{2}$, the vibration transmitted from the excitation sources to the protected devices can be significantly decreased.

In many cases, the relative motions ($z(t)$) between the responses and base excitations are more interesting. Therefore, one can calculate the ratio between the steady-state amplitude (relative motion) and the excitation amplitude as:

$$\left|\frac{Z_p}{Y}\right| = \sqrt{\frac{(\Omega/\omega_n)^4}{\left[1-(\Omega/\omega_n)^2\right]^2+[2\zeta(\Omega/\omega_n)]^2}} \quad (11.22)$$

11.2 Forced Vibrations Under Complex Periodical Excitations

Many types of excitation functions may be periodical, but not harmonic (Fig. 8.2a). This means that even though they may repeat themselves as shown in Fig. 8.2b–d, their responses cannot be described by vibrations at a single frequency. If the excitations extend over a wide range of frequencies, many resonant modes may be excited. It is more difficult to determine responses under periodical excitations than under harmonic excitations, especially if one needs to obtain the time domain responses.

However, for a convenient solution of responses, those excitations can be represented using a Fourier spectrum. This applies to, for example, ground excitations on running vehicles and dynamic loading due to wave and fluctuating wind velocity. As discussed in Sect. 12.4, although some of the motion records, such as seismic ground motions, are not perfectly periodical, in many applications they can still be modeled by using the Fourier transform techniques.

Moreover, compared to a time-domain method that requires the solutions of differential equations, the frequency-domain method only requires the solutions of algebraic equations. In addition, it often gives information regarding characteristics of frequency contents on the dynamic responses, which make it seldom necessary to transform the frequency-domain results back to the time-domain results through Fourier transformation.

Similar to the responses due to harmonic excitations, the complementary solution of vibrations under periodical force excitations normally dies out due to the presence of damping. Therefore, the steady-state solution is more interesting. This can be obtained, for a linear system, through combining the responses due to individual excitations in terms of the Fourier series [125].

The responses of an SDOF system subjected to periodical force excitations $(F(t) = c_0 + \sum_{i=1}^{N} c_i \sin(\Omega_i t + \gamma_i))$ can be expressed as:

$$x(t) = \frac{1}{m\omega_n^2} \left[c_0 + \sum_{i=1}^{\infty} c_i M(r_i, \zeta) \sin(\Omega_i t + \gamma_i - \phi_i) \right] \tag{11.23}$$

where:

c_0 and c_i are the average and the ith harmonic component of the Fourier series in terms of excitations, respectively (see Sect. 9.1);

$r_i = \frac{\Omega_i}{\omega_n}$ is the ratio between the excitation frequency and the natural frequency of the system;

$\phi_i = \tan^{-1}\left(\frac{2\zeta r_i}{1-r_i^2}\right)$;

ζ is the damping ratio of the system;

m is the equivalent mass of the system;

$M(r_i, \zeta)$ is the magnification factor of the system, i.e., $M(r_i, \zeta) = \dfrac{1}{\sqrt{\left(1-r_i^2\right)^2+4\zeta^2 r_i^2}}$;

ω_n is the natural frequency of the SDOF.

It is noted that the responses are essentially influenced by c_i and the ratio between the excitation frequency and the system's natural frequency. Further, c_i is influenced by the harmonic amplitude of sinuasoidal and cosinsoidal excitations. The responses are dominated by the harmonic amplitude with a frequency close to the system's natural frequency.

The upper bound of the maximum steady-state responses is:

$$x_{max} \leq \frac{1}{m\omega_n^2} \left[c_0 + \sum_{n=1}^{\infty} c_i M(r_i, \zeta) \right] \qquad (11.24)$$

11.3 Forced Vibrations Under Non-periodical Excitations

In the previous sections we have learned how to calculate the dynamic responses due to harmonic or periodical excitations. In particular, we notice that the total dynamic responses comprise both the complementary solution and the steady-state solution, of which the former is of a harmonic type with a frequency equal to the system's natural frequency, and the latter is of a periodical form with a frequency equal to the excitation's frequency Ω.

In this section, we will study the responses due to non-periodical excitations. Obviously, steady-state responses do not appear any more (the responses normally possess a non-zero steady-state) and the entire solution is transient [158]. In many cases, the homogeneous solution (almost equal to free vibrations) interacts with the forced responses even after the excitations have ceased.

To solve this problem, one can treat the excitations as the superposition of impulses of rather short duration, as will be discussed in Sect. 11.3.1, or represent the excitations by Fourier integral presented in Sect. 11.4, so that the excitations are essentially periodical.

Examples of non-periodical excitations are wave slaming impacts on ships or offshore platform topsides and transient pressure forces on structures due to explosion. Ground motion excitations due to earthquake are also non-periodical with short duration, while the maximum structural response normally occurs before excitations cease.

11.3 Forced Vibrations Under Non-periodical Excitations

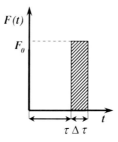

Fig. 11.12 Impulsive force excitations

11.3.1 Transient Responses to Force Excitation with Short Duration

Let's consider a structure excited by suddenly applied excitations $F(t)$ that are neither harmonic nor periodical. The general form of the governing equation of motions is:

$$m\ddot{x}(t) + c\dot{x}(t) + kx(t) = F(t) \tag{11.25}$$

By dividing the equation above with m and rearranging it, one obtains:

$$\ddot{x}(t) + 2\zeta\omega_n \dot{x}(t) + \omega_n^2 x(t) = \frac{1}{m}F(t) \tag{11.26}$$

With the initial conditions ($x(0)$ and $\dot{x}(0)$), the complete formulation can be established.

Again, in the case with damping presented, the complementary solution dies out sooner or later, leaving the particular solutions alone. However, compared to that of the maximum responses under periodical loading, the effects of damping are much less significant for decreasing the maximum response due to short duration excitations. This is primarily because the maximum responses due to the short duration excitations will appear in a short time, before the damping forces can efficiently absorb significant energy from the structure [161]. Therefore, in many cases, the damping can be neglected for calculating the transient responses due to short duration excitations.

However, the initial conditions and complementary solutions to the homogeneous equation significantly affect the short-term transient responses, and it is therefore advisable to obtain the complementary solutions and particular solutions simultaneously, with the initial conditions incorporated [162].

11.3.1.1 Responses Due to Arbitrary Force Excitations Using Convolution Integral

First we consider a system subjected to a force with a short duration as shown in Fig. 11.12. For a general close-form solution of Eq. (11.26), the convolution integral method can be used to obtain the responses. This method is derived using the equilibrium of momentum:

$$\int_{t_1}^{t_2} F(t)dt = m\left[\dot{x}(t_2) - \dot{x}(t_1)\right] \tag{11.27}$$

The time integral of force is designated by the symbol $\hat{\hat{F}}$:

$$\hat{\hat{F}} = \int_{\tau}^{\tau+\Delta\tau} F_0 dt \tag{11.28}$$

We here define an impulsive force with the amplitude of $\hat{\hat{F}}/\Delta\tau$ and the time duration of $\Delta\tau$. When $\hat{\hat{F}}$ is equal to unity, the force in the limiting case $\Delta\tau \to 0$ is called the unit impulse or Dirac delta function ($\delta(t - \tau)$), which has the following properties:

$$\int_0^{+\infty} \delta(t - \tau)d\tau = 1 \tag{11.29}$$

$$\delta(t - \tau) = 0 \text{ for } t \neq \tau \tag{11.30}$$

Therefore, the impulsive force applied at time τ is:

$$F(t) = F_0\delta(t - \tau) \tag{11.31}$$

The responses to an unit impulse applied at $t = 0$ with initial conditions equal to zero are called impulsive responses and are denoted by $h(t)$. For any time later than τ, the impulsive responses $h(t - \tau)$ can be obtained by shifting $h(t)$ to the right along the scale by $t = \tau$.

Thereafter, at $t = 0$, a radical change in the system motions takes place when the short duration and high amplitude forces excite an initial motion of the system, followed by free vibrations. For a unit impulse at $t = 0$, i.e., $F_0 = 1$, the velocity and displacement of the mass immediately after the initial impulse at $t = 0^+$ are therefore:

$$\dot{h}(0^+) = \frac{1}{m} \tag{11.32}$$

$$h(0^+) = 0 \tag{11.33}$$

11.3 Forced Vibrations Under Non-periodical Excitations

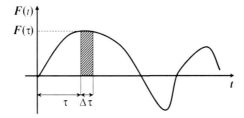

Fig. 11.13 Arbitrary force histories applied on a system

The velocity and displacement due to an applied step force $F(0)$ are:

$$\dot{x}(0^+) = \frac{1}{m} \tag{11.34}$$

$$x(0^+) = 0 \tag{11.35}$$

By realizing the initial condition $h(0) = 0$, one can derive the impulsive responses [158] of the undamped system:

$$h(t) = \begin{cases} \frac{1}{m\omega_n} \sin(\omega_n t) & \text{for } t > 0 \\ 0 & \text{for } t < 0 \end{cases} \tag{11.36}$$

Or

$$x(t) = \begin{cases} \frac{F_0}{m\omega_n} \sin(\omega_n t) = \frac{F_0}{k}[1 - \cos(\omega_n t)] & \text{for } t > 0 \\ 0 & \text{for } t < 0 \end{cases} \tag{11.37}$$

From the equation above it is noticed that the maximum displacement of the system due to the step excitations is twice the quasi-static displacement $\frac{F_0}{k}$.

The responses of the damped system are:

$$h(t) = \begin{cases} \frac{1}{m\omega_d} e^{-\zeta\omega_n t} \sin(\omega_d t) & \text{for } t > 0 \\ 0 & \text{for } t < 0 \end{cases} \tag{11.38}$$

Or

$$x(t) = \begin{cases} \frac{F_0}{m\omega_d} e^{-\zeta\omega_n t} \sin(\omega_d t) & \text{for } t > 0 \\ 0 & \text{for } t < 0 \end{cases} \tag{11.39}$$

The derivation above can be extended to calculate the responses under arbitrary excitation histories as shown in Fig. 11.13. The excitations $F(t)$ can be regarded as a series of impulses with different amplitudes. We here examine one impulse starting at time τ. Again, in the limiting case $\Delta\tau \to 0$, its contribution to the total responses at time t is:

$$\Delta x(t, \tau) = F(\tau)\Delta\tau h(t - \tau) \tag{11.40}$$

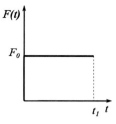

Fig. 11.14 Step force excitation

For a linear system, the principle of superposition is applicable. Therefore, the response at time t is the sum of responses due to a sequence of individual impulses, which are known as convolution integral:

$$x(t) = \int_0^t F(\tau)h(t-\tau)d\tau \qquad (11.41)$$

Or

$$x(t) = \sum F(\tau)h(t-\tau)\Delta\tau \qquad (11.42)$$

$h(t-\tau)$ is obtained from Eq. (11.38) by replacing t by $t-\tau$. Therefore, the damped responses in Eq. (11.38) can be rewritten as:

$$x(t) = \frac{1}{m\omega_d} \int_0^t F(\tau) e^{-\zeta\omega_n(t-\tau)} \sin[\omega_d(t-\tau)] d\tau \qquad (11.43)$$

It is obvious that the initial condition is not accounted for in the equation above.

In many cases, a closed-form solution of Eq. (11.26) of excitations does not exist, nor does the convolution integral. Therefore, numerical methods have to be used to either evaluate the convolution integral or directly solve Eq. (11.26).

11.3.1.2 Impulsive Responses Due to a Step/Rectangular Force Excitation with Short Duration

For a lightly damped (damping ratio $\zeta < 1$) system subjected to a force with a short duration as shown in Fig. 11.14, the entire solution for the dynamic responses is the sum of complementary and particular solutions $\left(\frac{F_0}{m\omega_n^2}\right)$:

$$x(t) = Xe^{-\zeta\omega t} \sin\left(\sqrt{1-\zeta^2}\omega_n t - \phi\right) + \frac{F_0}{m\omega_n^2} \qquad (11.44)$$

Strictly speaking, the homogeneous solution is only equal to the free vibration responses in cases in which there are no excitations. However, the

11.3 Forced Vibrations Under Non-periodical Excitations

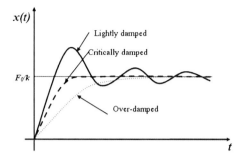

Fig. 11.15 Responses of an SDOF system due to the step force excitations

particular solution is equal to the forced responses when the responses reach a steady-state.

Because the duration of force is rather short, one can then assume the initial condition to be $x(0) = \dot{x}(0) = 0$. Therefore, the final solutions for the lightly damped system ($\zeta < 1$) are:

$$x(t) = \frac{F_0}{k}\left[1 - \frac{e^{-\zeta\omega_n t}}{\sqrt{1-\zeta^2}}\sin\left(\sqrt{1-\zeta^2}\omega_n t + \varphi\right)\right] \quad (11.45)$$

where $\varphi = \cos^{-1}(\zeta)$.

From the equation above it is noticed that the maximum displacement of the system with the damping presented is less than twice the quasi-static displacement $\left(\frac{F_0}{k}\right)$.

For a critically damped system ($\zeta = 1$), the responses are [165]:

$$x(t) = \frac{F_0}{k}[1 - (\omega_n + 1)e^{-\omega_n t}] \quad (11.46)$$

For an overdamped system ($\zeta > 1$), the responses are [165]:

$$x(t) = \frac{F_0}{k}\left\{1 - \frac{e^{-\zeta\omega_n t}}{2\omega_n\sqrt{\zeta^2-1}}\left[\lambda_1 e^{-\lambda_2 t} - \lambda_2 e^{-\lambda_1 t}\right]\right\} \quad (11.47)$$

where λ_1 and λ_2 are eigenvalues of the system.

The responses due to step force excitations are plotted in Fig. 11.15.

11.3.2 Responses Due to Arbitrary Base Excitations Using Convolution Integral

As discussed in Sect. 11.1.2, the equation of motions can be rewritten in terms of relative displacement $z(t) = x(t) - y(t)$ (Fig. 11.10) between the base and the mass m:

$$m\,\ddot{z}(t) + c\,\dot{z}(t) + kz(t) = -m\,\ddot{y}(t) \tag{11.48}$$

Dividing the equation above by m, one obtains:

$$\ddot{z}(t) + 2\zeta\omega_n\,\dot{z}(t) + \omega_n^2 z(t) = -\ddot{y}(t) \tag{11.49}$$

For systems without damping and initially at rest, the relative displacement is:

$$z(t) = -\frac{1}{\omega_n}\int_0^t \ddot{y}(\tau)\sin[\omega_n(t-\tau)]d\tau \tag{11.50}$$

Rewritten, the equation above gives [18]

$$z(t) = -\frac{1}{\omega_n}\left\{ \sin\omega_n t \int_0^t \ddot{y}(\tau)\cos(\omega_n\tau)d\tau - \cos(\omega_n t)\int_0^t \ddot{y}(\tau)\sin(\omega_n\tau)d\tau \right\} \tag{11.51}$$

11.3.3 Responses to Non-Periodical Excitations with Fourier Integral

The mathematical formulation in frequency domain is applicable to both periodical and non-periodical responses. In Sect. 11.2 we illustrated that, for a linear system, any periodical function in the time domain that contains a wide range of frequency (harmonic) components has an equivalent counterpart in the frequency domain, which can be represented by a convergent series of independent harmonic functions at the integral of the function's frequency as a Fourier series. This is because the system under study is stable before it is excited and for which motions die away after the excitations.

Similarly, the $h(t)$ characterizing a system in the time domain also has an equivalent counterpart: the transfer function $H(\omega)$ in the frequency domain. Their relationships can be clearly found by making the frequency approach zero, so that the first time interval stretches without bound, the function then becomes non-periodical. In this process, the values of each two adjacent discrete frequencies become closer and closer until they become continuous, leading to the Fourier series becoming a Fourier integral [158].

Let $F(t) = e^{i\omega t}$, the responses to non-periodical excitations can then be written in the form of Fourier transform pairs (inverse Fourier and Fourier transform) between $x(t)$ and $X(\omega)$:

$$x(t) = \frac{1}{2\pi}\int_{-\infty}^{+\infty} X(\omega)e^{i\omega t}d\omega = \frac{1}{2\pi}\int_{-\infty}^{+\infty} H(\omega)F(\omega)e^{i\omega t}d\omega \tag{11.52}$$

11.3 Forced Vibrations Under Non-periodical Excitations

Table 11.1 Relationship of responses between time and frequency domains

	Time domain	Frequency domain
Excitations	$F(t)$	Transformed excitation: $F(\omega)$
System characteristics (filter)	Impulsive responses: $h(t)$	Transfer function: $H(\omega)$
Responses	$x(t)$	Transformed response: $X(\omega)$

$$X(\omega) = \int_{-\infty}^{+\infty} x(t)e^{i\omega t}\,dt \tag{11.53}$$

where $F(\omega)$ is the Fourier transform of $F(t)$.

Similarly, the impulsive responses (under unit excitation $F(\omega) = 1$) are given by:

$$x(t) = \frac{1}{2\pi} \int_{-\infty}^{+\infty} X(\omega)e^{i\omega t}\,d\omega = \frac{1}{2\pi} \int_{-\infty}^{+\infty} H(\omega)e^{i\omega t}\,d\omega \tag{11.54}$$

$$H(\omega) = \int_{-\infty}^{+\infty} h(t)e^{i\omega t}\,dt \tag{11.55}$$

The relationship between the responses and force in the frequency domain can be clearly expressed as the Fourier transform of the responses:

$$X(\omega) = F(\omega)H(\omega) \tag{11.56}$$

From the equation above, it is noticed that the information regarding phase in responses $X(\omega)$ are included as the transfer function $H(\omega)$ contains the phase information.

Interested readers may conduct the proof of the relationship above as an exercise.

This expression is essential for calculating the dynamic responses due to random vibrations, in which the frequency composition (random) rather than time dependence (deterministic) of the responses is of interest (as will be elaborated in Sect. 11.4). The relationship between each item in the frequency domain expressed in Eq. (11.56) and its counterpart in the time domain in Eqs. (11.36) or (11.38) is summarized in Table 11.1. The essential part of this relationship is the time ($h(t)$) and frequency ($H(\omega)$) domain transfer functions, which are related through a Fourier transform pair.

Due to the computation efficiency and for a straight disclosure of a system's dynamic characteristics, the responses in the frequency domain are more popular than their time domain counterpart. However, the latter cherishes a better physical meaning since the former one only assumes a fully steady oscillatory state, whereas the real physical behaviors of many loadings and structural dynamic responses are essentially non-steady [166].

In addition, it should also be noted that, for calculating responses due to transient excitations, the use of Fourier transform often leads to contour

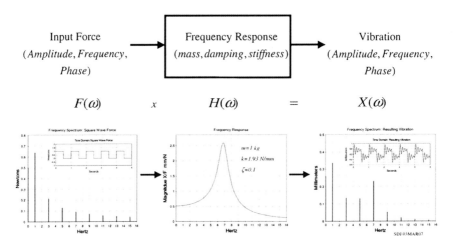

Fig. 11.16 Responses obtained by Fourier transfer (Figure by Lzyvzl)

integrations in the complex plane due to the definite integral in Eq. (11.52) [158]. In this case, the Laplace transform method, which is a modification of Fourier transform, is more appealing. Readers who are interested in this topic may read references by Meirovitch [158] and Thomson [18].

11.4 Forced Vibrations Under Random Excitations

11.4.1 Method

In Sect. 8.3, random motions are presented, which are the counterpart of deterministic motions. The major difference between the two motions is that the instantaneous value of random motions cannot be explicitly predicted at any time instant or be reproduced, while deterministic motions can. We characterize random motions with probability distributions, such as their mean and variance.

In the previous Sects. 11.1, 11.2 and 11.3, the responses to harmonic, periodical and transient excitations have been discussed. In this section, we are going to combine knowledge on random motions and forced vibrations previously learned to treat random vibration problems. Since it is not possible to explicitly predict values of random vibrations at a time instant, we have to treat them as obtaining the statistics of responses (unknown) from the given statistics of the loading.

To achieve this, we will present a direct method to calculate responses due to random excitations. As elaborated in Sect. 8.3, random excitations should be characterized by both their probabilistic properties and their frequency contents. They normally have components at multiple frequencies. For simplicity, it is assumed that the input is force, denoted as F, and the output is displacement, denoted as X. Readers should take both the input and output in a generic way and apply mathematical intuition together with the knowledge learned in the previous sections.

11.4 Forced Vibrations Under Random Excitations 165

Section 11.3.3 presents that, for random vibrations, the Fourier transform of response $X(\omega)$ in the frequency domain can be obtained as the product of the Fourier transform of forcing function $F(\omega)$ and the transfer function (frequency response function) $H(\omega)$. This is illustrated in Fig. 11.16.

It is noted that, in many cases, the random excitations cover a wide range of frequencies. Therefore, rather than being proportional to the amplitude of the individual responses at excitation frequencies, as the harmonic excitations are, the response amplitude primarily depends on the transfer function, which is a parameter representing the dynamic amplification properties of the excited structure/system itself, and the transfer function is independent of the amplitude of loading.

For an SDOF damped system with force as excitations, the transfer function is expressed in Sect. 11.1.1 as:

$$H(\omega) = \frac{1}{(k - m\omega^2) + ic\omega} \tag{11.57}$$

Here, one should notice that the transfer function is normalized by dividing the Fourier transform of response $X(\omega)$ with the input force amplitude,

---

Example: Derive the mean square response $E(x^2(t))$ under a harmonic excitation force $F(t) = F_0 \sin(\omega t)$.

Solution: The responses can be expressed as:

$$x(t) = \frac{(Xe^{i\omega t} + X^* e^{-i\omega t})}{2}$$

where X^* is the complex conjugate of X.

Therefore, the mean square responses can be written as:

$$E(x^2(t)) = \lim_{T \to \infty} \frac{1}{T} \int_0^T \frac{(X^2 e^{2i\omega t} + 2XX^* + X^{*2} e^{-2i\omega t})}{4} dt = \frac{XX^*}{2} = \frac{|X|^2}{2}$$

The mean square force can be expressed as:

$$E(F^2(t)) = \lim_{T \to \infty} \frac{1}{T} \int_0^T \frac{F_0^2}{2}[1 - \cos(2\omega t)] dt = \frac{F_0^2}{2}$$

From Eq. (11.56), one has:

$$E(x^2(t)) = E(F^2(t))H(\omega)H^*(\omega) = |H(\omega)|^2 E(F^2(t))$$

where $H^*(\omega)$ is the complex conjugate of $H(\omega)$.

i.e., $X(\omega)/F(\omega)$. We will soon show that, to be consistent with the definition of the power spectrum density that has been elaborated in Sect. 9.2, the response spectrum $X(\omega)$ can have a unit of square of the response amplitude per Hz or rad/s, which requires that the transfer function be squared during the calculation of the response spectrum in the form of power spectrum density.

In random vibrations, the phase angle has little meaning. Instead, the mean square value $(E(x^2(t)))$ of responses associated with the average energy over a time interval T is our main concern:

$$E(x^2(t)) = \lim_{T \to \infty} \frac{1}{T} \int_0^T x^2(t)dt \qquad (11.58)$$

The equation in the example above indicates that the mean square responses are equal to the mean square excitation multiplied by the square of the modulus of transfer function [18]. This conclusion can be extended to study the responses due to random loadings induced by wave, wind, earthquakes, etc.

In general, the mean square responses of a system are of engineering interest. Their contribution in each frequency interval $\Delta\omega$ due to the increment of frequency from between $\omega_i - \Delta\omega/2$ and $\omega_i + \Delta\omega/2$ is:

$$x(\omega_i) = H(\omega_i)\sqrt{\frac{\Delta\omega S_F(\omega_i)}{2\pi}} \qquad (11.59)$$

where $S_F(\omega_i)$ is the contribution from power spectra density of the excitation force between $\omega_i - \Delta\omega/2$ and $\omega_i + \Delta\omega/2$:

$$S_F(\omega_i) = \lim_{\Delta\omega \to 0} \frac{E(F^2(t) \text{ between } \omega_i - \Delta\omega/2 \text{ and } \omega_i + \Delta\omega/2)}{\Delta\omega} \qquad (11.60)$$

The process of the equation above is actually to make the $F(t)$ pass a band-pass filter with $\Delta\omega$ band and within the frequency interval $\omega_i - \Delta\omega/2$ and $\omega_i + \Delta\omega/2$, the output is squared, averaged and divided by $\Delta\omega$ [18]. Provided that $\Delta\omega$ is constant, the time history generated by power spectra density will be repeated with a period of $2\pi/\Delta\omega$. In order to increase this period, one may adopt a small value of $\Delta\omega$ or give frequency spacing uneven intervals [53].

Since the excitations and responses at M different frequencies ω_i are assumed to be randomly phased, the variance of the total response is the sum of the variances of components. The root of mean square responses, i.e., the square root of the variance, is written as:

$$x_{rms} = \sqrt{\sum_{i=1}^{M} x^2(\omega_i)} \qquad (11.61)$$

The power spectra density (mean square response spectrum) of the responses—which is also called the auto-spectra density function, indicating that only one

11.4 Forced Vibrations Under Random Excitations

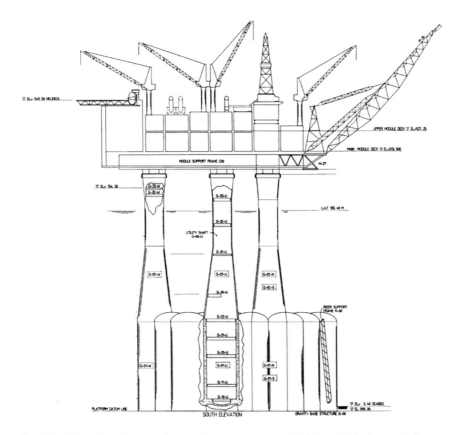

Fig. 11.17 Elevation drawing of a gravity-based structure (GBS) located in the North Sea

process is involved (in contrast to the cross-spectra density function, involving two processes)—introduced in Sect. 11.5, can then be calculated as:

$$S_X(\omega_i) = |H(\omega_i)|^2 S_F(\omega_i) \tag{11.62}$$

The above equation is widely used in many fields of engineering applications. For example, the response spectrum $S_X(\omega_i)$ for linear displacement responses of offshore structures or linear ship motions is calculated by filtering the wave energy spectrum (here $S_F(\omega_i) = S_\zeta(\omega_i)$, see Sect. 9.2) with an appropriate motion transfer function. This is achieved by multiplying each ordinate of the wave spectrum with the square of the modulus of motion transfer function at the corresponding frequencies, provided that the transfer function is normalized by dividing the wave amplitude [91].

Again, in the equation above, the information regarding phase in power spectra density responses $S_X(\omega_i)$ gets lost as the phase information is included in the

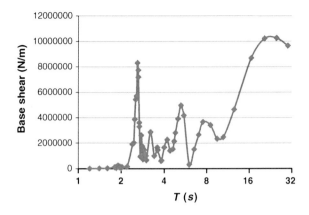

Fig. 11.18 The modulus of transfer function of base shear force for the GBS structure

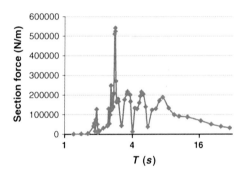

Fig. 11.19 The modulus of transfer function of section force at the GBS shaft

transfer function $H(\omega_i)$ (Eq. (11.57)), but not in $|H(\omega_i)|$, in which only the information for amplitude amplification can be found Fig. 11.17.

The transfer function can be expressed using various response measures, such as motions, forces, stresses or strains etc. Figures 11.18 and 11.19 show the modulus of transfer functions of base shear and section force on one shaft of a GBS structure shown in Fig. 11.17. It is clearly shown that the highest peak of the responses occurs at around the period ranges from 2.6 to 2.8 s, which is in the range between the second and the first eigenperiod of the GBS structure.

In many cases, the power spectra density is also expressed in terms of frequency in Hz. We then have:

$$S_X(f_i) = 2\pi S_X(\omega_i) \qquad (11.63)$$

11.4 Forced Vibrations Under Random Excitations 169

The mean square or the variance of the fluctuating responses is the area under the mean square response spectrum:

$$\sigma_X^2 = \int_{-\infty}^{\infty} S_X(\omega)d\omega \tag{11.64}$$

The two equations above are rather important for the random dynamic analysis.

11.4.2 White Noise Approximation

For structures subjected to loadings with frequencies far from the structures' natural frequency, and if the damping of the structure is also low, by revisiting Eq. (11.62) it is noticed that $H(\omega_i)$ is narrow banded and the responses of the structure are concentrated at the structure's natural frequency. In addition, if the excitations have similar energy content in all frequencies, the influence from the excitation spectrum $S_F(\omega_i)$ is less significant than that from the structure's transfer function $H(\omega_i)$. In this situation, the excitations can be assumed to be white noise, with its power spectrum density $W_0(\omega)$ defined as a constant covering the entire frequency range, known as white noise approximation:

$$W_0(\omega) = S_F(\omega) = C \quad -\infty < \omega < \infty \tag{11.65}$$

where C is a constant value.

The equation above indicates a flat power spectrum. Strictly speaking, it does not exist physically. However, it is a rather useful assumption in many fields related to vibration and control.

Table 11.2 Mean square responses to white noise excitations for an SDOF with stiffness k and the damping ratio ζ[167]

W_0, power spectra density of the excitation input[a]	Measure of the mean square responses[a]	Formula
Force excited system		
$F(t)$	$X(t)$	$\sigma_X^2 = 0.785 f_n W_0 / \zeta k^2$
$F(t)$	$\dot{X}(t)$	Responses are not finite
$F(t)$	$F_{TR}(t)$	$\sigma_{FT}^2 = 0.785 f_n W_0 (1 + 4\zeta^2) / \zeta$
Base excited system		
$Y(t)$	$X(t)$	$\sigma_X^2 = 0.785 f_n W_0 (1 + 4\zeta^2) / \zeta$
$\ddot{Y}(t)$	$\ddot{X}(t)$	$\sigma_{\ddot{X}}^2 = 0.785 f_n W_0 (1 + 4\zeta^2) / \zeta$
$\ddot{Y}(t)$	$Z(t)$	$\sigma_Z^2 = W_0 / (1984 \zeta f_n^3)$
$\ddot{Y}(t)$	$\ddot{Z}(t)$	Responses are not finite

[a] *Note* $F(t)$ is the force applied on the system; $X(t)$ is the absolute displacement of the mass; $Z(t)$ is the displacement of the mass relative to the ground; $Y(t)$ is the base displacement; $F_{TR}(t)$ is the force transmitted to the base

170   11 Forced Vibrations

With the aid of white noise excitation, Table 11.2 shows the closed form of expression of mean square of response σ_X^2.

The white noise approximation is widely adopted in the modern vibration analysis and measurement. For example, when the environmental dynamic loadings are unknown, in order to measure the eigenfrequencies of a structure, the unknown loadings may be assumed to follow white noise, which significantly simplifies the modal testing task and the subsequent analysis.

11.5 Cross-Covariance, Cross-Spectra Density Function and Coherence Function

11.5.1 Cross-Covariance in Time Domain

In Sect. 11.4, it is noted that the power spectra density (mean square response spectrum) of the responses does not contain any information regarding phase between the response and the loading. This information can be conveniently included by introducing cross-covariance or cross-correlation function:

$$C_{XY}(\tau) = E[X(t)Y(t+\tau)] \qquad (11.66)$$

where τ is the time lag, $X(t)$ and $Y(t)$ are two stationary processes.

The time lag τ at which the maximum occurs often has physical significance [46].

It is possible to reverse the order of the subscripts in the equation above, so that the following relationship also holds:

$$C_{YX}(\tau) = E[Y(t)X(t+\tau)] \qquad (11.67)$$

The cross-covariance possesses the property of symmetry, i.e., the two equations above are reflections of one another about the origin:

$$C_{XY}(\tau) = C_{YX}(-\tau) \qquad (11.68)$$

The cross-covariance has an upper and lower bound:

$$-\sigma_X\sigma_Y + \mu_X\mu_Y \le C_{XY}(\tau) \le \sigma_X\sigma_Y + \mu_X\mu_Y \qquad (11.69)$$

Cross-covariance has its limitations when studying dispersive propagation problems in which the time delay is frequency dependent, such as the propagation of seismic wave from bedrock to ground, which travels faster and farther at low frequencies than at high frequencies [46]. In this case, the cross-spectra density, as will be introduced in Sect. 11.5.2, need to be used.

11.5 Cross-Covariance, Cross-Spectra Density Function

11.5.2 Cross-Spectra Density in the Frequency Domain

The cross-spectra density function is defined as the Fourier transform of the cross-covariance:

$$S_{XY}(\omega) = \frac{1}{2\pi} \int_{-\infty}^{\infty} C_{XY}(\tau)e^{-i\omega\tau}d\tau \tag{11.70}$$

$$S_{YX}(\omega) = \frac{1}{2\pi} \int_{-\infty}^{\infty} C_{YX}(\tau)e^{-i\omega\tau}d\tau \tag{11.71}$$

As briefly mentioned in Sect. 11.4.1, different from the power spectra density (auto-spectra density), which includes only one process and is real valued, the cross-spectra density function includes two processes, and is therefore complex so that the information regarding the phase shift between the two processes is accounted for.

Similar to cross-covariance, symmetry also holds for the cross-spectra density function, i.e., $S_{XY}(\omega)$ and $S_{YX}(\omega)$ are complex conjugates of each other:

$$S_{XY}(\omega) = S_{YX}(-\omega) = S_{YX}^{*}(\omega) \tag{11.72}$$

Note that the area under spectrum density function $S_X(\omega)$ (discussed in Sect. 11.4) is the mean-square (variance) of the process $X(t)$, the integral of the cross-spectral density is equal to the expected value of the product $X(t)Y(t)$, which is their covariance.

11.5.3 Coherence Function in the Frequency Domain

The linear dependence of two processes $X(t)$ and $Y(t)$ can be expressed with non-dimensional coherence function:

$$Coh_{XY}(\omega) = \gamma_{XY}^2(\omega) = \frac{|S_{XY}(\omega)|^2}{S_X(\omega)S_Y(\omega)} = \frac{S_{XY}(\omega)S_{XY}^{*}(\omega)}{S_X(\omega)S_Y(\omega)} \tag{11.73}$$

Since $|S_{XY}(\omega)|^2 \leq S_X(\omega)S_Y(\omega)$, the coherence function is bounded by:

$$0 \leq \gamma_{XY}^2(\omega) \leq 1 \tag{11.74}$$

$\gamma_{XY}^2(\omega) = 1$ indicates a purely linear dependence between $X(t)$ and $Y(t)$. On the other hand, $\gamma_{XY}^2(\omega) = 0$ shows that there is no linear relationship between $X(t)$ and $Y(t)$ at all. However, the value of coherence function normally lies between these two extreme cases, indicating that systems are either not perfectly linear or

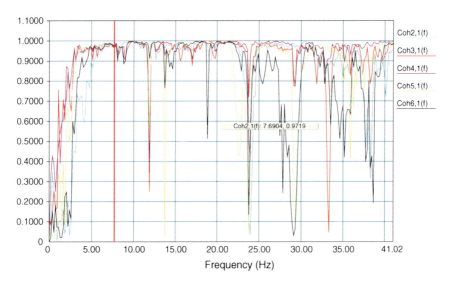

Fig. 11.20 Coherence functions at various locations on a ship deck during a modal testing

unconsidered simultaneous loads/noises other than $F(t)$ exist to corrupt the measurements.

In modal testing to obtain the eigenfrequencies and mode shapes, engineering environments always involve measurement noise, which comes from many sources, such as the environment (non-source related vibrations), quantization noise and electrical components etc. The obtained coherence function can be seen as a signal-to-noise ratio of the measurements. A coherence function equal to 1 indicates zero noise in the measurement, while when the value of the coherence function is 0, a pure noise situation is indicated.

Figure 11.20 shows coherence functions from a modal testing of a ship deck. It is clearly shown that, at the eigenfrequencies that are of interest (5–20 Hz) for structural engineers, the coherence function is close to 1 for almost all measurement positions (2–6). This indicates that the proportion of noise energy in the measured signal is in the order of a few percent. In general, the best signal-to-noise ratio was obtained at those measurement positions close to the excitation position.

In Sections 12.2.2.2 and 12.4.2, the coherence function will be applied to simulate the spatial variation of wind speeds and earthquake-induced ground motions.

Chapter 12
Calculation of Environmental Loading Based on Power Spectra

Wave, wind, ice and earthquake loadings are by their nature dynamic and are normally the major environmental loading to which structures are subjected.

As discussed in Sect. 11.1.1, when the frequencies of excitations due to these environmental loadings are well below the natural frequencies of the structure or the system, the responses are controlled by the stiffness. The dynamic responses will not exhibit resonance and can normally be neglected. A quasi-static analysis approach is routinely adopted to evaluate the behavior in this case. However, this may not always be adequate when excitation frequencies approach resonance condition or when there are transients (e.g. explosions, wave slamming on offshore structures or ships, ice floes impacting on ships and offshore structures), as discussed in Sect. 11.3. Furthermore, for structural responses that are sensitive to their self-vibrations, even if resonance does not occur, the self-vibration itself can cause problems such as fatigue. Dynamic effects are therefore also important to consider. An example of this is the responses of slender structures under both strong and mild wind loading.

As a convenient way of expressing the loading, the power spectra of environmental loads can give a direct feeling of the frequency contents of loads and responses, and are therefore often applied in structural dynamic analysis. They can be used primarily in the random vibration analysis (Sect. 11.4) or to generate the loading time histories used in time domain analysis (a way that is often called indirect time domain method).

12.1 Wave Loads

12.1.1 Calculation of Hydrodynamic Wave Loads

The hydrodynamic forces on a structural member such as a tubular component per unit length are calculated by Morison's equation [79]:

$$F = \rho \cdot A \cdot a + \rho \cdot C_m \cdot A_r \cdot a_r + \frac{1}{2} \rho \cdot C_D \cdot v_r |v_r| \cdot d \tag{12.1}$$

J. Jia, *Essentials of Applied Dynamic Analysis*, Risk Engineering,
DOI: 10.1007/978-3-642-37003-8_12, © Springer-Verlag Berlin Heidelberg 2014

where ρ is the density of the fluid; A is the cross-section area of the body; a is the component of the water particle acceleration normal to the member axis; C_M is the added mass coefficient; A_r is the reference area normal to the structural member axis; a_r is the relative acceleration between water particle and member normal to member axis; C_D is the drag coefficient; v_r is the water particle velocity relative to the member normal to the member axis; d is diameter of the member exposed to the water.

Morison's equation is only applicable when the diameter of the structural member d is less than 1/5 of the wave length, which is the case for many offshore structures such as jacket or jack-up structures.

The first item $\rho \cdot A \cdot a$ on the right hand side of the equation is the wave potential related Froude Krylov excitation force, which is the sum of the hydrodynamic pressures acting on the surface of the body.

The pressure disturbance due to the presence of the body is taken into account in the second item $\rho \cdot C_m \cdot A_r \cdot a_r$, which is the added mass ($\rho \cdot C_m \cdot A_r$) related force due to the relative acceleration (a_r) between the body and the fluid. In general, this depends on the flow condition as well as the location of the body. The wave frequency-dependent characteristics of the added mass may be disregarded for deeply submerged bodies provided that the dimensions of the body are smaller than the wave length.

Note that both the Froude Krylov force and added mass force are due to the inertia of the structures and the surrounding fluid. The viscous effects are then accounted for in the third item (drag force) $\frac{1}{2}\rho \cdot C_D \cdot v_r|v_r| \cdot d$. This item also indicates a nonlinear relationship between the resultant forces on structural members and the wave particle velocity or current speed. This nonlinearity will affect the determination of both statistical properties (non-Gaussian trend) and frequency contents of loading. For a submerged horizontal member subjected to regular wave (presented in Sect. 12.1.2.1) with horizontal water particle velocity $u \propto \sin(\omega t)$, the drag force can be studied based on the following relationship [24, 80]:

$$F_d \propto v_r|v_r| \propto \sin(\omega t)|\sin(\omega t)| \approx 0.85 \sin(\omega t) - 0.17 \sin(3\omega t) - 0.02 \sin(5\omega t)\ldots \tag{12.2}$$

From the equation above, it is obvious that the higher-order harmonic components will appear in the total wave loading, for example, at a wave frequency 3 ω the Fourier component is 20 % of that at a frequency of ω. With the presence of the current, the component at 3 ω will be reduced while a component at 2 ω occurs. These two components are roughly identical when the current velocity is 20 % of the wave particle velocity (18.5 and 17 % at 2 ω and 3 ω, respectively) [81]. This effect is difficult to address in any frequency domain approach, while it can be taken into account by a time domain analysis, as will be discussed in Sects. 17.3.5.4 and 17.3.1.

12.1 Wave Loads

12.1.2 Power Spectrum Density for Ocean Wave Kinematics

12.1.2.1 Long Crest Waves

Ocean waves are, in general, the most important phenomenon to consider among all environmental conditions affecting marine structures. The most typical waves are the wind-driven type: they appear as the wind starts to below, grow into mountainous swells in storms and completely disappear after the wind ceases blowing [82].

If one stands on a ship or an offshore platform, the sea waves can be observed as being made up of large and small waves essentially at various frequencies moving in many directions. This irregularity is an essential feature of sea waves. The wave environment comprises sea states, which are random processes described by a random wave model using wave amplitude energy density spectrum, often being abbreviated as wave energy spectrum $S(\omega_i)$. Discretizing the wave energy spectrum into N number of components $S(\omega_i) \cdot \Delta\omega$, the average wave energy per square meter of the sea surface for the wave component a_i at ω_i is:

$$E(\omega_i) = \frac{\rho g a_i^2}{2} \text{ kJ/m}^2 = \rho g S(\omega_i)\Delta\omega \qquad (12.3)$$

where ρ is the density of the sea water, g is the acceleration of Earth's gravity, a_i is the amplitude of wave components at frequency ω_i. $\Delta\omega$ is the difference between successive frequencies $\omega_i - \Delta\omega/2$ and $\omega_i + \Delta\omega/2$.

From the equation above, it is obvious that the spectrum density is:

$$S(\omega_i) = \frac{a_i^2}{2\Delta\omega} \text{ kJ/m}^2 \qquad (12.4)$$

The model may be visualized as a linear summation of a large number of periodical wavelets. Each of these wavelets has its own amplitude (a_i) and frequency (ω_i), as illustrated in Fig. 12.1. It is noted in the figure that all waves are assumed to travel in the same direction, i.e., the wave crests remain straight and parallel. Such waves normally referred to as long crested.

To calculate the irregular wave in one dimensional space, the linearized (Airy) wave theory [83] can normally be adopted to model the regular wave elevation at each frequency ω_i, leading to a long-crested sea or unidirectional random wave, which may be represented by linear superposition of N harmonic wave trains with random phase angles:

$$\zeta(x, t) = \sum_{i=1}^{N} a_i \cos(k_i x - \omega_i t + \gamma_i) \qquad (12.5)$$

where a_i is the amplitude of component i; ω_i is the angular frequency of component i with respect to a fixed position; k_i is the wave number; and γ_i is the random wave phase with uniform distribution at the interval ranging from $-\pi$ to π, which ensures that the wave elevation is stationary [84].

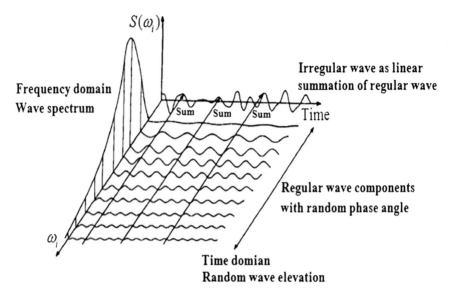

Fig. 12.1 Relation between the frequency and time domain representation of long crest wave elevation [87]

In the absence of current, the wave number is related to the wave frequency by:

$$\omega_i^2 = g \cdot k_i \tanh(k_i \cdot d) \tag{12.6}$$

where d and g are the water depth and acceleration of the Earth's gravity, respectively.

For infinite water depth, the wave number is:

$$k_i = \frac{\omega_i^2}{g} \tag{12.7}$$

For the irregular wave in two dimensional space, at location (x, y) with respect to an inertial reference frame, the irregular wave elevation can be calculated as:

$$\zeta(x, y, t) = \sum_{i=1}^{N} a_i \cos(k_i x \cos \phi + k_i y \sin \phi - \omega_i t + \gamma_i) \tag{12.8}$$

where ϕ is angle with respect to the inertial frame.

Within a short time interval, i.e., within the time frame of 20 min [86] to perhaps 10 h [87], the instantaneous irregular wave elevations can be assumed to follow Gaussian distribution and are stationary with the mean value equal to zero ($c_0 = \overline{\zeta(t)} = 0$). The elevations can be calculated by a linear superposition of the wave components. According to the stationary stochastic process (Chap. 10) and linear random wave theory, the water surfaces elevations and particle kinematics (the velocity and the acceleration) are all Gaussian distributed and are therefore

12.1 Wave Loads

fully defined by the variances of their associated probability distributions, which are equal to the area under their energy spectra. The variance of the wave elevation is then expressed as:

$$\sigma_{\zeta(t)}^2 = E\left[\left(\zeta(t) - \overline{\zeta(t)}\right)^2\right] = \frac{1}{T_z}\int_0^T \delta^2(t)dt = \int_0^\infty S_\zeta(\omega_i)d\omega \tag{12.9}$$

where T_z is the zero crossing wave period and E denotes the expected value.

The relation between the wave energy spectrum and the amplitude of wave components can be approximated as:

$$a_i = \sqrt{2S_\zeta(\omega_i)\Delta\omega} \tag{12.10}$$

where $i = 1, 2,..., N$.

Provided that $\Delta\omega$ is constant, the wave elevation will be repeated with a period of $2\pi/\Delta\omega$. In order to increase this period, a large number of N of discrete frequencies as well as giving the frequency spacing uneven distances may be used for the energy inputs to calculate the wave elevations and other items regarding wave kinematics. Depending on the frequency characteristics of the waves and structures, the frequency range of the wave energy spectrum is chosen slightly differently, while it normally ranges from 0.05–2.5 rad/s.

Associated with the water surface elevation in the equation above, at a position (x, z) and time t, with a water depth of d, the horizontal and vertical water particle velocity u and w, respectively, can be expressed as:

$$u(x, z, t) = \sum_{i=1}^N a_i \cdot \omega_i \cdot \chi_{hi} \cdot \cos(k_i x - \omega_i t + \gamma_i) \tag{12.11}$$

$$w(x, z, t) = \sum_{i=1}^N a_i \cdot \omega_i \cdot \chi_{vi} \cdot \cos(k_i x - \omega_i t + \gamma_i) \tag{12.12}$$

where the horizontal and vertical depth attenuation χ_{hi} and χ_{vi} can be expressed as:

$$\chi_{hi} = \frac{\cosh[k_i(z+d)]}{\sinh(k_i d)} \tag{12.13}$$

$$\chi_{vi} = \frac{\sinh[k_i(z+d)]}{\sinh(k_i d)} \tag{12.14}$$

It is noticed that, due to the depth attenuation, the base shear force of an offshore structure subjected to wave loadings generally increases with the decrease of water depth.

By differentiating the velocity equations (Eqs. (12.11) and (12.12)) with respect to time t, the horizontal and vertical water particle accelerations can then be written as:

$$\dot{u}(x,z,t) = \sum_{i=1}^{N} a_i \cdot \omega_i^2 \cdot \chi_{hi} \cdot \sin(k_i x - \omega_i t + \gamma_i) \tag{12.15}$$

$$\dot{w}(x,z,t) = -\sum_{i=1}^{N} a_i \cdot \omega_i^2 \cdot \chi_{vi} \cdot \cos(k_i x - \omega_i t + \gamma_i) \tag{12.16}$$

The spectral ordinates in terms of velocity and acceleration have a relation with their elevation (displacement) counterpart:

$$S_u(\omega_i) = \omega_i^2 \chi_{hi} S_\zeta(\omega_i) \tag{12.17}$$

$$S_v(\omega_i) = \omega_i^2 \chi_{vi} S_\zeta(\omega_i) \tag{12.18}$$

$$S_{\dot{u}}(\omega_i) = \omega_i^4 \chi_{hi} S_\zeta(\omega_i) \tag{12.19}$$

$$S_{\dot{v}}(\omega_i) = \omega_i^4 \chi_{vi} S_\zeta(\omega_i) \tag{12.20}$$

By assuming that waves are narrow banded, i.e., $\psi = 0$ so that $\overline{T}_z = \overline{T}_p$ (Sect. 10), indicating that all waves have more or less the same period, one can obtain the significant wave height by integrating the wave energy spectrum (Sect. 10.3) [89]:

$$\overline{H}_{1/3} = 4.0\sqrt{\lambda_0} \tag{12.21}$$

Typical wave energy expression in a frequency domain has the form:

$$S_{PM}(\omega_i) = \frac{A}{\omega_i^5} e^{-\frac{D}{\omega^4}} \tag{12.22}$$

A and D are two coefficients determined by wave characteristics. ω is the angular frequency of the wave.

The Pierson-Moskowits (PM) spectrum [90] can be used for the fully developed sea wave condition, i.e., the fetch and the duration are large and there is no disturbance from other sea areas. This is because, after a certain period of wind blowing, the sea elevation becomes statistically stable. The PM spectrum is mainly developed for the description of waves in the North Atlantic Ocean. It can be expressed using the mean significant wave height $\overline{H}_{1/3}$ and mean zero crossing mean period $\overline{T}_z$ as independent parameters and be input into the typical wave energy equation as:

$$A = 123.95 \overline{H}_{1/3}^2 \Big/ \overline{T}_z^4 \tag{12.23}$$

$$D = 495.8 \Big/ \overline{T}_z^4 \tag{12.24}$$

12.1 Wave Loads

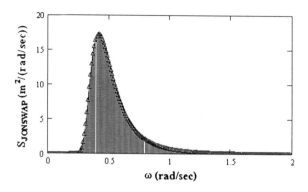

Fig. 12.2 Illustration of JONSWAP spectrum and the sampling frequency with 64 samples (△) per rad/s

In shallow waters with limited fetch due to geographical boundaries, and for extreme wave conditions, the JONSWAP spectrum, developed by the Joint North Sea Wave Project [92, 93], is recommended. Compared to the PM spectrum, the JONSWAP spectrum is narrow banded and extensively adopted by offshore industry. It is expressed by enhancing the peak of the PM spectrum, as shown in the following equation and Fig. 12.2:

$$S_{JONSWAP}(\omega_i) = \frac{A}{\omega_i^5} e^{-\frac{D}{\omega^4}} \cdot \gamma^\delta \qquad (12.25)$$

where $A = a \cdot g^2$, $D = 1.25 \cdot \omega_m^2$, $\delta = e^{-\frac{(\omega-\omega_m)^2}{2\sigma^2 \omega_m^2}}$, $g = 9.8$ m/s^2.

A, D and γ are functions of $\overline{H}_{1/3}$ and $\overline{T}_z$. a represents the level of high frequency tail, g is the acceleration of Earth's gravity. The JONSWAP spectrum has five parameters: a, ω_m, γ, σ_a, and σ_b. a can often be assumed to be 1.0. ω_m is the peak angular frequency of the wave spectrum.

σ represents the narrowness of the peak, and has a different value for frequencies lower (σ_a) and higher (σ_b) than the peak frequency ω_m as expressed as:

$$\sigma = \begin{cases} \sigma_a = 0.07, \omega < \omega_m \\ \sigma_b = 0.09, \omega \geq \omega_m \end{cases} \qquad (12.26)$$

The γ value indicates the enhancement of the spectrum peak. It normally ranges from 1.0 (flat) to 7.0 (very peaked) with 3.3 as an average value. For example, for most of the storms in the North Sea, γ can be assumed to be approximately 3.0 or less [93]. When γ is not specified, the following value may be taken [24, 94]:

$$\sigma = \begin{cases} 5, \text{ for } \frac{T_p}{\sqrt{H_s}} \leq 3.6 \\ e^{5.75 - 1.15\frac{T_p}{\sqrt{H_s}}}, \text{ for } \frac{T_p}{\sqrt{H_s}} > 3.6 \end{cases} \qquad (12.27)$$

Figure 12.3 shows the power spectrum axial force responses for joint 10,754 in the upper part of a jacket structure (Fig. 12.4) located in the North Sea. Since all the modal wave periods in the three adopted sea states are higher than the jacket's fundamental and second eigenperiod, generally two spectrum peaks can be identified at each sea state. One is close to the second eigenperiod of the jacket structure, indicating the free vibrations of the jacket. The other is close to the modal period of the wave loads. In addition, the magnitude of this part of power spectrum increases with the increase of the significant wave height H_s. It is noticed that the waves for all three adopted sea states come from the south, which explains why the global flexural vibration mode along the north–south (second eigenmode) is more relevant to the free vibrations of the jacket structure (for each sea state, one of the two spectrum peaks occurs close to the second eigenperiod).

12.1.2.2 Short Crest Waves

Strictly speaking, the long crest sea on which Fig. 12.1 is based only occurs in a laboratory or towing tank. In an actual ocean environment, the waves travel in many different directions, with the majority of energy contained along a "primary" direction v. The corresponding sea condition is referred to as a short crested irregular sea or confused sea [90]. Due to the angular dispersion or spreading of many wave systems coming from different directions, short-crestedness is especially apparent at locations closer to storm areas or coast lines with irregular sea bottom topology, and at conditions if the dominant wind direction is changing [91]. To account for this wave spreading effects, the average wave energy per square meter of the sea surface, for the wave component a_{ij} and in the direction band $\Delta \beta$ at ω_i is:

$$E(\omega_i) = \frac{\rho g a_{ij}^2}{2} \text{ kJ/m}^2 = \rho g S(\omega_i) \Delta \omega \Delta \beta \qquad (12.28)$$

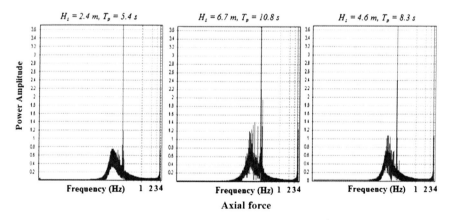

Fig. 12.3 The power spectrum of the axial forces at node 10,754 in the upper part of a jacket structure (shown in Fig. 12.4) due to the wave loads at three individual sea states (H_s is the significant wave height, T_p is the modal wave period.)

12.1 Wave Loads

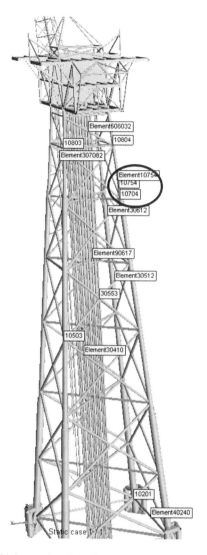

Fig. 12.4 The position of joint (node) 10,754 for the fatigue damage calculation

where $\beta = v_0 - v$, is the angle between the direction under consideration v and the primary wave direction v_0, ranging from $-\frac{\pi}{2}$ to $\frac{\pi}{2}$ (see Fig. 12.5 for an illustration); a_{ij} is the component of wave appropriate to the ith frequency and jth direction.

Therefore, the directional wave energy spectrum is written as:

$$S_\zeta(\omega_i, \beta) = \frac{a_{ij}^2}{2\Delta\omega\Delta\beta} \qquad (12.29)$$

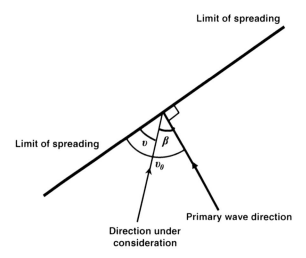

Fig. 12.5 Definition of directions for short crested wave

The variance of the wave elevation is then expressed as:

$$\sigma^2_{\zeta(t)} = \int_0^\infty \int_{-\pi}^{\pi} S_\zeta(\omega_i, \beta) d\beta d\omega \qquad (12.30)$$

The irregular wave elevation for the short crested sea can be calculated as:

$$\zeta(x,t) = \sum_{i=1}^{N} \sum_{j=1}^{M} a_{ij} \cos(-\omega_i t + \gamma_i) \qquad (12.31)$$

Note that the measurement of directional wave spectrum due to dispersion or spreading is rather difficult. In engineering practice, the short crested wave spectrum is expressed as a simple scaling of the long crested wave spectrum $S_\zeta(\omega_i)$, i.e., the scaling factor $A(\beta)$ is only a function of β without accounting for the influence from the frequency :

$$S_\zeta(\omega_i, \beta) = A(\beta) S_\zeta(\omega_i) \qquad (12.32)$$

where the scaling factor $A(\beta)$ by definition must have its peak at $\beta = 0$ and satisfy the following relationship :

$$\int_{-\pi}^{\pi} A(\beta) d\beta = 1.0 \qquad (12.33)$$

Therefore, the directional wave spectrum normally takes the form of:

$$S_\zeta(\omega_i, \beta) = C(n) \cos^n(\beta) S_\zeta(\omega_i) \qquad (12.34)$$

where n indicates the significance of short crestedness, it ranges from 2.0 to 4.0 for wind seas, and 6.0 or higher for swells (in many cases, swell seas are considered as being long crested, i.e., n is infinitely large), and $C(n)$ is a normalizing constant to satisfy Eq. (12.33):

$$C(n) = \frac{\Gamma(n/2 + 1)}{\sqrt{\pi}\Gamma(n/2 + 1/2)} \qquad (12.35)$$

For ship design purposes, n is often taken to be 2.0, which gives:

$$S_\zeta(\omega_i, \beta) = \frac{2}{\pi}\cos^2(\beta)S_\zeta(\omega_i) \qquad (12.36)$$

For most cases, the long crest sea is more conservative to use for extreme value predictions. The only exception may be the ship rolling for a weather vaning ship.

For final design of a new installation it is recommended that long crest sea be used when it is a conservative approach. However, if the short crested sea is to be used, n should be taken as the most conservative value in the range 2.0–10.0 for the wind sea. When utilizing short crest sea, it should as far as possible be verified that the modeling of short crest sea is representative of the wave events causing the governing loads on the structure.

It should also be noticed that, in severe sea conditions, water surface elevations and particle kinematics will no longer follow Gaussian distributions [95]. Nevertheless, this nonlinear non-Gaussian effect of wave kinematics is of less importance than other nonlinear effects such as nonlinear drag force effects (drag is proportional to the square of the wave particle velocity) [96] and can normally be neglected.

12.2 Wind Loads

12.2.1 Calculation of Aerodynamic Wind Load

The aerodynamic forces applied on a structural component per unit length are calculated by:

$$F = \frac{1}{2}\rho \cdot C_L \cdot v_r \cdot |v_r| \cdot A + \frac{1}{2}\rho \cdot C_D \cdot v_r|v_r| \cdot d \qquad (12.37)$$

where ρ is the density of the air (normally taken to be 1.25 kg/m^3). However, it should be noticed that from the sea surface to a height of 20–30 m above, the water spray may cause an increase of air density in this area; C_L is the lift coefficient; C_D is the drag coefficient or wind pressure coefficient; v_r is the wind velocity relative to the member normal to the member axis (note that if the wind load causes the structure to oscillate, the oscillation velocity of the structure should also be

Table 12.1 The drag coefficients specified by DnV RP-C205 [106] and NORSOK N-003 [71] for smooth tubular members

DnV RP-C205		NORSOK N-003	
Reynolds number[a]	C_D	Reynolds number[a]	C_D
$Re < 3.7 \cdot 10^5$	1.2	$Re < 5.0 \cdot 10^5$	1.2
$3.7 \cdot 10^5 < Re < 5.0 \cdot 10^5$	1.0		
$Re > 5.0 \cdot 10^5$	0.6	$Re > 5.0 \cdot 10^5$	0.65

[a] Reynolds number $Re = \frac{UL}{\nu}$, where U is the mean velocity of the object relative to the fluid (for wind, it is the mean wind speed); L is the characteristic linear dimension (for circular cylinders (tubular members), it is the diameter of the member); ν is the kinematic viscosity of the fluid (for air, it can be taken as $1.45 \cdot 10^5$ m^2 /s at 15 °C and standard atmospheric pressure)

accounted for when calculating v_r); A is the projected area on a plane normal to the coming wind direction; and d is the diameter of the member exposed to the air.

For tubular members, lift force does not exist. The first term on the right-hand side of the equation above then becomes zero. Thus, the forces on a member only include the viscous effects as included in the second term (drag force) $\frac{1}{2}\rho \cdot C_D \cdot v_r|v_r| \cdot d$. Similar to the wave-induced drag force, this second term also indicates a nonlinear relationship between the drag forces and the wind velocity. Unlike the wave drag load nonlinearities discussed at the end of Sect. 12.1.1, the effects of wind drag nonlinearities are generally insignificant for mild wind conditions (fatigue-dominated wind condition). However, for extreme wind loading conditions with strong wind or high wind turbulence intensity, these nonlinear effects may be relevant.

From the equation above, it is also noted that, due to the deformation or rotation of structures, the drag and lift forces may also change with time even if the wind velocity may not change over a short time duration.

The drag coefficients are a function of Reynolds number [97], Table 12.1 shows drag coefficient values for smooth tubular members.

Obviously the wind speed and the drag coefficient are the environmental parameters determining the wind load on structures.

Similar to wave-induced force presented in Sect. 12.1.1, inertia force also contributes to wind-induced force. However, since this inertia force is about two orders of magnitude smaller than the drag force [24], it is therefore neglected in almost all engineering calculations.

12.2.2 Power Spectrum Density for Wind Velocity Fields

Winds are generated due to the variation of temperature. Figure 12.6 shows the spectrum of the wind speed over a broad range of frequency from 1 s up to more than a year. The frequency content on the left side represents the yearly changes of seasons, the meteorological cycle of around 4 days' duration, and daily changes.

12.2 Wind Loads

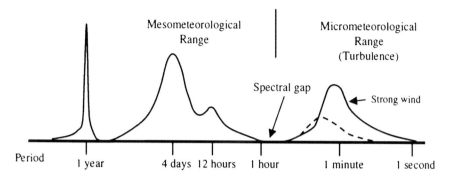

Fig. 12.6 Wind speed spectrum over a broad range of frequencies. On the *right side*, the *solid line* represents high turbulence during a period of high wind speeds, while the *dotted line* represents the reduced turbulence at lower wind speeds [98]

On the other hand, the wind turbulence with a period from tens of seconds to a couple of minutes can be clearly seen on the right side of the figure.

The gap around the 1 h period is also referred to as the "spectral gap." This separates the slowly changing wind (extremely low frequency) and the turbulent wind. Since at this period range the wind speed is more or less constant, it is defined as the constant mean wind speed, while the instantaneous wind speed changes are only due to the turbulence part, which can normally be assumed to be Gaussian, stationary and homogeneous over a short period of time (e.g. 10 min). Even though recent research on the spectral gap reveals that the gap is more a coincidental feature of the analysis technique [69, 99], the assumption that in the 10 min to 1 h range the mean wind speed is constant has proved to be an effective model for the relevant design of most types of engineering structures [69].

Therefore, the instantaneous wind speed is modeled as the summation of a slow varying mean wind part and a high frequency turbulent part as shown in Fig. 12.7. The mean wind speed $U'(z)$ is flowing horizontally and varies with height above ground or sea surface. The turbulent part varies in both time and space. By denoting the turbulence wind components u, v, and w in the along-wind (x), horizontal across-wind (y) and vertical across-wind (z) direction and the mean wind component as U', a wind velocity vector at time t can then be expressed in a Cartesian coordinate system:

$$V\{x, y, z, t\} = [U' + u, v, w]^T \tag{12.38}$$

Within the atmospheric boundary layer, which is the lowest 2 km of the Earth's atmosphere, the wind speed is strongly affected by the friction with the Earth's surface, known as wind shear. This reduces the wind speed from its undisturbed value at 2 km above to nearly zero at the ground or sea surface. The mean wind speed can be calculated by assuming a reasonable wind profile (Fig. 12.7) together with reference wind speed for a specified return period and time duration.

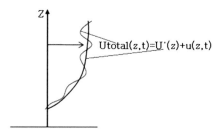

Fig. 12.7 Typical wind profile composed of mean wind speed components ($U'(z)$) and fluctuating wind speed component u(z, t) varying with height z and time t

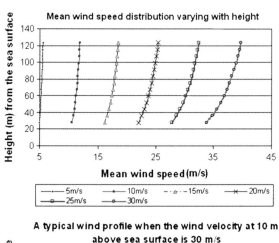

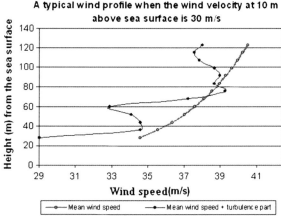

Fig. 12.8 Typical mean wind speed distribution (*upper*) varying with height for each reference wind speed ranging from 5 to 30 m/s, and an example of wind speed distribution including its turbulence part (*lower*)

12.2 Wind Loads

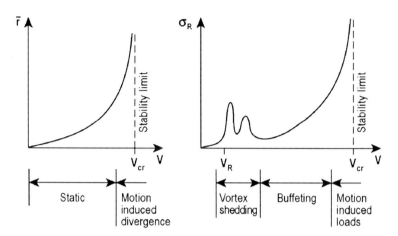

Fig. 12.9 Response characteristics varied according to the mean wind speed V. The *left* and *right figures* show the static and dynamic part of the responses, respectively ($\bar{r}$ is the mean value of the responses, σ_R is the standard deviation of the fluctuating part of the responses, V_{cr} is the wind speed corresponding to upper stability limit) [100]

Figure 12.8 shows the typical mean wind speed profiles and the turbulent part of wind speed.

As illustrated in Fig. 12.9, the mean wind speed is normally assumed to govern the static loads on structures (left figure). It is the fluctuating (gust or turbulent) part of wind that causes dynamic wind loading on structures. As shown in the right figure of Fig. 12.9, the dynamic part of the responses can be separated into three mean wind speed regions [100]. At low mean wind speed region, vortex-induced vibrations (Sect.1.1) can occur; at intermediate mean wind speed region, buffeting effects (due to the pressure fluctuation in the oncoming flow) may govern; in high mean wind speed regions, the motion-induced load effects, which are due to the interaction between the wind flow and the oscillating structural members, may be dominant [100]. The calculation of wind induced structural responses in these three regions is usually treated separately using different methods [100].

For both static and dynamic responses, when the wind speed is increased and approaches to a limiting value V_{cr}, a slight increase of mean wind speed will cause a dramatic increase of either static or dynamic responses. V_{cr} is here referred to as the upper statibility limit.

12.2.2.1 Calculation of Mean Wind Speed U'(z) for Along Wind

To describe the wind shear effects on the mean speed, two wind profile models are normally used, the logarithmic profile and the power law profile. Both profiles are widely used, and there is no general preference for either profile.

Table 12.2 Surface roughness length for various types of terrain

Type of terrain	z_0 (m)
City centers	1–10
Cities, forests	0.7
Suburbs, wooded countryside	0.3
Villages, countryside with trees and hedges	0.1
Open farmland, few trees and buildings	0.03
Flat grassy plains	0.01
Flat desert, rough sea	0.001
Calm sea	0.0002

By using the logarithmic profile, the mean along-wind speed at height z above the ground or sea surface can be expressed as:

$$U(z) = U_{10} \frac{\ln\left(\frac{z}{z_0}\right)}{\ln\left(\frac{10}{z_0}\right)} \tag{12.39}$$

where U_{10} is the reference wind speed with an average period from 10 to 60 min, and z_0 is the surface roughness length. Typical values of z_0 can be taken from Table 12.2.

In NORSOK standard N-003 [71], the mean along-wind speed at a height z above the sea surface and the corresponding mean period t [s] of not more than 3,600 s can be calculated by a mean wind profile:

$$U'(z,t) = U(z) \cdot \left[1 - 0.41I_u(z) \cdot \ln\left(\frac{t}{3600}\right)\right] \tag{12.40}$$

where the one-hour mean wind speed $U(z)$ [m/s] is given by:

$$U(z) = U_0 \cdot \left[1 + C \cdot \ln\left(\frac{z}{10}\right)\right] \tag{12.41}$$

where $C = 5.73 \cdot 10^{-2} \cdot [1 + 0.15U_0]^{0.5}$

$I_u(z)$ is the turbulence intensity, which is defined as the ratio between the standard deviation of the wind speed and the mean wind speed, which typically ranges from 0.1 to 0.4. For offshore sites, it is given in NORSOK standard N-003 [71] as:

$$I_u(z) = 0.06 \cdot [1 + 0.043U_0] \cdot \left(\frac{z}{10}\right)^{-0.22} \tag{12.42}$$

where U_0 (m/s) is the one-hour mean wind speed at $z = 10$ m above ground or sea surface, the information for U_0 and its probability of occurrence are normally given in the relevant wind-specification or metocean (meteorological and oceanographic) document.

From the equation above, it is shown that the greater the height, the lower the turbulence intensity will be. However, it should be clarified that the turbulent

12.2 Wind Loads

intensity will also be influenced by the topology and surface condition of the ground, which are not explicitly indicated by the equation above, which is only valid for offshore sites. Moreover, it should be noted that the standard deviation of the wind speed has little variation with increases in height. This is due to the increase of the wind speed with height that results in a decreased turbulence intensity with height.

In many engineering applications, the mean wind speed can also be simply taken with a power law profile:

$$U(z) = U_{10} \left(\frac{z}{10} \right)^{\alpha} \qquad (12.43)$$

where U_{10} is the reference wind speed with an average period from 10 to 60 min, α is the parameter depending on the topology of the terrain [101]. For flat desert or rough sea, its value ranges from 0.12 as 0.14; for open terrain with very few obstacles (e.g. open farm or grass land), it is normally taken to be 0.16; for terrain uniformly covered with obstacles 10–15 m in height (e.g. residential suburbs, small towns, small fields with bushes, tress and hedges), α can be taken to be 0.28; if the terrain has large and irregular objects (e.g. city centers), α can be taken to be 0.40.

12.2.2.2 Calculating the Turbulence (Fluctuating) Wind Components u, v, and w Using a Wind Spectrum

Along-Wind Spectrum

The fluctuating part of wind is simulated from wind spectra and coherence functions, so that the spatial statistical properties are maintained.

The fluctuating part of the along-wind speed component u is approximately constant in space but varies with time. It can be defined as 1 point spectrum $S_u(f_i)$ ($[\mathrm{m^2 s^{-2}/Hz}]$) for all locations in space.

The most widely used wind energy spectra are the Harris and Davenport spectra.

The Harris spectrum [101] is expressed as:

$$S_u(f_i) = \sigma_u^2 \frac{4 \frac{L_u}{U_{10}}}{\left[1 + 70.8 \cdot \left(\frac{f_i L_u}{U_{10}} \right)^2 \right]^{\frac{5}{6}}} \qquad (12.44)$$

where f_i denotes the frequency in Hz, L_u is the integral length scale, typically between 60 and 400 m with a mean value of 180 m; U_{10} is the 10 min mean wind speed; σ_U is the standard deviation of the wind speed.

The Davenport spectrum [102] is expressed as:

$$S_u(f_i) = \sigma_u^2 \frac{\frac{2}{3}\left(\frac{L_u}{U_{10}}\right)^2 f_i}{\left[1 + \left(\frac{f_i L_u}{U_{10}}\right)^2\right]^{\frac{4}{3}}}$$

(12.45)

where the L_u is specially referred to as the Davenport integral length scale, which can normally be taken to be 1200 m.

Both Harris and Davenport spectra are developed for wind over land, and are not recommended for application for wind fields with significant components within the low frequency range ($f_i < 0.01$ Hz).

For situations in which excitations in the low-frequency range are of importance, such as wind over water, the Frøya (NPD) spectral density proposed by Andersen and Løvseth [103, 104] is widely used:

$$S_u(f_i) = \frac{320 \cdot \left(\frac{U_0}{10}\right)^2 \cdot \left(\frac{z}{10}\right)^{0.45}}{\left(1 + \tilde{f}^n\right)^{\frac{5}{3n}}}$$

(12.46)

where $n = 0.468$.

$$\tilde{f} = 172 \cdot f_i \cdot \left(\frac{z}{10}\right)^{0.667} \cdot \left(\frac{U_0}{10}\right)^{-0.75}$$

(12.47)

where z is the height above the sea surface and U_0 is the one hour mean wind speed at 10 m above the sea surface.

The Frøya spectrum was originally developed for neutral conditions over water in the Norwegian Sea. The use of the Frøya spectrum can therefore not be recommended in areas where stability effects are important. A frequency of 1/ 2400 Hz defines the lower bound for the application of the Frøya spectrum [106]. Whenever it is important to estimate the energy in the low frequency range of the wind spectrum over water, the Frøya spectrum is considerably better than other wind spectra such as Davenport or Harris spectrum [101, 106].

The two upper figures in Fig. 12.10 show typical Frøya wind spectra at two locations with a distance of 36.6 m. They illustrate that the target (solid lines) and the simulated (dots) values for both wind power spectra and the coherence function between two points (series) agree very well.

By observing Fig. 12.10, it is clearly shown that the majority of wind spectra energy is below 0.1 Hz, meaning that the wind loads' peak frequency is far below the natural frequencies of typical engineering structures (above 0.3 Hz), i.e. the structural resonance is not likely to occur due to wind flutter effects. However, strong winds can excite structures exhibiting a large amplitude of forced vibrations, which may cause integrity problems due to plasticity or stability. Moreover, the self-vibrations of structure itself can also cause fatigue failure of structural joints or components. Figure 12.11 shows the stress history and response spectra

12.2 Wind Loads

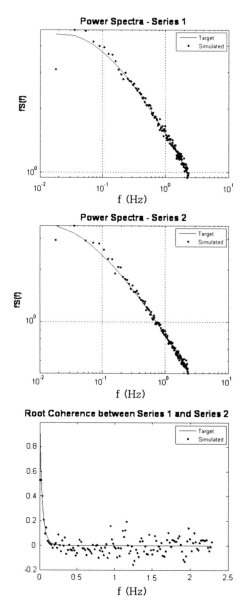

Fig. 12.10 A verification of wind simulation (*dots* in the figure) in the frequency domain between the two locations (series) in space with the separation of $\Delta = 36.6$ m. The *solid lines* represent the target wind spectra

of the axial, in-plane and out-of-plane forces in a conical joint at the upper part of a flare boom, under the wind speed of 30 m/s from the direction perpendicular to the longitudinal direction of the flare boom. Axial stress makes the most significant

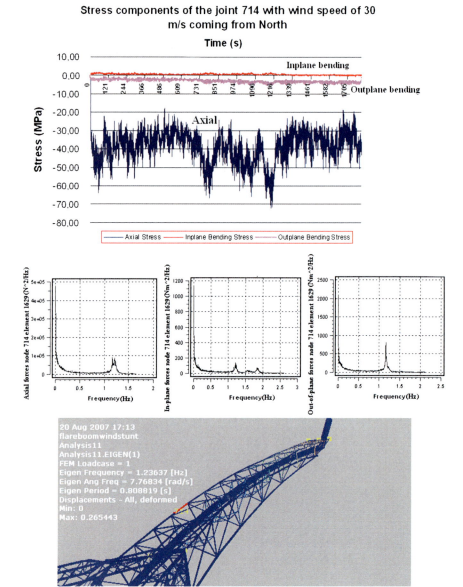

Fig. 12.11 The stress history (*upper figure*) of joint 714 and response spectra (*middle figure*) at a wind speed of 30 m/s from the north (outward from the paper plane and perpendicular to the longitudinal direction of the flare boom); the response peak frequency clearly indicates the contribution from the fundamental eigenmode (*lower figure*)

contribution to fatigue damage. Except for large responses close to a rather low frequency (0.1 Hz) due to the quasi-static responses of the flare boom subjected to wind, response peaks appear around the frequency of 1.2 Hz, which is close to,

12.2 Wind Loads

and slightly lower than, the global fundamental frequency (vertical flexural vibration of the flare tip) of the flare boom. This indicates that the free vibrations of the flare boom top significantly contribute to the dynamic responses of this joint. This response peak frequency is slightly lower than the fundamental eigenfrequency of the flare boom mainly because of the damping and stress-softening effects (a reduction in initial elastic stiffness due to axial compressive force, discussed in Sect. 6.3) of the structure. By varying the reference wind speed and calculating the flare boom's responses, it is discovered that the higher the wind speed, the more significant dynamic effects (as compared with quasi-static response) the structure will exhibit.

Several other wind spectra are also widely used, such as the Kaimal spectrum [108], Simiu and Leigh spectrum [109], Ochi and Shin spectrum [110]. Interested readers may consult the relevant references.

Across-Wind Spectrum

For structures that are not vertical, such as an obliquely-oriented flare boom, the across-wind turbulence may make a more significant contribution to the total structural responses. Based on the calculation of nonlinear wind-induced responses of a high-rise tubular structures, Jia [17] performed a fatigue assessment by rainflow counting (Sect. 17.3.5.1) the nonlinear stress responses and applying the Miner summation rule (Sect. 17.2.5). It is found that, in many structural joints, the fatigue life without including the across-wind components is significantly higher than that including the across-wind contribution. Generally, the vertical acrosswind components make a slightly more significant contribution on the fatigue damage of the structure than that of the horizontal across-wind components.

Different from the way of multiplying a scaling factor to the spectrum of the primary direction, as a short crest sea wave spectrum does, the one point spectra for across fluctuating wind components v and w are specified in a separated spectrum [105]:

$$S_{vorw}(f) = \left(\frac{\sigma_\varepsilon^2}{f_i}\right) \cdot \frac{A_\varepsilon \cdot f_\varepsilon}{(1 + 1.5A_\varepsilon \cdot f_\varepsilon)^{1.667}} \tag{12.48}$$

where σ is the standard deviation; ε is a symbolic index indicating the direction in y–z plane (horizontal across wind direction or vertical across wind direction); and $f_\varepsilon = \frac{f_i \cdot {}^x L_\varepsilon}{\varepsilon}$, where ${}^x L_\varepsilon$ is the integral scale length of the relevant turbulence component, which may be determined from ESDU 86,010 [111]. It can be assumed that $A_\varepsilon = 9.4$ if it is not specified.

Spatial Variation of Wind Speed by Coherence Function

Since wind gusts vary in space, i.e., wind speeds vary at different locations, but the speed at any two arbitrary locations are correlated, the correlation decreases with

194   12 Calculation of Environmental Loading Based on Power Spectra

the increase of the distance between the two locations. This correlation can be expressed in terms of single-point spectra S_h and S_j (with the separation Δ between locations h and j) with the corresponding coherence spectra as expressed by:

$$S_{hj}\{\Delta, \omega_i\} = \sqrt{S_h(\omega_i) \cdot S_j(\omega_i)} \sqrt{Coh_{hj}(\omega_i, \Delta)} e^{\sqrt{-1}\varphi_{ij}\{\omega_i\}} \tag{12.49}$$

where $\phi_{hj}\{\omega_i\}$ is the phase spectrum, which, for simplicity, can be assumed to be 0.

For wind over water, the Frøya coherence spectrum $Coh(\Delta, \omega_i)$ for both either the wind components v or w between any two points (x_1, y_1, z_1) and (x_2, y_2, z_2) can be expressed by:

$$Coh(\omega_i, \Delta) = e^{\left[-\frac{1}{U_0} \cdot \left(\sum_{h=1}^{3} A_h^2\right)^{0.5}\right]} \tag{12.50}$$

where:

- U_0 is the 1 h mean wind speed
- Δ is the separation between the two points (x_1, y_1, z_1) and (x_2, y_2, z_2)
- $A_h = \alpha_h \cdot \left(\frac{\omega_i}{2\pi}\right)^{r_h} \cdot \Delta_h^{q_h} \left(\frac{\sqrt{z_1 \cdot z_2}}{10}\right)^{-p_h} [m/s]$

The coefficients α_i, r_i, q_i, p_i and the separations Δ_h can be taken as illustrated in Table 12.3.

Since the wind velocity is assumed to be stationary and homogeneous, in some cases, the along-wind separation is disregarded. Thus, the along wind may be represented by the one-point wind spectra without accounting for the spatial variation.

Simulation of Turbulent Wind Speed Time Series

In order to reduce the computation efforts, the cross-covariance (Sect. 11.5.1) between u, v and w may be assumed to be insignificant, indicating that the u, v and w components can be obtained independently. The fluctuating parts of the wind spectra can then be expressed by:

$$S\{\Delta, \omega_n\} \approx diag[S_{uu}, S_{vv}, S_{ww}] \tag{12.51}$$

By subdividing the frequency range into N segments ($i = 1, 2,..., N$) and the wind fields into M points in space grids, an M by M cross-spectral density matrix can then be expressed for each sample frequency ω_i and each flow component as:

$$S\{\omega_i\} = \begin{bmatrix} S_{11}(\omega_i) & S_{12}(\omega_i) & \cdots & S_{1M}(\omega_i) \\ S_{21}(\omega_i) & S_{22}(\omega_i) & \cdots & S_{2M}(\omega_i) \\ \cdots & \cdots & \cdots & \cdots \\ S_{M1}(\omega_i) & S_{M2}(\omega_i) & \cdots & S_{MM}(\omega_i) \end{bmatrix} \tag{12.52}$$

12.2 Wind Loads

Table 12.3 Coefficients and separation for the 3 ($h = 1, 2, 3$)-dimensional coherence spectrum

h	α_h	r_h	Δ_h(m)	q_h	p_h
1	2.9	0.92	$|x_2 - x_1|$	1.00	0.4
2	45.0	0.92	$|y_2 - y_1|$	1.00	0.4
3	13.0	0.85	$|z_2 - z_1|$	1.25	0.5

If one wants to simulate the wind speed time series from the wind spectrum, a Cholesky decomposition can first be performed on the matrix above so that it can be rewritten as the product of lower triangular matrix $H(\omega_i)$ and its transposed complex conjugate:

$$S(\omega_i) = H\{\omega_i\} \cdot H^{*T}\{\omega_i\} \tag{12.53}$$

where:

$$H\{\omega_i\} = \begin{bmatrix} H_{11}(\omega_i) & 0 & \dots & 0 \\ H_{21}(\omega_i) & H_{22}(\omega_i) & \dots & 0 \\ \dots & \dots & \dots & \dots \\ H_{m1}(\omega_i) & H_{m2}(\omega_i) & \dots & H_{mm}(\omega_i) \end{bmatrix} \tag{12.54}$$

Aas-Jakobsen and Strømmen [112] present a criterion for selecting the element length in the finite element-based structure model as:

$$\Delta l \leq \begin{cases} \min\{L_{tot}/3, 2\Psi\} : \text{symmetric modes} \\ \min\{L_{tot}/3, \Psi\} : \text{anti-metric modes} \end{cases} \tag{12.55}$$

where Δl is the element length; L_{tot} is the total length of the structure (or the active mode); and Ψ is the span-wise integral of the coherence function at a dominating eigenfrequency [112].

By realizing the fact that the wind grid size and element length are interchangeable quantities, this criterion may provide practical recommendations for selecting a decent wind grid size.

Based on the assumption of Gaussian process, the simulated wind velocity components u, v, and w in space (at point m) at time t can finally be expressed as:

$$v_m(t) = \sum_{k=1}^{m} \sum_{i=1}^{N-1} |H_{mk}(\omega_i)| \cdot \sqrt{2\Delta\omega} \cdot \cos[\omega_i t + \psi_{mk}(\omega_i) + \theta_{ki}] \tag{12.56}$$

where:

- $\Delta\omega$ is the sample density in the frequency domain
- $\psi_{mk}\{\omega_i\} = \arctan\left(\frac{\text{Im}[H_{mk}\{\omega_i\}]}{\text{Re}[H_{mk}\{\omega_i\}]}\right)$ is the phase angle between two points in space. If Eq. (12.51) is fulfilled, then $\psi_{mk}\{\omega_n\} = 0$
- θ_{ki} is a random phase angle uniformly distributed between 0 and 2π

In order to optimize computer storage and computation speed, Eq. (12.56) can be rewritten in an exponential format expressed by:

$$v_m(t) = \sum_{k=1}^{m} \sum_{i=1}^{N-1} \left(|H_{mk}(\omega_i)| \cdot \sqrt{2\Delta\omega} \cdot e^{\left[\sqrt{-1}(\psi_{mk}\{\omega_i\}+\theta_{ki})\right]} \right) \cdot e^{\sqrt{-1}\omega_i t} \qquad (12.57)$$

It should be noticed that the calculation of the wind turbulence in the current section is based on the assumption that the wind occurs on an open landscape with free flow condition. The blocking and speed-up effects of the nearby structures or topology may modify the wind field to a great extent [17].

12.3 Ice Loads on Narrow Conical Structures

Our knowledge on the ice load is rather limited even if it has attracted extensive research efforts in recent years. Ice load differs significantly between static and dynamic loading [113]. For dynamic ice loading, the initial contact conditions are invariably irregular and non-uniform. The duration of the ice impact is generally determined by kinetic energy of the impacting ice feature, which may come to rest during the impact process.

The ice load on marine structures can be divided into two types: total or global load, and local loads (pressures). The global load affects the overall motion and stability of structures, while local load affects areas from 1 m^2 to as much as 100 m^2 [24]. For both types of loads, the ice thickness and velocity are two essential parameters to determine the load level.

With respect to local ice load, close observations show that, when ice crushes a structure, local ice pressure on small patches or a narrow line-like area can reach rather high values [114]. Such areas are termed high pressure/critical zones, as discussed in reference [115]. This pressure may be well beyond the normal uni-axial crushing strength of ice [115]. The explanation is that the stress field is in fact multi-axial. The process of ice-structure interaction is characterized by fracture and damage processes, which play a key role in the appearance and disappearance of high pressure zones. Further, the high pressure has an important effect on local ice actions. However, the global ice pressure is significantly lower than the local ice pressure because in local regions of ice the intense pressure occurs over a short time.

It is noted that the strength of ice due to bending is less than half of that due to crushing, and the use of conical or sloping structures will induce the bending failure of the ice sheet when the ice crushes structures. Therefore, they are often a preferred design option for structures located in ice-covered waters. Recently, field measurements also indicate that the adoption of conical structures can avoid severe steady-state vibrations [116]. Figure 12.12 shows that a cone (right) is installed on an offshore monopod structure originally with a vertical leg (left) at China Bohai Sea.

12.3 Ice Loads on Narrow Conical Structures

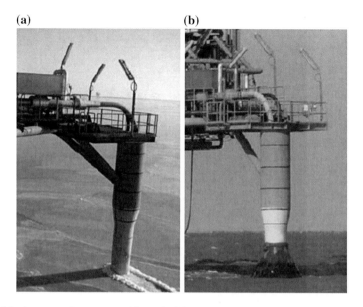

Fig. 12.12 A monopod structure with vertical leg structure (*left*) during the first winter, later modified with an ice-breaking cone (*right*) installed during the second winter [116]

Even though the ice loads on offshore structures are generally limited by the failure of the ice itself [117], the ice-induced vibrations in offshore structures can be rather severe when the ice is moving driven by winds or current. Dynamic loading scenarios include both transient impact loads and continuous ice failure loads. Resonance of fixed offshore structures due to ice loading may occur, as discussed later in this section.

Kärnä et al. [118] derived the spectrum density function of the crushing-level ice force on vertical structures.

Ice loads on conical and sloped structures are generally narrow banded with a long tail toward higher loads in the probability distribution. Several researchers [119, 120, 121] have proposed the power spectrum density of level ice forces on narrow conical structures, all of which are in the form of a Neumann spectrum [122]:

$$S_{ice-cone}(f) = \frac{A\overline{F}_0^2 \overline{T}^{-\chi}}{f^p} e^{\left(-B\frac{1}{\overline{T}^\alpha f^\beta}\right)} \qquad (12.58)$$

where A, B, χ, p, α, and β are parameters obtained from experiments. By applying curve fitting from observation data from the Bohai Sea to the Neumann spectrum, the values of A, B, χ, p, α, and β are 10, 5.47, 2.5, 3.5, 0.64 and 0.64, respectively [121]; $\overline{F}_0$ is the force amplitude on the target structure; $\overline{T} = \frac{L_b}{V}$ is the mean period of ice force on the structure; L_b the ice breaking length; and V is the ice velocity.

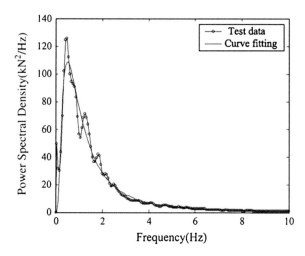

Fig. 12.13 One-sided power spectrum density functions of ice force obtained from load panel data (*test data*) and Neumann's formula (*curve fitting*) [121]

Based on the panel test to measure the ice force time history on a jacket leg at China Bohai Bay [121], Fig. 12.13 shows a comparison between the power spectrum density functions of ice forces obtained from test data and results of curve fitting using Neumann's formula presented above. The two curves match each other well. By observing this figure, it is also noticed that the local peak frequencies and the majority of the spectra energy are concentrated in a frequency range between 0.3 and 2.5 Hz, which is in the frequency range of most of the offshore jacket platforms, i.e. a noticeable dynamic response amplification due to ice loading may be expected, which can interfere with the serviceability due to excessive vibrations on platforms or even structural integrity with regard to fatigue and plasticity development.

Note that the formula above only applies for calculating level ice-induced loading, even though this type of forces are generally higher than that caused by ice floes impacting.

In addition, the spatial variability of ice load exists for offshore structures with multiple legs [123]. This further complicates the prediction of the ice loading and the subsequent structural responses.

12.4 Earthquake Ground Motions

12.4.1 Power Spectrum of Seismic Ground Motions

Even though many excitations are nearly periodical and stationary, such as wave and wind loadings on structures, ships' propeller excitation forces etc. [124], earthquakes' ground motions are neither periodical (e.g., strong ground motions

12.4 Earthquake Ground Motions

have not repeated themselves during any earthquake event), nor stationary, i.e., the intensity builds up to a maximum value in the early part of the motions, then remains constant for a period of time, and finally decreases near the end of the motions [125].

However, as a modification of the traditional power spectrum density function, Kanai [126] and Tajimi [127] presented that, for both engineering and research purposes, one may still assume that strong ground accelerations are a stationary stochastic process by passing a white noise process (the power spectral density is constant over the entire frequency range, see Sect. 11.4.2) through a filter, i.e., the actual excitations are regarded as a function of output from a series of filters (usually a linear second-order system) subjected to white noise input S_0. In terms of acceleration amplitude, the Kanai-Tajimi model can be expressed as:

$$S_g(\omega) = \frac{\omega_g^4 + 4\xi_g^2\omega_g^2\omega^2}{\left(\omega_g^2 - \omega^2\right)^2 + 4\xi_g^2\omega_g^2\omega^2} S_0 \qquad (12.59)$$

where ω_g and ξ_g are characteristic ground frequency and damping ratio, respectively. S_0 is a scaling factor to define the white noise intensity level. The power spectral density is often filtered twice in order to remove the singularities at $\omega = 0$, i.e., the non-zero power spectrum density occurs for zero frequency.

The spectral density has its maximum value when $\omega = \omega_g$. By a proper selection of ω_g and ξ_g, Eq. (12.59) can be used to represent different spectral density shapes [128]. Kanai [126] and Tajimi [127] reported that ξ_g varies from 0.2 (relatively narrow banded) for soft soil to 0.6 (relatively wide banded) for hard rock sites. While many researchers and engineers tend to use $\xi_g = 0.6$, from a geotechnical engineering point of view, it is reasonable to assume that the power spectra of the horizontal ground motions have a similar shape, while the vertical motion component is more wide banded. Based on this assumption, Kubo and Penzien [129] simulated the San Fernando Earthquake with $\xi_g = 0.2$ and $\xi_g = 0.3$ for two horizontal ground motion components and $\xi_g = 0.6$ for the vertical component. Figure 12.14 [130] illustrates the Kanai-Tajimi power spectral density functions for soft, medium and stiff soil conditions. It is obvious that the spectrum under soft soil conditions is more narrow banded than that of the stiff soil conditions. Therefore, resonance of structures is more likely to occur under soft soil conditions than under stiff soil conditions.

Compared to the selection of ξ_g, the determination of ω_g is more important. This is because even if the wide-band power spectral density tends to overestimate the contribution from high frequencies, this normally does not result in a significant change of ground motions. However, the ω_g determines the dominant frequency of ground motion input [131]. When the dominant frequency is close to the natural frequency of a structure subjected to the ground motions, resonance of structural responses would occur. A typical value of $\omega_g = 5\pi$ can be assigned for rock sites.

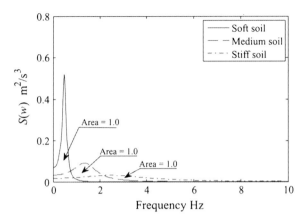

Fig. 12.14 The Kanai-Tajimi power spectral density functions for different types of soil (ξ_g is assumed to be 0.2 for soft soil, 0.4 for medium soil and 0.6 for stiff soil) [130]

A few previously adopted values of ξ_g and ω_g are listed as follows [132]: for El Centro 1940 N–S component, $\omega_g = 12$ and $\xi_g = 0.6$; for Kobe 1995 N–S component, $\omega_g = 12$ and $\xi_g = 0.3$; for Uemachi, the simulated ground motion using fault rupture model gives $\omega_g = 3$ and $\xi_g = 0.3$. Those selections of values are widely used in the research of tuned mass dampers [133, 134].

It should be noted that models expressed with the power spectral density elaborated above can only provide the excitation information phenomenally, they give no information on how spectra amplitudes are scaled with earthquake source and distance (the attenuation effects [23]). This drawback may be eliminated by calibrating the model to the measured ground motions. In cases when no such data is available, an alternative is to fit them into physical power spectra density models based on seismological description of source and wave propagation [135]. Interested readers may read references by Hanks and McGuire [136], Boore [137] and Herman [138].

12.4.2 Spatial Variation of Ground Motions by Coherence Function

Similar to that of the wind field simulation, the spatial variation of earthquake-induced ground motions can have a significant effect on the responses of extended structures. Section 16.1.4 briefly discusses the effects of spatial variation of ground motions. The spatial variation can be described mathematically either by auto-covariance and cross-covariance (in the time domain), or by coherence functions (in the frequency domain). In engineering practice, the ground motions at significant different locations can normally be defined by a homogeneous and

12.4 Earthquake Ground Motions

isotropic Gaussian stochastic model, with its spatial variability then expressed by its coherency spectrum, or coherence function. In establishing the coherence functions, it is normally sufficient to account for the effects of seismic wave passage and the loss of coherence; the function depends on the frequency and separation distance. The most widely used model to represent the spatial variation of ground motions is based on the Luco and Wong coherency function [139]. This establishes the coherence of any pairs of locations, which can be expressed as:

$$Coh(\xi, \omega_n) = e^{\left[-\left(\frac{\alpha \omega_n \xi}{v_s}\right)^2\right]} \cdot e^{\left[i\frac{\omega_n \xi^L}{v_{app}}\right]} \tag{12.60}$$

where α indicates the mechanical characteristics of the soil; a low value of α (e.g. $2 \cdot 10^{-4}$) represents a slow exponential decay in the coherency as the frequency ω_n and separation distance ξ increase. On the other hand, a high value of α (e.g. 10^{-3}) represents a sharp exponential decay in the coherency as frequency and separation distance increase.

The first term in the equation above also shows an exponential decay of coherence due to the variation of separation distance ξ between two locations, shear wave velocity v_s and frequency ω_n. This item controls the geometric incoherence of ground motions, which decreases as soil becomes stiffer. The second term in the equation above represents the seismic wave passage effect. It gives the longer signal arrival delay when the projected horizontal distance ξ^L and the frequency ω_n increase, and the apparent velocity v_{app} decreases. The coherence level is mainly governed by shear wave velocity and apparent surface wave velocity. When the surface wave travels at infinite speed ($v_{app} = \infty$), the second term equals to 1, the seismic waves arrives at all locations simultaneously, and the loss of coherence is only due to the geometric incoherence (first term). The shear wave velocity is generally much lower than the apparent wave velocity. For soft soil, v_s is in a range of 200–300 m/s [140].

It should be noted that several other models are available to simulate the coherence of asynchronous ground motion, such as the Harichandran-Vanmarcke model [141], Loh-Yeh model [142], Feng-Hu model [143], Oliveira-Hao-Penzien model [144], and Qu-Wang-Wang model [145] etc. For more details of those models, the reader may read the references cited above.

A distinction needs to be made between the power spectrum and response spectrum (Sect. 16.2) in earthquake engineering applications. Even if both of them are expressed in the frequency domain, the former one is of a stochastic nature while the latter is based on a calculation of maximum responses under ground motion history, i.e., the response spectrum is essentially a deterministic type. For details of these two types of spectra, see reference [23].

Chapter 13
Vibration of Multi-Degrees-of-Freedom Systems

In most cases, a system or a structure possesses more than a SDOF, which indicates that the vibrations on each degree-of-freedom are likely to be different from the others but also coupled. The number of degrees-of-freedom is equal to the number of independent coordinates necessary and sufficient to describe complete motions of the system. Therefore, the system will have the same number of eigenfrequencies and mode shapes as degrees-of-freedom. The free vibrations of the system involve several simultaneous oscillations at various eigenfrequencies. What distinguishes the free vibrations of multi-degree-of-freedom (MDOF) from that of SDOF is that, when all degrees-of-freedoms (all parts of the system) move harmonically at the same eigenfrequency as the system, a certain displacement configuration or shape, called the principal mode or normal mode of vibrations, is formed. Moreover, the number of normal modes is equal to the number of degrees-of-freedom. This means that the system can be modeled as a series of masses. The number of normal modes is equal to the number of masses multiplied by the number of directions that masses can move translationally and rotationally. Through the superposition of the normal mode vibrations, more general types of vibrations for the system can be obtained.

Generally, three methods are used for solving the vibration problem for systems possessing more than one DOF: direct/exact method, mode superposition method and direct time integration method.

Table 13.1 shows the applicability of three methods for solving the equations of motions. It is noticed that, compared to the direct integration method, the mode superposition is more efficient for solving the linear dynamic problem, as will be discussed in Sect. 13.4. Practically, only the latter two methods are suitable for the analysis of systems with MDOF.

13.1 Equations of Motions

We start with the vibrations of two degrees-of-freedom-system as shown in Fig. 13.1a, and then extend the method to an MDOF system. The governing equations of motions are formulated by the following two equations, each of which

---

J. Jia, *Essentials of Applied Dynamic Analysis*, Risk Engineering,
DOI: 10.1007/978-3-642-37003-8_13, © Springer-Verlag Berlin Heidelberg 2014

203

Table 13.1 Applicability of three methods for solving the equation of motions [125]

Method	MDOF	Nonlinearity	Non-classical damping[a]	Solutions type
Direct/exact/stiffness	Not good	No	No	Closed form
Mode superposition	Yes	Not efficient	No	Closed form for periodical excitations, numerical for complex excitations
Direct integration	Yes (but can be computationally demanding)	Yes	Yes	Numerical

[a] Note: non-classical damping means that the damping matrix is not diagonal

13.1 Equations of Motions

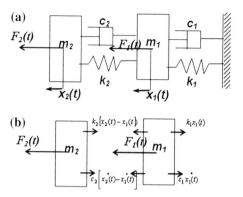

Fig. 13.1 A two degrees-of-freedom-system under external force excitations $F(t)$: **a** spring-mass-damper system; **b** force equilibrium for each mass

corresponds to an equilibrium state (free body diagram shown in Fig. 13.1b) for one mass:

$$m_1 \ddot{x}_1(t) + c_1 \dot{x}_1(t) - c_2 \left[\dot{x}_2(t) - \dot{x}_1(t) \right] + k_1 x_1(t) - k_2 [x_2(t) - x_1(t)] = F_1(t) \tag{13.1}$$

$$m_2 \ddot{x}_2(t) + c_2 \left[\dot{x}_2(t) - \dot{x}_1(t) \right] + k_2 [x_2(t) - x_1(t)] = F_2(t) \tag{13.2}$$

Note that the two equations above are coupled in that the first one contains $x_2(t)$ and the second one contains $x_1(t)$. Further, it is realized that the number of differential equations is equal to the number of degrees-of-freedom.

The equations above can also be expressed in a matrix form as:

$$\begin{bmatrix} m_1 & 0 \\ 0 & m_2 \end{bmatrix} \begin{Bmatrix} \ddot{x}_1(t) \\ \ddot{x}_2(t) \end{Bmatrix} + \begin{bmatrix} (c_1 + c_2) & -c_2 \\ -c_2 & c_2 \end{bmatrix} \begin{Bmatrix} \dot{x}_1(t) \\ \dot{x}_2(t) \end{Bmatrix}$$
$$+ \begin{bmatrix} (k_1 + k_2) & -k_2 \\ -k_2 & k_2 \end{bmatrix} \begin{Bmatrix} x_1(t) \\ x_2(t) \end{Bmatrix}$$
$$= \begin{Bmatrix} F_1(t) \\ F_2(t) \end{Bmatrix} \tag{13.3}$$

The more general form of the equation above can be written as:

$$[M_{n \times n}] \{\ddot{X}_{n \times 1}\} + [C_{n \times n}] \{\dot{X}_{n \times 1}\} + [K_{n \times n}] \{X_{n \times 1}\} = \{F_{n \times 1}\} \tag{13.4}$$

where n is the number of degrees-of-freedom.

We extend our study from the spring-mass system to a cantilever beam model fixed at one end as shown in Fig. 13.2a with five degrees-of-freedom: three translations and two rotations. The beam is divided into three segments. At node 1

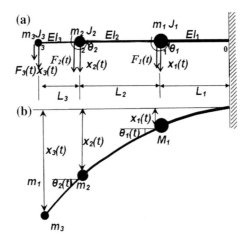

Fig. 13.2 A beam with three segments and five degrees-of-freedom ($x_1(t), x_2(t), x_3(t), \theta_1(t), \theta_2(t)$) under external force $F(t)$ at each segment's end: **a** beam model; **b** space diagram

of the two adjacent segments, half of the mass sum (m_1) for the two adjacent segments and rotation inertia (J_1) associated with node 1 are concentrated together with the moment of inertia I_1. At node 2 of the beam, half of the mass sum (m_2) for the two adjacent segments and rotation inertia (J_2) associated with node 2 are concentrated together with the moment of inertia I_2. At node 3 of the beam, half of the mass of the third segment (between node 2 and node 3) and any additional mass attached to node 3 (m_3) are concentrated together with the moment of inertia (I_3) of the third segment. This approach of assigning mass and rotation inertia properties lumped at the nodal coordinates is called the lumped mass method. It is normally applicable to represent the essential inertia feature of the mass distribution for a typical structure. The major advantage of the lumped mass method is that, as will be discussed in Sect. 13.6, a diagonal mass matrix can be used, which reduces the computational effort. Therefore, it is widely adopted as an efficient way of formulating mass matrix in both analytical and finite element methods.

For segment 1, the relationship between the elastic forces (F)/moments (M) and deflection (x) and rotation (θ) at nodes 0 and 1 is:

$$\begin{bmatrix} m_0 & 0 & 0 & 0 \\ 0 & J_0 & 0 & 0 \\ 0 & 0 & m_1 & 0 \\ 0 & 0 & 0 & J_1 \end{bmatrix} \begin{Bmatrix} \ddot{x}_0(t) \\ \ddot{\theta}_0(t) \\ \ddot{x}_1(t) \\ \ddot{\theta}_1(t) \end{Bmatrix} + \begin{bmatrix} \frac{12EI_1}{L_1^3} & -\frac{6EI_1}{L_1^2} & -\frac{12EI_1}{L_1^3} & -\frac{6EI_1}{L_1^2} \\ -\frac{6EI_1}{L_1^2} & \frac{4EI_1}{L_1} & \frac{6EI_1}{L_1^2} & \frac{2EI_1}{L_1} \\ -\frac{12EI_1}{L_1^3} & \frac{6EI_1}{L_1^2} & \frac{12EI_1}{L_1^3} & \frac{6EI_1}{L_1^2} \\ -\frac{6EI_1}{L_1^2} & \frac{2EI_1}{L_1} & \frac{6EI_1}{L_1^2} & \frac{4EI_1}{L_1} \end{bmatrix} \begin{Bmatrix} x_0(t) \\ \theta_0(t) \\ x_1(t) \\ \theta_1(t) \end{Bmatrix} = \begin{Bmatrix} F_0(t) \\ M_0(t) \\ F_1(t) \\ M_1(t) \end{Bmatrix}$$

(13.5)

13.1 Equations of Motions

Note that the fixed support condition at node 0 indicates that all the displacements and rotations at node 0 are zero. The equation above is then simplified as:

$$
\begin{bmatrix} m_1 & 0 \\ 0 & J_1 \end{bmatrix} \begin{Bmatrix} \ddot{x}_1(t) \\ \ddot{\theta}_1(t) \end{Bmatrix} + \begin{bmatrix} \frac{12EI_1}{L_1^3} & \frac{6EI_1}{L_1^2} \\ \frac{6EI_1}{L_1^2} & \frac{4EI_1}{L_1} \end{bmatrix} \begin{Bmatrix} x_1(t) \\ \theta_1(t) \end{Bmatrix} = \begin{Bmatrix} F_1(t) \\ M_1(t) \end{Bmatrix} \tag{13.6}
$$

Similarly, for segment 2, the relationship between the elastic forces (F)/ moments (M) and deflection (x) and rotation (θ) at nodes 1 and 2 is:

$$
\begin{bmatrix} m_1 & 0 & 0 & 0 \\ 0 & J_1 & 0 & 0 \\ 0 & 0 & m_2 & 0 \\ 0 & 0 & 0 & J_2 \end{bmatrix} \begin{Bmatrix} \ddot{x}_1(t) \\ \ddot{\theta}_1(t) \\ \ddot{x}_2(t) \\ \ddot{\theta}_2(t) \end{Bmatrix} + \begin{bmatrix} \frac{12EI_1}{L_2^3} & -\frac{6EI_1}{L_2^2} & -\frac{12EI_1}{L_2^3} & -\frac{6EI_1}{L_2^2} \\ -\frac{6EI_1}{L_2^2} & \frac{4EI_1}{L_2} & \frac{6EI_1}{L_2^2} & \frac{2EI_1}{L_2} \\ -\frac{12EI_1}{L_2^3} & \frac{6EI_1}{L_2^2} & \frac{12EI_1}{L_2^3} & \frac{6EI_1}{L_2^2} \\ -\frac{6EI_1}{L_2^2} & \frac{2EI_1}{L_2} & \frac{6EI_1}{L_2^2} & \frac{4EI_1}{L_2} \end{bmatrix} \begin{Bmatrix} x_1(t) \\ \theta_1(t) \\ x_2(t) \\ \theta_2(t) \end{Bmatrix}
$$

$$
= \begin{Bmatrix} F_1(t) \\ M_1(t) \\ F_2(t) \\ M_2(t) \end{Bmatrix}
$$

$$\tag{13.7}$$

Note that any stiffness term in the assembled stiffness matrix may be obtained by adding together the corresponding stiffness associated with those nodal coordinates. Similarly, the force and moment contribution at a node can also be obtained by adding the forces and moments at this node and the adjacent nodes. We can then assemble the stiffness at node 1 by calculating the contribution of forces and moments from both segments 1 and 2:

$$
[K] = \begin{bmatrix} k_{11}^{(1)} & k_{12}^{(1)} & k_{13}^{(1)} & k_{14}^{(1)} & 0 & 0 \\ k_{21}^{(1)} & k_{22}^{(1)} & k_{23}^{(1)} & k_{24}^{(1)} & 0 & 0 \\ k_{31}^{(1)} & k_{32}^{(1)} & \left(k_{33}^{(1)} + k_{11}^{(2)}\right) & \left(k_{34}^{(1)} + k_{12}^{(2)}\right) & k_{13}^{(2)} & k_{14}^{(2)} \\ k_{41}^{(1)} & k_{42}^{(1)} & \left(k_{43}^{(1)} + k_{21}^{(2)}\right) & \left(k_{44}^{(1)} + k_{22}^{(2)}\right) & k_{23}^{(2)} & k_{24}^{(2)} \\ 0 & 0 & k_{31}^{(2)} & k_{32}^{(2)} & k_{33}^{(2)} & k_{34}^{(2)} \\ 0 & 0 & k_{41}^{(2)} & k_{42}^{(2)} & k_{43}^{(2)} & k_{44}^{(2)} \end{bmatrix} \tag{13.8}
$$

where $k_{ij}^{(m)}$ is the stiffness coefficient. Its lower indices serve to locate the appropriate stiffness coefficients in the corresponding stiffness matrix, while the upper indices are used to identify the beam segment.

From the equation above, it is noticed that the stiffness matrix of two adjacent elements overlap by a 2 by 2 matrix. In addition, the assembling of the mass matrix

208   13   Vibration of Multi-Degrees-of-Freedom Systems

for the entire beam model is exactly the same as described for assembling the stiffness matrix.

We therefore conclude a more general form of the equation of equilibrium for beams with N degrees-of-freedom, which is expressed in 2×2 N mass and stiffness matrix form as:

$$
\begin{bmatrix} m_1 & & & & \\ & J_1 & & & \\ & & \bullet & & \\ & & & \bullet & \\ & & & & m_N \\ & & & & & J_N \end{bmatrix} \begin{Bmatrix} \ddot{x}_1(t) \\ \ddot{\theta}_1(t) \\ \bullet \\ \bullet \\ \ddot{x}_N(t) \\ \ddot{\theta}_N(t) \end{Bmatrix} + \underset{2N \times 2N}{[K]} \begin{Bmatrix} x_1(t) \\ \theta_1(t) \\ \bullet \\ \bullet \\ x_N(t) \\ \theta_N(t) \end{Bmatrix} = \begin{Bmatrix} F_1(t) \\ M_1(t) \\ \bullet \\ \bullet \\ F_N(t) \\ M_N(t) \end{Bmatrix}
$$

(13.9)

The equation above is the basic matrix formulation of the equations of a lumped mass (Sect. 13.6) system. The knowledge of the matrix theory is required to solve this equation. It is noted that, for large degrees-of-freedom, the solutions using the direct/exact method can be extremely tricky and therefore impractical. The solution for a two-degrees-of-freedom system using the direct/exact method will be presented in Sect. 13.2.

13.2 Free Vibrations of the Two-Degrees-of-Freedom System: Direct/Exact Method

By neglecting the damping in Fig. 13.1, and further by assuming that if the system is vibrating in a normal mode, the displacements of two masses are in phase and harmonic, the solutions of the motions can then be written as:

$$x_1(t) = X_1 \cos(\omega t) \tag{13.10}$$

$$x_2(t) = X_2 \cos(\omega t) \tag{13.11}$$

By substituting the equations above into Eq. (13.3) and also setting the external force as zero for free vibrations, one obtains:

$$
\begin{bmatrix} k_1 + k_2 - \omega^2 m_1 & -k_2 \\ -k_2 & k_2 - \omega^2 m_2 \end{bmatrix} \begin{Bmatrix} X_1 \\ X_2 \end{Bmatrix} = \begin{Bmatrix} 0 \\ 0 \end{Bmatrix} \tag{13.12}
$$

Again, the equation above indicates that the vibrations of two masses are coupled. It can only be valid for nontrivial X_1 and X_2, leading to the determinant of the coefficient matrix being vanished:

$$
\begin{vmatrix} k_1 + k_2 - \omega^2 m_1 & -k_2 \\ -k_2 & k_2 - \omega^2 m_2 \end{vmatrix} = 0 \tag{13.13}
$$

13.2 Free Vibrations of the Two-Degrees-of-Freedom System

The equation above results in the characteristic, or frequency equation:

$$m_1 m_2 \omega^4 - [m_1 k_1 + m_2(k_1 + k_2)]\omega^2 + k_1 k_2 = 0 \tag{13.14}$$

The solution of the two eigenfrequencies for the system can then be calculated as the real positive roots of ω:

$$\omega_{1,2} = \sqrt{\frac{[m_1 k_2 + m_2(k_1 + k_2)] \pm \sqrt{[m_1 k_2 + m_2(k_1 + k_2)]^2 - 4m_1 m_2 k_1 k_2}}{2m_1 m_2}}$$

$$\tag{13.15}$$

In free vibrations, the relative deflections of x_1 and x_2 define the mode shape. At eigenfrequency ω_i:

$$x_1 = \frac{k_2}{k_1 + k_2 - m_1 \omega_i^2} x_2 \tag{13.16}$$

Or

$$x_1 = \frac{k_2 - m_2 \omega_i^2}{k_2} x_2 \tag{13.17}$$

where $i = 1$ or 2.

The two equations above indicate that the relative value between x_1 and x_2 forms a unique "shape" but not a unique value, i.e., they do not express the real responses until known excitations are applied on the system. By arbitrarily choosing one of these two (x_1 and x_2), the other can be calculated as well. This also means that any comparison between x_1 and x_2 only makes qualitative sense within the same vibration mode, i.e., the comparison between different modes of vibrations does not make any sense unless the associated modal mass (as will be presented later in this section) and the excitations are known.

By applying the equations presented above, the two vibration mode shapes ($\{\phi\}_1$ and $\{\phi\}_2$) at the two eigenfrequencies (ω_1 and ω_2) of the beam (without rotational degrees-of-freedom) are shown in Fig. 13.3. These resemble the vibrations of a system with two degrees-of-freedoms, and have a unique shape at each eigenfrequency even though the amplitude is arbitrary. By observing the figure, it is also found that at the second eigenmode, the vibration amplitude x_1 and x_2 is out of phase.

It should be noted that one can also assume that the solutions of the motions ($x_1(t)$ and $x_2(t)$) are a function of $\cos(\omega t)$, which will finally reach the same conclusion.

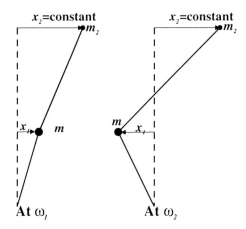

Fig. 13.3 Mode shapes at two eigenfrequencies of a beam with two translational degrees-of-freedoms

13.3 Forced Vibrations of Two Degrees-of-Freedom Systems: Direct Method

This time, we still assume that the external forces are harmonic but expressed in another format:

$$F_1(t) = F_{10}e^{i\omega t} \tag{13.18}$$

$$F_2(t) = F_{20}e^{i\omega t} \tag{13.19}$$

And the steady-state solution is:

$$x_1(t) = X_1 e^{i\omega t} \tag{13.20}$$

$$x_2(t) = X_2 e^{i\omega t} \tag{13.21}$$

The equilibrium equation (Eq. (13.3)) can be rewritten as:

$$\begin{bmatrix} (-\omega^2 m_{11} + i\omega c_{11} + k_{11}) & (-\omega^2 m_{12} + i\omega c_{12} + k_{12}) \\ (-\omega^2 m_{12} + i\omega c_{12} + k_{12}) & (-\omega^2 m_{22} + i\omega c_{22} + k_{22}) \end{bmatrix} \begin{Bmatrix} X_1 \\ X_2 \end{Bmatrix} = \begin{Bmatrix} F_{10} \\ F_{20} \end{Bmatrix} \tag{13.22}$$

Let $Z_{ij}(\omega) = \left(-\omega^2 m_{ij} + i\omega c_{ij} + k_{ij}\right)$, which is called impedance. We then define:

$$[Z(\omega)] = \begin{bmatrix} Z_{11}(\omega) & Z_{12}(\omega) \\ Z_{21}(\omega) & Z_{22}(\omega) \end{bmatrix} \tag{13.23}$$

13.3 Forced Vibrations of Two Degrees-of-Freedom Systems 211

The system of equation can be expressed as:

$$\{X(\omega)\} = [Z(\omega)]^{-1}\{F_0(\omega)\} = [H(\omega)]\{F_0(\omega)\} \tag{13.24}$$

where $[H(\omega)] = [Z(\omega)]^{-1}$ is the transfer function.

The responses in the frequency domain are then expressed as:

$$x_1(\omega) = H_{11}(\omega)F_1(\omega) + H_{12}(\omega)F_2(\omega) \tag{13.25}$$

$$x_2(\omega) = H_{21}(\omega)F_1(\omega) + H_{22}(\omega)F_2(\omega) \tag{13.26}$$

where $H_{ij}(\omega)$ is complex that represents the relationship between responses at ith degree-of-freedom and forces acting at jth degree-of-freedom [168].

13.4 Forced Vibrations of MDOF: Modal Superposition Method

We recall the general expression of the equation of motions in a matrix form:

$$[M_{n\times n}]\left\{\overset{\bullet\bullet}{X}\right\}_{n\times 1} + [C_{n\times n}]\left\{\overset{\bullet}{X}\right\}_{n\times 1} + [K_{n\times n}]\{X_{n\times 1}\} = \{F_{n\times 1}\} \tag{13.27}$$

Regardless of which types of excitation force $\{F_{n\times 1}\}$ is, the equation above can always be solved either directly or by transforming it into a simpler form.

We have already, by using an N-segment beam model, presented the relevant knowledge on how to assemble the stiffness and mass matrix and implement them into the equation of motions, i.e., direct/matrix method. It is noticed that for a rather large degrees-of-freedom system, not only the solution of the assembled equation in the matrix form, but also the process of matrix assembling itself is complicated. This difficulty promotes the application of modal analysis, which is a convenient method for solving vibration problem for systems with a large degrees-of-freedoms, providing that the stiffness matrix is constant, i.e., the structure/system is linear (see Sect. 15.2 for the definition of nonlinearities).

In the modal analysis, the coupled equations of motions are transformed into a series of uncoupled/independent equations. Each of these equations is analogous to the equation of motions for an SDOF system, and can be solved in the same manner. The equivalent single degree-of-freedom-system is the one for which the kinetic energy, internal strain energy, and work done by all external forces are at all times equal to the same quantities for the complete MDOF system when vibrating in this normal mode alone [169].

By excluding the eigenfrequencies and mode shapes associated with rigid motions, in which the system moves as a solid part with a zero eigenfrequency (Sect. 4), the number of eigenfrequencies and mode shapes is equal to the number of degrees-of-freedom of the system. It is noticed that the number of uncoupled equations needed to be solved is equal to the number of eigenmodes to be accounted for. For structures with the dynamic responses dominated by the first few eigenmodes, great computational efficiency can be achieved.

From a physical point of view, an initial excitation of a structure/system will cause it to vibrate and the system responses will be a combination of eigenmodes, where each eigenmode oscillates at its associated eigenfrequency.

Let's assume that a structure has n degrees-of-freedoms and is vibrating at an eigenfrequency ω with the mode shape $\{\phi_{n\times1}\}$. The mode shape does not vary with time. The solution of time variation of the responses can then be expressed as a harmonic function multiplied by the mode shape:

$$\{x(t)\}_n = \{\phi\}_n \cos(\omega t) \tag{13.28}$$

where the lower indices serve to identify the order of the vibration mode.

By neglecting the damping and inserting the equation above into Eq. (13.27), one has:

$$-[M_{n\times n}]\{\phi\}_n \omega^2 + [K_{n\times n}]\{\phi\}_n = \{F_{n\times1}\} \tag{13.29}$$

For free vibrations, set $\{F_{n\times1}\} = \{0\}$. The equation above will yield n eigen-pairs (eigenfrequencies and mode shapes). It is necessary to solve this linear eigenproblem by rearranging the equation above:

$$\left([K_{n\times n}] - \omega^2[M_{n\times n}]\right)\{\phi\}_n = 0 \tag{13.30}$$

This equation is called the matrix eigenvalue problem, and actually includes n linear homogenous equations. The non-trivial solutions of $\{\phi\}_n$ exist if all eigenvalues are different, this can be obtained by vanishing the determinant of the coefficient matrix as expressed in the characteristic equation:

$$\left|[K_{n\times n}] - \omega^2[M_{n\times n}]\right| = 0 \tag{13.31}$$

To avoid the solutions for rigid body vibrations, all elements in $[K_{n\times n}]$ must be positively defined, as is the case for structures with supports to the ground.

It is noted that the equation above is of a polynomial order in terms of the number of degrees-of-freedom. It is not easy to solve this equation because no explicit formulas are available for directly obtaining the eigenvalues (roots) of the equations when the degrees-of-freedom are higher than four. Therefore, iteration methods must first be adopted to find one of the eigenpairs (eigenfrequencies or mode shapes), and the other one can then be calculated directly. For the detailed technique for solving the eigenvalue, the textbook by Chopra [124] is recommended.

An eigenvector is arbitrary to the extent that a scalar multiple of it is still a solution of the equation above, i.e., it only indicates the vibration shape in space given by the relative values as aforementioned in Sect. 13.2. However, it is con-venient to choose this multiplier in such a way that $\{\phi\}_n$ has some desirable property. Such $\{\phi\}_n$ is called the normalized eigenvector [170]. The most common procedures are to either scale $\{\phi\}_n$ such that its largest component is unity or so

13.4 Forced Vibrations of MDOF 213

that the modal mass (also called the generalized mass as described later on in this section) is unity:

$$\{\phi\}_n^T[M_{n\times n}]\{\phi\}_n = 1 \tag{13.32}$$

Up to now, we have learned the method for how to obtain eigenpairs (eigenfrequencies and mode shapes). But the objective of this section is to find the solution of the equation of motions (Eq. 13.27). The difficulty of solving the equation lies in the fact that this equation is coupled through the response terms by the virtue of the stiffness, damping and mass matrix. In order to solve this problem, one has to find a way to uncouple the equations so that each normal mode can be determined separately as an SDOF system. For this purpose, we need to be familiarized with the property of mode shape orthogonality.

We first present this property mathematically through studying the matrix eigenvalue Eq. (13.30) with two separate eigenfrequencies, ω_r and ω_s:

$$\left([K_{n\times n}] - \omega_r^2[M_{n\times n}]\right)\{\phi\}_r = 0 \tag{13.33}$$

$$\left([K_{n\times n}] - \omega_s^2[M_{n\times n}]\right)\{\phi\}_s = 0 \tag{13.34}$$

Pre-multiplying the first equation above by $\{\phi\}_s^T$ and the second equation by ϕ_r^T gives:

$$\{\phi\}_s^T \left([K_{n\times n}] - \omega_r^2[M_{n\times n}]\right)\{\phi\}_r = 0 \tag{13.35}$$

$$\{\phi\}_r^T \left([K_{n\times n}] - \omega_s^2[M_{n\times n}]\right)\{\phi\}_s = 0 \tag{13.36}$$

where the upper indices ^{T} means the transpose of the mode shape vector.

Since both $[K_{n\times n}]$ and $[M_{n\times n}]$ are symmetric, we can transpose the second equation (eq. (13.36)) above:

$$\{\varphi\}_s^T \left([K_{n\times n}] - \omega_s^2[M_{n\times n}]\right)\{\varphi\}_r = 0 \tag{13.37}$$

Subtracting Eq. (13.37) from Eq. (13.35) and rearranging the result, the obtained equation is:

$$\left(\omega_s^2 - \omega_r^2\right)\{\phi\}_s^T[M_{n\times n}]\{\phi\}_r = 0 \tag{13.38}$$

For different eigenpairs, i.e., $r \neq s$ and $\omega_r \neq \omega_s$, we have:

$$\{\phi\}_s^T[M_{n\times n}]\{\phi\}_r = 0 \tag{13.39}$$

Substituting the equation above into Eq. (13.35) and rearranging it, one obtains:

$$\{\phi\}_s^T[K_{n\times n}]\{\phi\}_r = 0 \tag{13.40}$$

Physically, the mode shape orthogonality means that the work done by the sth mode inertia forces in going through the rth mode displacements is zero. Another implication is that the work done by the equivalent static forces associated with the

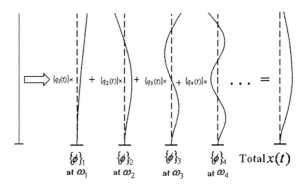

Fig. 13.4 Illustration of modal superposition in which the responses ($x(t)$) in "real" coordinates are the linear sum of product between the eigenvectors (constant in time) and the generalized/modal coordinates (varied with time) for each eigenmode

displacement in the sth mode in going through the rth mode displacements is zero [125].

With the knowledge of mode shape orthogonality, we can now express the displacement response vectors of a structure/system due to the initial conditions or excitations by transforming the responses to a set of uncoupled equations, which is also called the modal superposition:

$$\begin{aligned}\{x^{(1)}(t)\} &= \phi_1^{(1)} q_1(t) + \phi_2^{(1)} q_2(t) + \cdots + \phi_n^{(1)} q_n(t) \\ \{x^{(2)}(t)\} &= \phi_1^{(2)} q_1(t) + \phi_2^{(2)} q_2(t) + \cdots + \phi_n^{(2)} q_n(t) \\ &\cdots \\ \{x^{(n)}(t)\} &= \phi_1^{(n)} q_1(t) + \phi_2^{(n)} q_2(t) + \cdots + \phi_n^{(n)} q_n(t) \end{aligned} \quad (13.41)$$

where $q_i(t)$ is the generalized/modal coordinates varied with time. The upper indices indicate the location of the deflection along the structure/system, while lower indices indicate the order of the eigenmode. The procedure is illustrated in Fig. 13.4.

The essential feature of modal superposition is to use eigenmode shapes to uncouple the equations of motions. The uncoupled equations are in terms of modal coordinates $q_i(t)$, which can be obtained by solving each equation independently.

With the absence of damping, $q_i(t)$ will be infinite when a forcing frequency reaches ω_i. Therefore, to avoid resonance, for a system with N ($n = N$) degrees-of-freedom, the forcing frequency has to be away from the N eigenfrequencies ω_i.

Rewrite the equations above in a compact form:

$$\{x(t)\} = [\phi]\{q(t)\} \quad (13.42)$$

Or conversely,

$$\{q(t)\} = [\phi]^T \{x(t)\} \quad (13.43)$$

13.4 Forced Vibrations of MDOF

Where $[\varphi]$ is called the modal matrix and is constructed from the n modal column vectors or mode shapes.

It is clearly seen that, rather than from the summation of Cartesian deflections, the transformation to the generalized coordinate is equivalent to the deflection from the summation of mode shapes. The solution for the generalized coordinate will be presented later in this section.

Inserting the equation above into the equation of motions Eq. (13.27), one gets:

$$[M_{n\times n}][\phi_{n\times n}]\left\{\ddot{q}(t)_{n\times 1}\right\} + [C_{n\times n}][\phi_{n\times n}]\left\{\dot{q}(t)_{n\times 1}\right\} + [K_{n\times n}][\phi_{n\times n}]\left\{q(t)_{n\times 1}\right\}$$
$$= \phi[\phi_{n\times n}]\left\{F(t)_{n\times 1}\right\}$$

$$(13.44)$$

For an ith order normal mode of vibrations, the equation above can be written in a compact form:

$$m_i\,\ddot{q}_i(t) + c_i\,\dot{q}_i(t) + k_i q_i(t) = F_i(t) \qquad (13.45)$$

where:

$m_i = \left[\{\phi_i\}^{(j)}\right]^{\mathrm{T}}[M]\left[\{\phi_i\}^{(j)}\right]$ is called the generalized mass or modal mass;

$c_i = \left[\{\phi_i\}^{(j)}\right]^{\mathrm{T}}[C]\left[\{\phi_i\}^{(j)}\right]$ is called the generalized damping or modal damping;

$k_i = \left[\{\phi_i\}^{(j)}\right]^{\mathrm{T}}[K]\left[\{\phi_i\}^{(j)}\right]$ is called the generalized stiffness or modal stiffness;

$F_i(t) = \left[\{\phi_i\}^{(j)}\right]\{F(t)\}$ is called the generalized excitation force or modal force.

In the mode superposition method, the modal mass is of special importance in determining the characteristics of dominated dynamic responses and how many eigenmodes need to be included in a dynamic analysis. Let r^{st} be the static displacement of the masses along a particular direction, resulting from the static application of a unit base displacement without damping. We hereby define a coefficient vector:

$$\{L_i\} = [\phi]^T[M]\{r^{st}\} \qquad (13.46)$$

The modal mass participation due to the ith eigenmode is then defined as:

$$P_i = \frac{L_i^2}{m_{ii}} \qquad (13.47)$$

where $m_{ii} = 1$ for each index if the eigenvectors have been normalized with respect to the mass matrix (Eq. 13.32). Due to the orthogonality of the eigenvectors, for both lumped mass and consistent mass matrix $[M]$ (as will be elaborated in section 13.6), m_{ij} $(i \neq j)$ is always zero. If all modes are used, these ratios will be equal to 1.0.

Engineering experiences indicate that, for most of the dynamic analysis practices, only the first few eigenmodes are required to obtain the responses with sufficient accuracy. As a rule of thumb, along each principal direction of a structure, more than 85 % of the participating mass should be included in the calculation of the dynamic responses. It should be noted that this criterion is intended to estimate the accuracy of a solution for the structure's base motion only. Strictly speaking, it cannot be used as an exact error estimator for other types of loading such as point loads acting on the structure [171], even though it is used in many cases as a judgment on the reliability of the set-up for performing dynamic analysis. Figure 13.5 shows the modal mass participation influenced by the number of eigenmodes included for a gravity-based structure (GBS). It is noticed that the mass participation in global X and Y reaches to more than 90 % when only the first five eigenmodes have been included. However, the mass participation in Z direction converges to more than 90 % after 11 eigenmodes have been included.

Table 13.2 describes a flare boom's eigenfrequencies, mode shapes and the corresponding modal mass along each direction. It is clearly shown that, instead of the fundamental eigenmode, the 2nd and 4th eigenmode (shown in Fig. 13.6) have the largest modal mass along the horizontal X and Y direction, respectively. By observing Fig. 13.7, which shows the power spectra of wind induced displacement responses at various locations of the flare boom, it is clearly shown that almost all response peaks occur at periods close to the 2nd and 4th eigenperiod of the flare boom, indicating the significant influence due to the high percent of modal mass participation by these two eigenmodes.

Since the mode shape is governed by the dynamic characteristics of a structure and has no absolute magnitude, the responses can only be obtained if the generalized coordinate $q_i(t)$ in Eq. (13.43) is known.

For free vibrations, the $q_i(t)$ can be obtained by setting the modal force and damping in Eq. (13.45) to zero:

$$\ddot{q}_i(t) + \omega_i^2 q_i(t) = 0 \tag{13.48}$$

It shows that the vibrations of the ith eigenmode are independent of vibrations in other modes. The general solution is then:

$$q_i(t) = A_i \cos(\omega_i t + \varphi_i) \tag{13.49}$$

where A_i and φ_i are the amplitude and the phase, respectively.

For forced vibrations, the total responses of the system are to solve $q_i(t)$ for a total of N equations with the form of Eq. (13.45) and by substituting them into Eq. (13.41). Convolution integral may be used for obtaining $q_i(t)$ as:

$$q_i(t) = \frac{1}{m_i \omega_i} \int_0^t F_i(\tau) \sin(\omega_i(t - \tau)) d\tau \tag{13.50}$$

where τ is the time when the loading has been applied. For a causal system, it affects the responses if $t > \tau$.

13.4 Forced Vibrations of MDOF

The equation above can be integrated numerically or by parts if the force is an analytical function. A number of textbooks address the solutions of this equation. See the detailed solution strategies in the reference by Bathe [173].

Table 13.2 A flare boom's eigenfrequencies, mode shapes and the associated modal masses along three perpendicular directions

Mode order	Eigenperiod (s)	Mode shape	Modal mass along X direction (tons)	Modal mass along Y direction (tons)	Modal mass along Z direction (tons)
1	1.462	Local sway vibration at the tip of the original high and lower pressure pipes	0.0	10.8	0.0
2	1.424	1st bending eigenmode of the main flareboom	132.0	0.0	158.0
3	1.070	Local coupled sway vibration at the tip of all pipes and flare boom	0.0	32.3	0.0
4	0.810	Global torsional vibration of the flare boom coupled with the sway vibration at the tip of all pipes and flare boom	0.0	103.0	0.0
5	0.751	2nd bending eigenmode of the main flare boom	51.7	0.0	78.5
6	0.734	1st torsional vibration of the wind struts coupled with the sway vibration at the tip of all pipes and flare boom	0.5	1.0	0.6

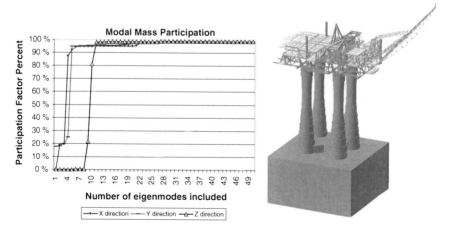

Fig. 13.5 Modal mass participation (*left*) influenced by the number of the eigenmodes selected for a GBS platform (*right*) [172]

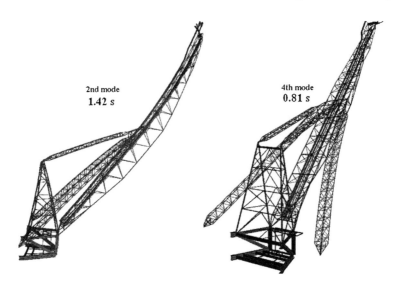

Fig. 13.6 The 2nd and 4th eigenmode shapes of the flare boom (courtesy of Aker Solutions)

Note that the modal damping matrix $[c_i]$ is diagonal, the usual way of implementing it into the equation of motions (Eq. 13.44) is through the damping ratio ζ_i in each eigenmode:

$$\ddot{q}_i(t) + 2\zeta_i \omega_i \dot{q}_i(t)_i + \omega_i^2 q_i(t) = \frac{F_i(t)}{m_i} \qquad (13.51)$$

Even though the implementation of damping ratio ζ_i is not mathematically correct, it provides a reasonable response calculation if ζ_i is small. In addition, for MDOF systems, this does not satisfy the dynamic equilibrium: modal damping introduces additional loads on the system, leading to the difference between the sum of inertia forces of all masses and base shear (sum of the forces at members attached to the base).

Significant computation cost can be reduced by using the modal superposition method. This is especially the case for structures with a large number of degrees-of-freedoms and with the dynamic responses dominated by the first few eigenmodes. Consider Eq. (13.41) applied to an structure with 9,000 degrees-of-freedom. If only the first six eigenmodes are significant in contributing to the dynamic responses, we then have:

$$\underset{9{,}000 \times 1}{\{x^i(t)\}} = \underset{9{,}000 \times 6}{\left[\{\phi^i\}_j\right]} \underset{6 \times 1}{\{q(t)\}_j} \qquad (13.52)$$

13.4 Forced Vibrations of MDOF

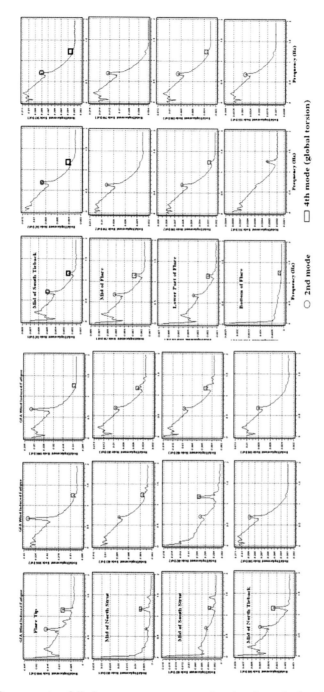

Fig. 13.7 Power spectra of displacement responses on various locations of a flare boom, which is subjected to a wind load with a reference mean wind speed of 24 m/s (courtesy of Aker Solutions)

220   13 Vibration of Multi-Degrees-of-Freedom Systems

Rather than solving 9,000 simultaneous coupled equations, only six uncoupled equations are needed to be solved for $q^1(t)$, $q^2(t) \ldots q^6(t)$.

However, the modal superposition technique is generally limited to the smooth solutions of a linear structure/system.

13.5 Forced Vibrations of MDOF: Direct Time Integration Method

13.5.1 Introduction to the Method

Exact/analytical solutions to equations of motions are usually not possible if the excitations vary arbitrarily with time or if the system is nonlinear [125]. The solutions can be obtained by using a numerical step-by-step procedure for the integration of the equations of motions, which is called the direct integration method.

The major difference between the modal superposition method and the direct integration method is that in the former, a transformation is always performed prior to the numerical integration.

Since the solutions are time histories, it is obvious that the excitations varying with time must be defined at every time step, i.e., the excitations must be a deterministic function.

Let's first set up the governing equations of motions for a system at two adjacent time instants t_i and t_{i+1} so that $\Delta t_i = t_{i+1} - t_i$:

$$m\ddot{x}(t_i) + c\dot{x}(t_i) + kx(t_i) = F(t_i) \tag{13.53}$$

$$m\ddot{x}(t_{i+1}) + c\dot{x}(t_{i+1}) + kx(t_{i+1}) = F(t_{i+1}) \tag{13.54}$$

Generally, three types of time-stepping procedures are available for obtaining the responses: interpolation (normally linearly) of the excitation input between two adjacent time instants t_i and t_{i+1}; finite difference expression of acceleration and velocity; and variation of accelerations.

The first type of method is only applicable for linear systems. The second and third type of methods are especially suited for solving equations of motions for nonlinear systems. The most commonly used methods are the central difference method, the Houbolt method, the Wilson-θ method, and the Newmark method [173]. Among them, the Newmark method is perhaps the most popular tool in many practical dynamic analyses for solving second-order differential equations with MDOF, mainly due to its accuracy. It is a type of the finite difference method and is a "single time step" method. Using this method, the motions at each

13.5 Forced Vibrations of MDOF 221

degree-of-freedom are updated step by step through iterative calculations, which enable the calculation of the responses in the time domain.

The Newmark time-stepping scheme is based on the following assumptions:

$$x_{t+\Delta t} = x_t + \Delta t \, \dot{x}_t + \frac{\Delta t^2}{2}(1 - 2\beta) \, \ddot{x}_t + \Delta t^2 \beta \, \ddot{x}_{t+\Delta t} \qquad (13.55)$$

$$\dot{x}_{t+\Delta t} = \dot{x}_t + \Delta t \,(1 - \delta)\ddot{x}_t + \Delta t \delta \ddot{x}_{t+\Delta t} \qquad (13.56)$$

where Δt is the time increment and β and δ are the coefficients that define the variation of acceleration over a time step, and are related to the integration accuracy and stability. The coefficient β denotes the variation in the acceleration during the time-incremental step from t to $t + \Delta t$. Different values of β indicate different schemes of interpolation of the acceleration over each time-step. For example, $\beta = 0$ indicates a scheme equivalent to the central difference method, and $\beta = 1/6$ together with $\delta = 1/2$ corresponds to the linear acceleration method; the latter can also be obtained if $\theta = 1$ in the Wilson-θ method (see Ref. [173]). In addition, the Newmark method is unconditionally stable, i.e., the time-step Δt can be chosen without requirements, since there is a guarantee of solution stability, provided that $\beta \geq 0.25(0.50 + \delta)^2$ and $\delta \geq 0.50$.

For a linear system, at each time step, the solutions can be obtained without iteration. However, for a nonlinear system, as will be presented in a Sect. 15, a decent iteration scheme has to be used.

Note that the number of operations in the direct integration method is directly proportional to the number of time steps in the analysis. Hence, the direct integration scheme is effective when the duration of an event to be analyzed is short (i.e., for a few time steps).

The direct integration may be performed by either explicit (conditionally stable, Sect. 13.5.2) or implicit (unconditionally stable, Sect. 13.5.3) schemes [173].

Example: A vehicle dynamics code based on the Newmark scheme for direct time integration.

In order to calculate a car's vibrations and tire reaction forces when the tire is subjected to base excitations, a code for calculating the vehicle dynamic responses subjected to the base excitations has been developed. In the code, the car is modeled with 27 degrees-of-freedom as shown in Fig. 13.8 [177]. It comprises the stiffness and damping between vehicle tire and deck (1–12) in longitudinal (x), transverse (y) and vertical (z) directions; stiffness and damping between vehicle body and tire (15, 18, 21, 24) in the vertical direction; stiffness between vehicle body and tire (13, 14, 16, 17, 19, 20, 22, 23,) in the longitudinal and transverse directions. The degrees-of-freedoms 25 and 26 represent the coupling of front and rear suspension, respectively. The deformation of these two stiffness components is proportional to the difference of deformation between 15 and 18 for the front and

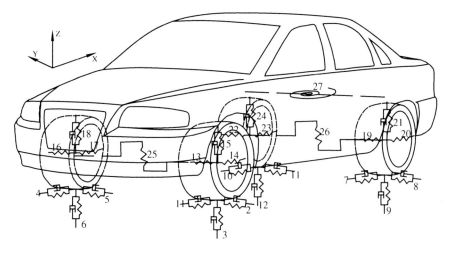

Fig. 13.8 Mechanical modeling of a full car with 27 degrees-of-freedoms

21 and 24 for the rear. Position 27 is an elastic component to take care of an eventual coupling in the horizontal plane. Its deformation is proportional to the angle between the rear axis (rear wheel centers) and the normal to the symmetry line of the body. The equations of vehicle motions were derived by using the Lagrangian principle for non-conservative systems from Eq. (2.6) (Sect. 2.3):

$$\frac{d}{dt}\left(\frac{\partial T}{\partial \dot{q}_j}\right) - \frac{\partial T}{\partial q_j} + \frac{\partial V}{\partial q_j} + \frac{\partial D}{\partial \dot{q}_j} = Q_j, j = 1, 2, \ldots, n$$

The equation above can also be written as the equilibrium of forces:

$$F^{kinetic}\left(\ddot{q}, \dot{q}, q, t\right) + F^{internal}\left(\ddot{q}, \dot{q}, q, t\right) = F^{external}\left(\ddot{q}, \dot{q}, q, t\right)$$

where $F^{kinetic}$, $F^{internal}$, $F^{external}$ are the vector-functions of generalized kinetic, internal and external forces, respectively.

For solving the equation above, the Newmark scheme is used to obtain the initial value, and the Newton–Raphson iteration method (Sect. 15.5.2) is used for reaching an equilibrium state (minimize the residual of the equation) for every time step as well as for solving the static problem (such as the displacement under a static load) of the vehicle system.

First the ship motions in all degrees-of-freedoms (Fig. 13.9) need to be obtained from either measurement or seakeeping calculations. The motions are then transferred to the position of the deck that is in contact with the car

13.5 Forced Vibrations of MDOF

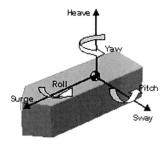

Fig. 13.9 A ship's fixed coordinate system (X: surge, Y: sway, Z: heave)

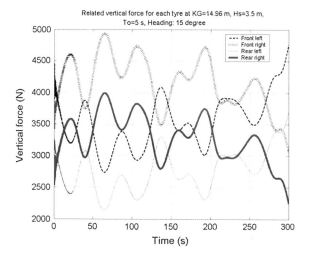

Fig. 13.10 An example of the vertical reaction force time series between the tire and the deck in the time domain at a position of X = 100 m, Y = 10 m and Z = 15 m from the center of ship motions, with the wave coming from the stern (back) and the ship speed being 20 knots (10.3 m/s)

tires and input into the model. The vehicle's dynamic responses can then be calculated. By comparing the simulation results with the real measurement of reaction forces for a real car on board a ship, it is concluded that the model can predict the motions of the vehicles with sufficient accuracy. It is therefore suggested that the code can also predict the reactions between the tire and decks with sufficient accuracy [178].

Figure 13.10 shows the vertical tire reaction forces for a sea stern wave with a wave height of 3.5 m, a wave period of 5 s and at a ship advancing speed of 20 knots. It is shown that the values of vertical forces are quite different among different tires. For the vertical force, the front right and rear right tires vary almost in phase, while the amplitude of the former one is larger than the latter. The front left and rear left tire also vary almost in phase, but the former is larger than the latter. The sum of vertical forces for all four tires is normally in the same scale as, but not exactly equal to, the gravity load of the vehicle. This small difference is mainly due to the roll motion, but is also influenced by motions at other degrees-of-freedom [179].

13.5.2 Explicit Integration Method

For physical problems that are short duration transient events, and which are often in combinations of nonlinearities such as large deformation or contact, the most popular explicit method is the central difference method, in which the acceleration and velocity can be evaluated at time t with displacements at time $t - \Delta t$, with Taylor series expansion:

$$\ddot{x}_t = \frac{1}{\Delta t^2}(x_{t-\Delta t} - 2x_t + x_{t+\Delta t}) \tag{13.57}$$

$$\dot{x}_t = \frac{1}{2\Delta t}(-x_{t-\Delta t} + x_{t+\Delta t}) \tag{13.58}$$

where Δt is the time step.

It is noted that both the expressions above have the same order of errors Δt^2. By combining the two equations above, the $x_{t-\Delta t}$ can be eliminated:

$$x_{t+\Delta t} = x_t + \Delta t \dot{x}_t + \frac{1}{2}(\Delta t)^2 \ddot{x}_t \tag{13.59}$$

By substituting the relations for $\ddot{x}_t$ and $\dot{x}_t$ in Eqs. (13.57) and (13.58) into the equations of motions, one can solve for $x_{t+\Delta t}$:

$$\left(\frac{m}{\Delta t^2} + \frac{c}{2\Delta t}\right)x_{t+\Delta t} = F_t - \left(k - \frac{2m}{\Delta t^2}\right)x_t - \left(\frac{m}{\Delta t^2} + \frac{c}{2\Delta t}\right)x_{t-\Delta t} \tag{13.60}$$

It should be emphasized that an explicit integration method such as the central difference method requires a time step Δt smaller than the critical time step $\Delta t_{critical}$. Under such conditions the integration is said to be conditionally stable. If a time step is larger than $\Delta t_{critical}$, erroneous unbounded time-history responses will occur. In a finite element analysis, the critical time step can be determined by the fact that a wave is not allowed to pass through two nodes in the same time increment:

$$\Delta t_{critical} < \frac{T_n}{\pi} = \frac{L}{C} \tag{13.61}$$

where T_n is the smallest period of the finite element assembly; L is the element length; C is the wave propagation speed, for isotropic elastic material; and $C = \sqrt{\frac{\rho}{E}}$ (square root of ratio between material density and Young's modulus).

13.5 Forced Vibrations of MDOF

Courant number is usually adopted to measure the relative size of time step when using an explicit integration method:

$$C_n = \frac{\Delta t}{\Delta t_{critical}}$$

(13.62)

Theoretically, the time step should be as small as possible to achieve a high numerical accuracy. However, when the explicit integration method is used in a finite element analysis, regardless of how much smaller than the critical one the time step is, the analysis only results in an accurate calculation of highest frequency responses that the finite element mesh size can handle, which is not of any engineering interest. Therefore, the time step close to the critical one is usually satisfactory to guarantee numerical accuracy. For a typical dynamic analysis, a maximum Courant number is recommended to be 0.95–0.98 [223].

Equation (13.61) indicates that the critical time step for an FE modeling with the same material is determined by the size of the smallest element in the model. In practice, this requires that the size among each element be as equal as possible. In many cases, this leads to an increase of time step. For structures with large deformations, some elements may deform to a shorter length, this should also be taken into account by assigning a relatively large element length for areas that can deform to a shorter length.

In addition, the critical time step can also be increased by increasing the density of the modeled material, which is often referred to as "mass scaling." Apparently, the drawback of the "mass scaling" is that the dynamic inertia forces are overestimated. Therefore, the "mass scaling" can only be applied to a small number of elements that are not critical with respect to strength and stability, this normally refers to the elements with smallest size on uncritical areas.

A step-by-step procedure for using the central difference method consists in:

(1) Form the stiffness (k), mass (m) and damping (c) term/matrix
(2) Solve the equations of motions to obtain the initial value of $x_{t=0}$
(3) Select a time step Δt ($\Delta t < \Delta t_{critical}$)
(4) Calculate $x_{t+\Delta t}$ using $x_{t+\Delta t} = x_t + \Delta t \dot{x}_t + \frac{1}{2}(\Delta t)^2 \ddot{x}_t$
(5) For each time step, repeat the solution:

$$\left(\frac{m}{\Delta t^2} + \frac{c}{2\Delta t}\right)x_{t+\Delta t} = F_t - \left(k - \frac{2m}{\Delta t^2}\right)x_t - \left(\frac{m}{\Delta t^2} + \frac{c}{2\Delta t}\right)x_{t-\Delta t}$$

From the procedure above, the advantage of explicit integration becomes apparent: no stiffness, mass and damping term/matrices are re-calculated in each time step, the solutions can be carried out on the element level, and relatively little high-speed storage is required. Furthermore, if the stiffness, mass and damping matrices for each element are the same, one only needs to calculate or read from back-up storage for the matrices of the first element in the series [173].

The disadvantage of the explicit integration method mainly lies in the fact that it requires a small time increment to make the solution stable. Therefore, the

Table 13.3 Numerical accuracy based on the choice of the α and β [42]

δ	β	Accuracy
$0.5 \leq \delta \leq 2\beta$		Unconditionally stable, but does not guarantee accuracy
$\delta \geq 0.5$	$\beta \geq 0.25(\alpha + 0.5)^2$	Improved accuracy with artificial damping in higher vibration modes
$\delta = 0.5$	$\beta = 0.25$	Improved accuracy, and the formulation corresponds to a trapezoidal rule with constant average acceleration
$\delta \geq 0.5$	$\beta \geq 0.25$	Improved accuracy with artificial damping in higher vibration modes. Performs better for small time step
$\delta = \frac{1}{6}$	$\beta = 0.5$	Improved accuracy, and the formulation corresponds to a linear acceleration. Performs better for small time step and tends to be unstable for large time step

method is not suitable for simulating long duration time events. In addition, the effectiveness of the method strongly depends on the use of diagonal stiffness and damping matrices (or neglects the damping for lightly damped system). In addition, explicit integration normally presents more accurate results with lumped mass modeling than that with consistent mass modeling (see Sect. 13.6). This sometimes also requires a fine mesh size in order to obtain a necessary accuracy due to the diagonal mass requirement.

There are also other types of explicit direct integration methods such as the Runge–Kutta method [174], but the central difference method is thus far the most efficient one and has been adopted by many commercial finite element analysis codes such as LS/DYNA, ABAQUS as the representative explicit integration technique.

13.5.3 Implicit Integration Method

Contrary to the explicit integration method, an implicit method is suited for dynamic analysis with relatively long time durations with regard to both loading and responses, regardless of whether nonlinearities are involved or not. This fits for dynamic analysis in the civil engineering field where typical dynamic loadings due to earthquake, wave and wind are represented by an event with a duration of more than 30 s. The implicit scheme requires a factorization of effective element matrices at each time step while not requiring a maximum (critical) time step, i.e., it is unconditionally stable. However, the computation accuracy cannot be guaranteed without a careful consideration of the time step.

In the implicit integration approach, the responses at each time $t + \Delta t$ are evaluated based on the equilibrium conditions at time $t + \Delta t$. The commonly used implicit integration methods are based on the following assumption:

$$x_{t+\Delta t} = x_t + \Delta t \, \dot{x}_t + \frac{\Delta t^2}{2} (1 - 2\beta) \, \ddot{x}_t + \Delta t^2 \beta \, \ddot{x}_{t+\Delta t} \qquad (13.63)$$

13.5 Forced Vibrations of MDOF

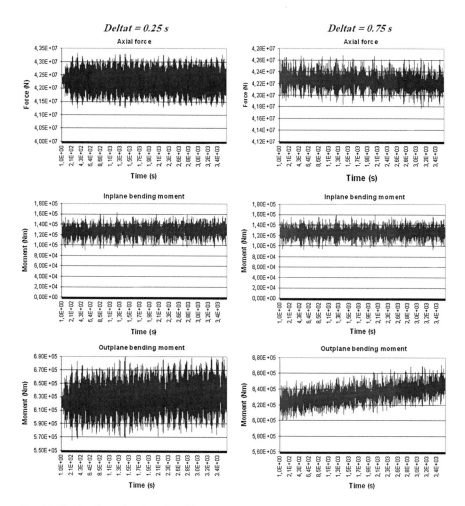

Fig. 13.11 One hour duration time history responses at a structural joint of a jacket structure subjected to wave loading, with $\Delta t = 0.25$ s (*left*) and 0.75 s (*right*), respectively

$$\dot{x}_{t+\Delta t} = \dot{x}_t + \Delta t (1 - \delta) \ddot{x}_t + \Delta t \delta \ddot{x}_{t+\Delta t} \tag{13.64}$$

It is noted that the parameters δ and β not only define the types of method, but also determine the stability and accuracy as described in Table 13.3. Typical implicit methods are the Houbolt method, the Wilson-θ method, and the Newmark method, which have been presented in Sect. 13.5.1.

Because the stiffness, mass and damping matrices have to be re-evaluated at each time step, the computational cost for implicit methods is directly related to the model size and time step.

As aforementioned, numerical stability and accuracy can be achieved if the time step length in the direct time integration algorithm is small enough to

Fig. 13.12 The Siri jack-up structure located in the North Sea (courtesy of Dong Energy Denmark)

accurately integrate the responses in the lowest period component. However, this may require a rather small time step and consequently lead to an increased computation cost. In reality, the vibrations in the lowest period (highest frequency) range may not be a necessary contributor to the dynamic responses.

Note that the implicit method is unconditionally stable, the choice of time step is only based on accuracy considerations. Compared to the explicit method, the computational effort at each time step is higher. Therefore, the implicit method is only competitive when a much larger time step than that of the explicit method is adopted. As mentioned above, while the implicit integration method is unconditionally stable, an overly large time step will result in an inaccurate calculation results. Basically, the time step should ensure an accurate representation of the excitations and forced and free vibration response components [124]. Therefore, it depends on both a structure's eigenfrequencies and loading frequencies. By presenting that structural vibration modes with frequency higher than 3 times of the highest frequency (ω_u) of interest only participate quasi-statically in the modes with frequency lower than 3 ω_u, Cook and his co-workers [233] recommended that 20 time steps per period of ($2\pi/\omega_u$) could provide sufficient accuracy, i.e., $\Delta t < (2\pi/\omega_u)/20 = 0.3/\omega_u$, unless a smaller Δt is required because of convergence difficulties caused by nonlinearities. It is noted that the similar recommendation is also given by Bathe [173]. On the other hand, too small a time step may gradually increase the responses with time due to the artificial vibration responses/noises related to higher vibration modes.

13.5 Forced Vibrations of MDOF

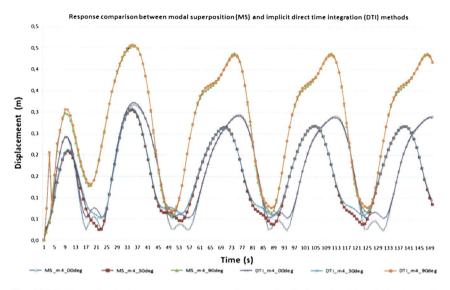

Fig. 13.13 Dynamic displacement responses at the top of a jack-up leg (the circle shown in Fig. 13.14) due to waves coming from different directions of 0, 30 and 90° (courtesy of Dong Energy Denmark)

Strictly speaking, the implicit time integration is only unconditionally stable for linear dynamic analysis. For dynamic analysis involving any types of nonlinearities, the unconditional stability cannot be guaranteed. Figure 13.11 shows the force response history at a joint of a jacket structure subject to dynamic wave loadings. It is observed that both the response amplitudes and cycles are significantly lower for $\Delta t = 0.75$ s than that for $\Delta t = 0.25$ s. Especially for the out of plane bending moments, the average value of the moments is increased with time, which shows a sorts of "unstable" phenomenon in the numerical computation. This may be caused by the increased numerical damping due to the large time step set-up [225].

13.5.4 Comparison between Modal Superposition and Direct Time Integration Method

As previously discussed, the major difference between direct time integration and modal superposition is that in the latter, a transformation is always performed prior to the numerical integration.

To calculate the dynamic response under wave load of the Siri jack-up platform shown in Fig. 13.12, both modal superposition and implicit direct time integration methods are used in the finite element analysis. Figure 13.13 shows the comparison of the displacement at the top of a jack-up leg (the circle shown in Fig. 13.14) due to the waves coming from different directions (0, 30 and 90°). The time step for both methods are set as 0.1 s, which gives more than 60 calculations for each

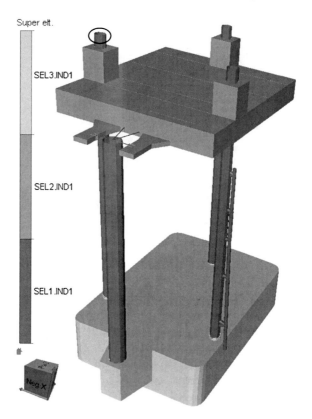

Fig. 13.14 The geometry model of Siri jack-up structure for finite element analysis for which the dynamic responses are calculated (courtesy of Dong Energy Denmark)

natural period (around 6.5 s). In the calculation using modal superposition method, 10 eigenmodes ranging from 6.5 to 0.7 s are included. From the figure, it is clearly shown that the calculated responses between the two methods agree quite well.

For problems where the responses are linear and are dominated by only up to a couple of "lowest" eigenmodes, the modal superposition method is more efficient than the direct integration method. Otherwise, the direct integration method is preferred.

In practice, the dynamic responses calculated using the modal superposition method require a pre-run to obtain the eigenpairs, which is followed by the superposition of the modal response.

13.6 Lumped and Consistent Mass

We recall that the equation of motions for a system in a matrix form is:

$$[m]\{\ddot{x}(t)\} + [c]\{\dot{x}(t)\} + [k]\{x(t)\} = \{F(t)\} \tag{13.65}$$

13.6 Lumped and Consistent Mass

where $[m]$, $[c]$, and $[k]$ are the global mass, damping and stiffness matrices obtained from the assembly of the individual element matrices. $\left\{\ddot{x}(t)\right\}$, $\left\{\dot{x}(t)\right\}$, and $\{x(t)\}$ are the nodal acceleration, velocity and displacement vectors, respectively. $\{F(t)\}$ is the force vector.

The global mass and damping matrices can be formulated by using the same element shape function matrix $[N]$:

$$[m] = \int \rho^e [N]^T [N] dV \tag{13.66}$$

$$[c] = \int c^e [N]^T [N] dV \tag{13.67}$$

where ρ^e and c^e are the mass density and viscous damping coefficient for element e.

The mass represented using the same shape functions as element stiffness and damping matrix is called the consistent mass. The consistent mass matrix is symmetric and generally full at the element level but has the same sparse topology as the system stiffness matrix on the global level.

Besides consistent mass modeling, a simple and historically prior formulation is lumped mass, which is formulated by placing discretized masses m_i at the nodal point of an element, such that the summation of all nodal masses $\sum m_i$ is the total element mass. Compared to that of the consistent mass, a lumped mass is diagonal and normally has no rotary inertia unless it is arbitrarily assigned.

The lumped mass matrix for a two-node truss or beam element can be expressed as:

$$[m] = \frac{\rho^e LA}{2} \begin{bmatrix} 1 & 0 \\ 0 & 1 \end{bmatrix} \tag{13.68}$$

$$[m] = \frac{\rho^e LA}{2} \begin{bmatrix} 1 & 0 & 0 & 0 \\ 0 & 1 & 0 & 0 \\ 0 & 0 & 1 & 0 \\ 0 & 0 & 0 & 1 \end{bmatrix} \tag{13.69}$$

where L and A is the length and cross-section area of the element.

Generally, consistent mass modeling gives a higher accuracy of dynamic responses than lumped mass modeling, especially for dynamic responses that are significantly influenced by the higher order of eigenfrequencies. This applies to structures possessing high redundancy and with loading in frequency range corresponding to higher-order vibration modes. However, for dynamic responses dominated by the first few eigenmodes, a lumped mass modeling normally gives sufficient calculation accuracy and requires less computational efforts and storage space than that of the consistent mass. For wave propagation problems, the lumped mass modeling may even be more accurate because there are fewer spurious oscillations.

Chapter 14
Damping

14.1 Types of Damping and its Effects

A structure subjected to oscillatory deformation contains a combination of kinetic and potential energy. During this process, energy will always be dissipated, leading to a response decrease in case of free oscillations. This dissipation of energy is caused by damping, which is essentially the conversion of the mechanical energy of a structure into thermal energy. Therefore, forces due to damping are non-conservative. For dynamic sensitive structures, the structural dynamic performance may highly depend on the initial assumption of damping, which is essential to decrease the dynamic responses through energy absorption and dissipation.

Damping that occurs on a structure can be categorized as either inherent damping, which occurs naturally within the structure or its environment, or external damping through installed apparatus such as a damper.

The inherent damping can be generated by various mechanisms such as viscosity, hysteresis, yielding and friction, and also externally by the actions at supports, radiation of energy (radiation damping) into the ground and fluid damping (aerodynamic and hydrodynamic damping). It is influenced by many factors such as the surface finish, lubrication, area of contact, normal load, damage or wear, temperature, humidity etc. [198–200].

Note that the amount of inherent damping cannot be estimated with certainty. A known level of damping may be introduced through the energy dissipation of devices added to structural systems, known as supplemental damping or artificial dampers [23]. For example, damping can be introduced by installing external dampers, such as base isolation bearings made of rubber, plastic or sliding material, or mechanical dampers such as dynamic absorber [18] and viscous dampers etc. Figure 14.1 shows hydraulic shock absorbers (a type of viscous dampers) that can decelerate the motions in a controlled manner.

Generally, for a lightly damped structure, the contribution from inherent damping is much less than that of inertia and stiffness. Frequency-dependent damping is normally only effective at or close to the lower order of eigenfrequencies, and normally most effective at or close to the natural frequency, i.e., when a structure

---

J. Jia, *Essentials of Applied Dynamic Analysis*, Risk Engineering, 233
DOI: 10.1007/978-3-642-37003-8_14, © Springer-Verlag Berlin Heidelberg 2014

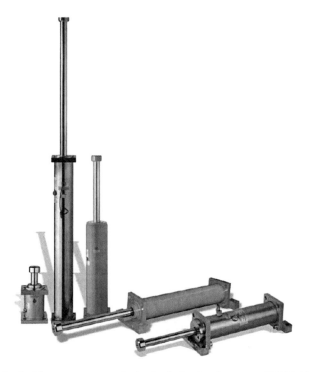

Fig. 14.1 Mechanical dampers to protect shock–shock absorber (courtesy of ITT Enidine Inc, USA)

is vibrating in resonance or is impulse-excited and the resulting vibrations are allowed to die out exponentially (in case of viscous dampers) at its natural frequency. In the latter case, damping is most effective when the time interval of impulsive forces allows the vibrations to decay down to the background level [203].

Despite the importance of damping on the dynamic response calculation, it is also noted that both the estimation (Sect. 14.3) and modeling (Sect. 14.2) of the damping are rather difficult tasks. The resulting uncertainty poses a great challenge for evaluating the calculated responses of dynamic sensitive structures.

14.2 Damping Modeling

Modeling of damping may take the form of viscous (proportional to velocity), dry friction/Coulomb (constant), hysteresis, yielding or fluid (proportional to the square of the velocity). The approaches of modeling damping are to some extent empirical, and aim to capture the essential energy dissipative effects relative to the dynamic behavior of interests [204]. Today, there is no unified damping model that can perfectly represent various types of energy dissipation. Figure 14.2 illustrates major types of damping modelings that will be discussed in the subsequent sections.

14.2 Damping Modeling

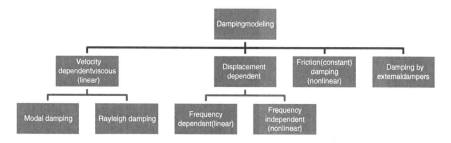

Fig. 14.2 Categories of damping modeling

14.2.1 Pure Viscous Damping

Viscous damping model (Figs. 14.3a and 14.4) is widely used for structures within the elastic deformation limits. When structures deform further to the inelastic range, the viscous damping coefficient varies depending on the deformation amplitude. However, in a dynamic analysis, this variation of damping coefficient is not usually explicitly accounted for. This situation can be handled by approximating a damping value corresponding to the expected deformation amplitude level at the deformation close to the linear elastic limit [125].

For viscous damping, under free vibration decay, the damping force is proportional to the velocity. It can be written as:

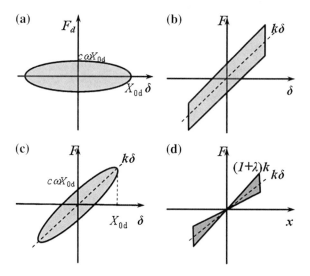

Fig. 14.3 Modeling of different types of damping (F_d: force due to damping only, F: force due to stiffness and damping, δ: extension), the shaded area indicates the energy dissipated per cycle

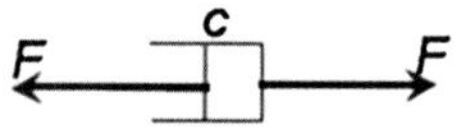

Fig. 14.4 Physical representation of viscous damping modeling force

$$
\begin{aligned}
F_{\mathrm{d}} = c\,\dot{x}(t) &= c\omega X_{0\mathrm{d}}\cos(\omega t - \phi) \\
&= c\omega\sqrt{X_{0\mathrm{d}}^2 - X_{0\mathrm{d}}^2\sin^2(\omega t - \phi)} \\
&= c\omega\sqrt{X_{0\mathrm{d}}^2 - x^2(t)}
\end{aligned}
\tag{14.1}
$$

where $X_{0\mathrm{d}} = \dfrac{F_0/k}{\frac{c\omega}{\omega_n}}$ is the steady-state displacement due to the harmonic force excitation with the maximum amplitude of F_0.

Rearranging the equation above, one obtains:

$$
\left(\frac{\delta}{X_{0\mathrm{d}}}\right)^2 + \left(\frac{F_{\mathrm{d}}}{c\omega X_{0\mathrm{d}}}\right)^2 = 1
\tag{14.2}
$$

where δ is the extension of the mass.

The equation above shows an ellipse shape in the force-extension diagram as shown in Fig. 14.3a. The area enveloped by the ellipse is the dissipated energy per cycle $\pi X_{0\mathrm{d}}c\omega X_{0\mathrm{d}}$.

14.2.2 Friction/Coulomb Damping

Due to its mechanical simplicity and convenience, friction/Coulomb damping exists in many mechanical and structural systems, such as the damping generated by the friction of structural joints, between members and connections, or between structural and non-structural components etc. It is typically constant within the static (before a system begins to move) or kinetic (after the system begins to move) regime. But at the boundary between these two regimes, the friction force dramatically changes. An example to illustrate a typical friction phenomenon is the friction between car tire and ground. Figure 14.5 shows a test set-up of a car tire under a horizontal displacement (transversely) with a speed of 1 mm/s. The friction coefficient is defined as:

$$
\mu = T/N
\tag{14.3}
$$

where T and N are contact forces tangent to the deck surface (horizontal force) and normal to the ground surface (normal force), respectively.

The friction coefficients during this moving are documented as shown in Fig. 14.6. It can be seen that immediately after the tire begins to slide at point C,

14.2 Damping Modeling

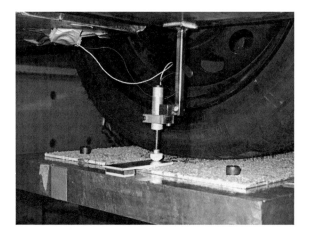

Fig. 14.5 Friction testing under the horizontal moving loads T and normal load N

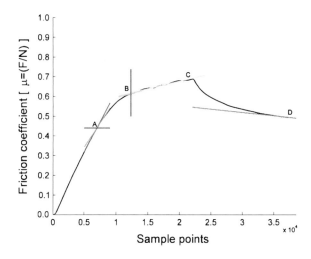

Fig. 14.6 Friction coefficients varied with displacement (proportional to sample points) from the original location. The displacement is applied at the center of the car tire

the friction coefficient significantly decreases, which illustrates a shift from a static to kinetic friction regime.

However, for simplicity, the friction is often modeled as inherently linear by assuming that the coefficient of kinetic friction is equivalent to its static counterpart.

For a realistic structure, the friction damping force (F_d) is always accompanied by the elastic force (F_e) due to stiffness, making the total resisting force (Fig. 14.3b) the sum of the two:

$$F = F_d + F_e = \mu N + k\delta \tag{14.4}$$

where μ is the coefficient of sliding or kinetic friction, and N is the normal force. The equation of motions due to Coulomb damping can be written as:

$$m\ddot{x}(t) + \mu N \text{sgn}(\dot{x}(t)) + kx(t) = 0 \tag{14.5}$$

where $\text{sgn}(\dot{x}(t))$ is a sign function depending on the instant velocity.

The equation above is nonlinear, and can be solved either numerically or analytically by breaking the time axis into segments with an altered sign of velocity. Interested readers may read references by Chopra [124] and Rao [205] for the solutions of the equation.

There are several characteristics that distinguish Coulomb damping from viscous damping:

- The equation of motions is nonlinear with Coulomb damping and linear with viscous damping.
- Coulomb damping is independent of velocity and displacement.
- The magnitude of Coulomb damping does not affect the frequency of motion, i.e., the eigenfrequency does not change with the addition of Coulomb damping.
- The transient responses for free vibrations decay linearly by a constant amount $\left(\frac{4\mu N}{k}\right)$ per cycle (i.e., in time $\frac{2\pi}{\omega_n}$) due to Coulomb damping and decay exponentially for viscous damping.
- Friction damping force can overcome restoring elastic force when the extension is small enough. This indicates that it can stop the motions typically with a permanent displacement from the neutral position. However, viscous damping can theoretically never stop motions.
- The responses are periodical with the addition of Coulomb damping, but it can be non-periodical for a viscously overdamped system.

Friction type of damping is widely used in engineering mechanism or structures. For example, to resist earthquake loading, friction dampers were installed in almost 100 land-based structures worldwide, for both concrete and steel buildings, elevated water towers, and for new construction and retrofit of existing structures. Examples are T Boeing Commercial Airplane Factory in Everett, WA, Sonic City Office Tower in Ohmiya, Asahi Beer Tower in Tokyo, and Moscone West Convention Center in San Francisco etc. [23].

14.2.3 Frequency-Dependent Hysteretic Damping

Pure viscous damping is mathematically convenient due to its linearity of the resulting equation of motions, and it can capture the essential damping behavior of the observed behavior in many practical circumstances. However, it is not necessarily the best damping model to represent an actual structure, in which the

14.2 Damping Modeling

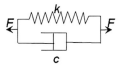

Fig. 14.7 Physical representation of frequency dependent hysteretic damping mechanism

viscous damping force (F_d) is always accompanied by the force (F_e) due to stiffness, making the total resisting force the sum of the two (Figs. 14.3c and 14.7):

$$F = F_d + F_e = c\omega\sqrt{X_{0d}^2 - \delta^2} + k\delta \tag{14.6}$$

This is called frequency-dependent hysteretic damping force. The word "frequency dependent" means that the damping is dependent on displacement amplitude instead of instantaneous velocity. The equation above represents the behavior for typical yielding materials. It essentially implies a rotation (Fig. 14.3c) of viscous damping force diagram (Fig. 14.3a) due to the presence of stiffness term ($k\delta$). However, since the stiffness term is elastic, the dissipated energy enveloped by the ellipse is still the same as that of the viscous damping model.

Readers need to understand the common and distinguishing features between the dynamic and static hysteresis loop. Both dissipate energy through deformation, but whereas the formal one is frequency dependent, the dissipated energy is proportional to the excitation frequency. For the latter, the excitations are statically applied, i.e., the excitation frequency is zero, the dissipated energy (area enveloped by the closed ellipse shape) is zero because the force-extension diagram becomes a single valued curve within an elastic limit; when a structure exhibits yielding and subsequent plastic deformation, the force-extension diagram becomes hysteresis loops again purely due to the plasticity and it is not related to the extension rate (excitation frequency is zero). This is known as static hysteresis.

The lightly damped system comprising frequency-dependent hysteretic damping can be measured from either a free decay (Sect. 14.3.1) or by reading the sharpness of frequency response curve (half power method, see Sect. 14.3.5).

In real structures, except for the forms of damping described above, many other mechanisms may also contribute to energy dissipation. Therefore, in a dynamic analysis, the damping is normally modeled in a highly idealized manner, and in most cases, by the modeling of viscous damping. The damping value is selected so that the dissipated energy in the modeled damping mechanism is identical or close to that of the realistic situation. This is also called equivalent viscous damping, as will be discussed in Sect. 14.2.6.

Under extreme dynamic loading such as seismic excitations, especially in dynamic time history analysis, the dominant damping contribution is likely to be due to the plastic deformation of a structure, i.e., the energy dissipates through the inelastic force and deformation loop. Therefore, a more accurate and direct modeling of damping is to define the plastic material properties of a target structure. There is then usually no need to use viscous damping modeling.

Note that with the modeling of the viscous or frequency dependent hysteretic damping, linear equation of motions can be formulated. However, frequency-independent hysteretic damping (Sect. 14.2.4), friction/Coulomb damping (Sect. 14.2.2), and hydrodynamic damping (Sect. 14.2.5) will result in nonlinear equations of motions.

14.2.4 Frequency-Independent Hysteretic Damping

For certain materials (such as sand), with the energy losses due to internal friction the damping force increases when the mass moves away from and decreases when it moves toward the origin, the resisting force can then be expressed as the sum of a force due to stiffness $k\delta$ and a force proportional to the instantaneous displacement but in the opposite direction of the instantaneous velocity, which is formulated as:

$$F = k\delta + \lambda k \frac{\dot{\delta}(t)}{\left|\dot{\delta}(t)\right|}\delta \tag{14.7}$$

Unlike the viscous damping or frequency dependent hysteretic damping, the equation of motions for frequency-independent hysteretic damping is nonlinear.

14.2.5 Fluid (Hydrodynamic or Aerodynamic) Damping

Fluid damping, also called drag damping, arises due to the relative motions between the material/member and the fluid (water or air) surrounding the material/member. The damping or drag force per unit length along a member is proportional to the square of the relative velocity normal to the member axis, which is expressed by the Morison equation [201]:

$$F = \frac{1}{2}\rho \cdot C_D \cdot v_r |v_r| \cdot d \tag{14.8}$$

where ρ is the density of the fluid; C_D is the drag coefficient; v_r is the relative fluid particle velocity relative to the member and normal to the member axis; and d is the cross-section dimensions (for tubular members, it is the diameter of the member exposed to the fluid).

The equation above also indicates a nonlinear relationship between the relative velocity and the damping force, which leads to mathematical difficulties for solving the nonlinear equation of motions. In many cases, it is solved by a linearization with respect to the maximum fluid velocity or an equivalent energy of the fluid particle kinematics.

14.2 Damping Modeling

14.2.6 Equivalent Viscous Damping

Even though viscous damping modeling has obvious advantages, the energy dissipation for an actual structure is more prone to be displacement proportional rather than velocity proportional, and is sometimes a combination of the two. This leads to the concept of equivalent viscous damping [202], which is to define the damping of a system using viscous damping based on the equivalent energy dissipation between the viscous damping and that of the actual system (elaborated in Sect. 14.2.6.2). In case of relatively low damping (less than 15 %), viscous, friction and hysteretic damping can be conveniently expressed by equivalent viscous damping, which will be presented in Sects. 14.2.6.2, 14.2.7 and 14.2.8.

14.2.6.1 Specific Damping Factor and Loss Factor

Consider an SDOF system with viscous or hysteretic damper subjected to harmonic loading $F(t) = F_0 \sin(\Omega t)$. It is noted that the work done by conservative forces such as elastic, inertia and gravitational forces in a complete loading cycle will be zero. Therefore, the net work will be dissipated by damping only. When the motions reach steady-state, the energy dissipation (E_d) during a complete cycle by viscous damping is illustrated in the left figure of Fig. 14.8, which can be expressed as:

$$E_d = \int F_d d\delta = \int_0^{2\pi/\omega} \left(c \dot{\delta} \right) \dot{\delta} \, dt = c \int_0^{2\pi/\omega} [\Omega X_{0d} \cos(\Omega t - \phi)]^2 dt = \pi \Omega c X_{0d}^2 \tag{14.9}$$

From the equation above, it is found that, rather than being a constant value, the energy dissipation is proportional to the excitation frequency Ω or the square of the motion amplitude X_{0d}.

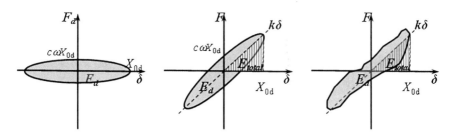

Fig. 14.8 Energy dissipation and strain energy by a viscous damper (*left*, strain energy is zero), hysteretic damper (*middle*) and measurement from real structures (*right*)

The equation above is only valid with the presence of spring stiffness k, as shown in the middle figure of Fig. 14.8, which gives:

$$E_d = \pi \Omega c X_{0d}^2 = 2\pi \zeta \frac{\Omega}{\omega_n} k X_{0d}^2 \qquad (14.10)$$

With total energy expressed as either the maximum potential/strain energy $\left(\frac{1}{2} k X_{0d}^2\right)$ or the maximum kinetic energy $\left(\frac{1}{2} m \Omega^2 X_{0d}^2\right)$, one can measure the dissipation as the fraction of the total energy, which is called specific damping capacity:

$$\frac{E_d}{E_{total}} = \frac{2\pi \zeta \frac{\Omega}{\omega_n} k X_{0d}^2}{\frac{1}{2} k X_{0d}^2} = 4\pi \zeta \frac{\Omega}{\omega_n} \qquad (14.11)$$

If the loss of energy due to damping is only supplied by the excitations, the steady-state responses can only be reached if the excitation frequency Ω is equal to the system's nature frequency ω_n. Therefore, the specific damping expressed by the equation above can be rewritten as:

$$\frac{E_d}{E_{total}} = 4\pi \zeta \qquad (14.12)$$

In a more common way, the energy dissipation can be investigated by a loss factor defined as:

$$\eta = \frac{E_d/2\pi}{E_{total}} = 2\zeta \qquad (14.13)$$

Realistic measurement of force-response diagram (right figure of Fig. 14.8) does not show a perfectly ellipse shape. However, damping level can be conveniently calculated by measuring the total energy (E_{total}) and energy dissipation (E_d) as shown in the right figure of Fig. 14.8.

14.2.6.2 Equivalent Damping Measured by the Hysteretic Loop

The most convenient determination of equivalent damping ζ_{eq} is by measuring the harmonic force and harmonic responses at $\Omega = \omega_n$:

$$E_d = E_{total} 4\pi \zeta_{eq} \frac{\Omega}{\omega_n} = E_{total} 4\pi \zeta_{eq} \qquad (14.14)$$

This gives:

$$\zeta_{eq} = \frac{E_d}{4\pi E_{total}} \qquad (14.15)$$

14.2 Damping Modeling

The concept and calculation procedure of equivalent viscous damping can be extended to MDOF systems, in which each eigenmode has an individual equivalent viscous damping so that the dissipated energy in viscous damping matches the energy loss when the system vibrates at that eigenfrequency and corresponding mode shape. This is assumed by dynamic analysis using the modal superposition method (Sect. 13.4). In many engineering structures, the damping level increases with the increase of eigenmode orders.

It should also be noted that, strictly speaking, equivalent viscous damping modeling is only valid for frequency domain analysis where the excitation is harmonic. However, as will be shown in Sect. 14.2.9, practically, the use of this concept has been extended to model the damping in the form of modal damping, Rayleigh damping, and even non-proportional damping, for both time and frequency domain analysis.

14.2.7 Equivalent Viscous Damping with Coulomb Damping

For an SDOF system under the harmonic force excitations $F(t) = F_0 \sin(\Omega t)$, the equivalent viscous damping with Coulomb damping is:

$$\zeta_{eq} = \frac{2F_c}{\pi \chi F_0} \frac{1}{\Omega/\omega_n} \tag{14.16}$$

where F_c is the constant friction/Coulomb damping force.

The magnification factor χ, which is a ratio between the displacement amplitude and the maximum quasi-static deflection $X_0 = F_0/k$ under the force F_0, is calculated as:

$$\chi = \frac{\sqrt{1 - \left(\frac{4F_c}{\pi F_0}\right)^2}}{1 - (\Omega/\omega_n)^2} \quad \text{for} \quad \frac{4F_c}{\pi F_0} < 1 \tag{14.17}$$

14.2.8 Equivalent Viscous Damping with Frequency Dependent Hysteretic Damping

A hysteretic damping coefficient can be obtained from the logarithmic decrement (Δx) test (Sect. 14.3.1):

$$h = \frac{\Delta x}{\pi} \tag{14.18}$$

For an SDOF system under harmonic force excitations at the frequency of Ω, the equivalent viscous damping is:

$$\zeta_{eq} = \frac{h}{2\Omega/\omega_n} \tag{14.19}$$

The magnification factor can then be calculated as:

$$\chi = \frac{1}{\sqrt{\left[1 - (\Omega/\omega_n)^2\right]^2 + h^2}} \tag{14.20}$$

14.2.9 Practical Damping Modeling for Dynamic Analysis

It is rather difficult to explicitly implement and formulate detailed damping forces for an entire structure. Therefore, for the sake of both mathematical convenience and the representativeness of energy dissipation, practical damping models have to be adopted for a dynamic analysis. This mainly includes proportional (modal damping and Rayleigh damping) and nonproportional damping.

14.2.9.1 Modal Damping

Physically, an actual structure comprises abundant damping mechanisms all along it. However, even with the most convenient viscous damping modeling, it is still impossible to model all those mechanisms by one by one individual dampers. Therefore, in most cases, modal damping (Sect. 13.4) is widely adopted in computer modeling to approximate the energy dissipation within the structure. In order for the modal equations to be uncoupled, the damping must fulfill the condition:

$$2\omega_i\zeta_i = \phi_i^T c_i \phi_i \tag{14.21}$$

where ζ_i is the viscous modal damping at ith eigenmode.

14.2.9.2 Rayleigh Damping

Damping effects can be conveniently accounted for in modal superposition analysis with the fulfillment of the equation above. However, in nonlinear dynamic analysis for which the mode shapes are changing with stiffness changes (especially for dynamic analysis with direct time integration methods), and with more realistic

14.2 Damping Modeling

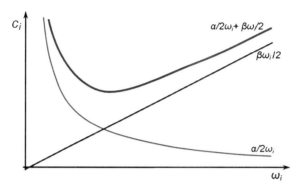

Fig. 14.9 Rayleigh damping as a function of frequency

damping that is varied with frequency, Rayleigh damping [173] is commonly used. It is a linear combination of the system's mass and stiffness as shown in Fig. 14.9 and the damping at frequency ω_i is:

$$c_i = \alpha m_i + \beta k_i \tag{14.22}$$

where α with the unit of s^{-1} and β with the unit of s are two coefficients to be determined from two given damping ratios at two specific frequencies of vibrations.

αm_i and βk_i, namely mass proportional and stiffness proportional damping, respectively, are the simplest way to formulate a proportional damping matrix because the undamped mode shapes are orthogonal with respect to each of these [161].

α and β can be evaluated by the solution of a pair of simultaneous equations at two separate frequencies as follows:

With the orthogonality properties of mass and stiffness matrix, the equation above can be rewritten by inserting it into Eq. (14.21) as:

$$2\omega_i \zeta_i = \alpha + \beta \omega_i^2 \tag{14.23}$$

Rearranging the equation above, the relationship between modal damping (ζ_i) and Rayleigh damping is finally expressed as (Fig. 14.9):

$$\zeta_i = \frac{\alpha}{2\omega_i} + \frac{\beta \omega_i}{2} \tag{14.24}$$

It is normally recommended that the two specific frequencies for determining Rayleigh damping should ensure reasonable damping values in all the modes significantly contributing to the vibrations. At the frequency outside the range of these two frequencies, the damping will dramatically increase and the modal responses at the corresponding frequency range will almost be eliminated. Practically, this can be used to damp out the high and low frequency vibrations/noises that are outside the frequency range of interests. In many cases, the variation of damping ratio with frequency is not available, and one can then assume the damping at the two specific frequencies to be identical.

Even though Rayleigh damping is very convenient for modeling, it cannot be physically justified: the mass proportional damping introduces externally supported dampers, which do not exist for a fixed structure. As illustrated in Fig. 14.9, the stiffness proportional damping increases the damping dramatically at a higher order of eigenmodes, which is not physically true either, even if it is numerically efficient.

Example: In a modal testing for an offshore structure, the two important eigenfrequencies are 0.12 and 0.23 Hz with the corresponding modal damping of 3 and 5 %. Establish the Rayleigh damping that will be used in a dynamic analysis using direct time integration methods.

Solution: Using Eq. (14.24): $\zeta_i = \frac{\alpha}{2\omega_i} + \frac{\beta\omega_i}{2}$

$$3\% = \frac{\alpha}{2(2\pi \times 0.12\text{Hz})} + \frac{\beta(2\pi \times 0.12\text{Hz})}{2} = 0.66\alpha + 0.38\beta$$

$$5\% = \frac{\alpha}{2(2\pi \times 0.23\text{Hz})} + \frac{\beta(2\pi \times 0.23\text{Hz})}{2} = 0.35\alpha + 0.72\beta$$

Therefore, we obtain $\alpha = 0.0076$ and $\beta = 0.0658$.

14.2.9.3 Caughey Damping

If one needs to specify damping ratios at more than two eigenmodes, instead of Rayleigh damping, an extended or more generalized form of Rayleigh damping called Caughey damping can be used:

$$c = m_i \sum_{i=0}^{\eta-1} \gamma_i \left(m_i^{-1} k_i\right)^i \tag{14.25}$$

where γ_i is a constant, and η is the number of modes one wants to specify damping.

The modal damping ratio ζ_r at modes r higher than η can then be expressed as:

$$\zeta_r = \frac{1}{2} \sum_{i=0}^{\eta-1} \gamma_i \omega_r^{2i-1} \tag{14.26}$$

Similar to modal damping, the mass and stiffness matrices adopted in formulating Caughey damping also satisfy the mode shape orthogonality condition. However, Caughey damping normally results in a full matrix, which is computationally demanding for solving the equations of motions. Therefore, it is generally not practical to use this type of damping.

14.2.9.4 Nonproportional Damping

Both modal and Rayleigh damping are proportional damping, the corresponding damping matrices can be diagonalized in the modal matrix of the undamped system [206]. This provides computational convenience, but only applies for lightly damped structures with uniformly distributed damping mechanism, where off-diagonal terms in the damping matrix can be neglected. However, for many types of structures such as an offshore jacket structure shown in Fig. 14.10, the majority of the structural damping (Sect. 14.5.2) is concentrated at the joints between structural members, which in many cases does not result in a proportional distribution in damping [207]. Moreover, with the advent of external artificial dampers to mitigate the dynamic structural responses, the damping force values may be in the same levels as that of stiffness or inertia forces. The locally-installed dampers will also make the distribution of damping disproportional to that of mass or stiffness, and the equations of motions will then be coupled by means of undamped mode shapes. Furthermore, if various parts of a structure are constructed with different materials, the energy dissipation mechanisms in the different parts also vary, and the

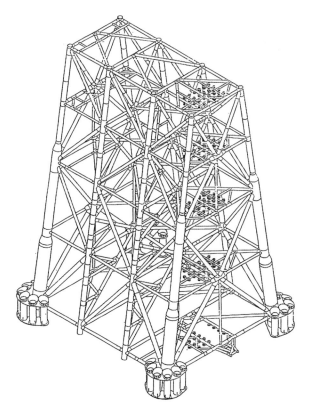

Fig. 14.10 A typical offshore jacket structure with braces and legs connected at joints (courtesy of Aker Solutions)

distribution of inertia and elastic forces will differ from one part to another. In addition, many structures have locally concentrated defects that also exhibit a non-uniformally distributed damping through the structures. All the factors above emphasize the significance of off-diagonal terms in the damping matrix and result in nonproportional damping matrices. In those conditions, both analytical and computer model must account for this effect in a more detailed and exact manner.

Similar to proportional damping matrices, nonproportional damping matrices can be obtained by a direct assembly of viscous damping matrices from different parts of a structure. This introduces a coupling between the undamped modal coordinate equations of motions. Note that the resulted modal coordinate damping matrix is a full matrix; therefore, the superposition method cannot be employed for structures/systems with nonproportional damping.

Generally, the equations of motions with nonproportional damping can be solved by using either a step-by-step integration of the geometric coordinate system or normal coordinates [208]. In addition, a so-called pseudo force iteration method, developed by Claret and Vinancio-Filho [209], can also be used to solve the equations by moving the off-diagonal coefficients to the right side (external force) of the equations of motions.

For computer modeling, the majority of commercial finite element analysis codes do not have the capability to solve the equations of motions with nonproportional damping. In order to overcome this challenge, user programd scripts have to be added to the codes. This also adds the physical requirement of data storage space for finite element based damping matrices and the combined system damping matrix.

By modeling the Tsing Ma Bridge (a 1,377 m span with a steel truss stiffening deck and two concrete towers) and the Humen Bridge (a 888 m span with a steel box deck and two concrete towers) with both nonproportional damping and proportional Rayleigh damping (2–5 %), and exciting them with three component acceleration time histories recorded from the El Centro earthquake, Qin and Lou [208] showed that the maximum relative response differences between the two types of damping modeling can be up to 36 % for bridge decks, 42 % for towers, and 16 % for main cables.

However, in many cases, the off-diagonal coupling coefficients can be neglected, yielding uncoupled equations that can finally be solved by linear methods such as modal superposition.

14.3 Measuring Damping

Nowadays, an accurate estimation of damping is still an extremely challenging task for dynamic analysis. The major challenge is caused by the difficulty of isolating various types of damping (e.g., material, structural, and hydrodynamic damping etc.) from an overall measurement. Furthermore, realistic damping can only be obtained from measurements conducted under actual operating conditions.

14.3 Measuring Damping

Table 14.1 Applicability of time and frequency methods for measuring damping

	Time methods	Frequency method
Damping level	Lightly to extremely damped system	Lightly damped system
Linearity	Linear	Linear
Modal interaction from other vibration modes	No	Yes

The methods for measuring damping are usually divided into two major groups according to whether the responses are expressed in the time domain or the frequency domain, i.e., time response or frequency response methods. Typical time response methods include the free decay method, the hysteretic loop method and the step response method [165]. The free decay method (Sect. 14.3.1) aims to measure the rate of response decay of transient oscillations in free vibrations, which show an exponential decay (logarithmic decrement, Fig. 3.9 in Sect. 3.3) for viscous damping and linear decay (Sect. 14.2.2) for Coulomb damping. The hysteretic loop method is performed by measuring the harmonic force and response at a resonance condition as introduced in Sect. 14.2.6. Readers may notice that the utilization of time response methods implies that the system's transient vibrations only contain one particular eigenmode that is of engineering interest (this is in practice performed by exciting a structure at the corresponding eigenfrequency in anti-node positions with initial impacts). However, strictly speaking, the modal interactions from other orders of eigenmodes will influence the measured data, which will introduce a certain amount of errors in the calculated damping. This dilemma can be solved by the frequency response methods.

The frequency response methods indirectly assess the energy loss balanced by external excitations. Typical examples of this type of method include the half-power (bandwidth) method (as will be discussed in Sect. 14.3.5) and the amplification-factor method [165] presented in Sect. 14.3.4.

Before performing the measurement of damping for a system or a structure, one needs to select a model that can sufficiently characterize the nature of the energy dissipation of the target system or structure, which is followed by choosing items to be measured in order to describe the damping model.

Before each measurement method is elaborated, one should notice that the damping obtained from the methods previously mentioned is based on the assumption of linearity for a system. In many cases, this assumption cannot be justified, because in a nonlinear system, the damping is affected by the response amplitude, especially when a system/structure vibrates violently. For example, significant damping can be produced when a structure sustains large plastic deformation. Table 14.1 briefly summarizes the applicability of time and frequency methods for measuring damping.

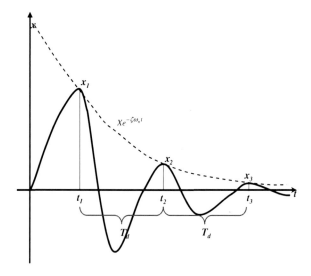

Fig. 14.11 Responses of an oscillator with viscous damping in a free decay test

14.3.1 Free Decay Method

The free decay method, also called the logarithmic decrement method, measures the free vibration responses of a system excited by an impulse (or an initial condition excitation).

For a system with viscous damping, the two consecutive free vibration response peaks shown in Fig. 14.11 are expressed as:

$$x_1 = Xe^{(-\zeta\omega_n t_1)} \sin(\omega_d t_1 + \phi) \tag{14.27}$$

$$x_2 = Xe^{(-\zeta\omega_n t_2)} \sin(\omega_d t_2 + \phi) = Xe^{(-\zeta\omega_n t_2)} \sin\left[\omega_d\left(t_1 + \frac{2\pi}{\omega_d}\right) + \phi\right]$$
$$= Xe^{(-\zeta\omega_n t_2)} \sin(\omega_d t_1 + \phi) \tag{14.28}$$

The ratio of the two consecutive response peaks is:

$$\frac{x_1}{x_2} = \frac{Xe^{(-\zeta\omega_n t_1)} \sin(\omega_d t_1 + \phi)}{Xe^{(-\zeta\omega_n t_2)} \sin(\omega_d t_1 + \phi)} = e^{(-\zeta\omega_n t_1 + \zeta\omega_n t_2)} = e^{\left[-\zeta\omega_n t_1 + \zeta\omega_n \left(t_1 + \frac{2\pi}{\omega_d}\right)\right]} = e^{\left(\zeta\omega_n \frac{2\pi}{\omega_d}\right)} \tag{14.29}$$

The logarithm decrement is then defined as the natural logarithm of the response peak ratio:

$$\delta_{1-2} = \ln\frac{x_1}{x_2} = \zeta\omega_n \frac{2\pi}{\omega_d} = \zeta\omega_n T_d \tag{14.30}$$

where T_d is the damped eigenperiod of the system or structure.

14.3 Measuring Damping 251

Therefore, the damping ratio can then be calculated as:

$$\zeta = \frac{\delta_{1-2}\omega_d}{2\pi\omega_n} \tag{14.31}$$

For calculating the damping ratio based on the ratio of any two response peaks x_i and x_j ($i < j$), a more general form of expression is:

$$\zeta = \frac{\delta_{i-j}\omega_d}{2\pi(j-i)\omega_n} \tag{14.32}$$

In case of low damping ($\zeta < 0.1$), one can assume $\omega_d = \omega_n$, and the equation above can be approximated as:

$$\zeta = \frac{\delta_{i-j}}{2\pi(j-i)} \tag{14.33}$$

It is worth mentioning that the free decay method is not only used in determining damping characteristics of a structure or a mechanical system, but also adopted to measure the stabilizing performance of floating structures (floating platforms or ships) from oscillations due to water wave excitations [163]. Since unexpected large motions can cause discomfort for people and breakdown of facilities on floating structures [164], this is important to improve the serviceability of floating structures.

> **Example:** Under an initial impulse excitation at the free end of a cantilever beam, the 3rd and 6th peaks of damped free vibration amplitudes measured are 0.16 and 0.09 m. Determine the system damping ratio of this cantilever beam.
>
> **Solution:** $\zeta = \frac{\delta_{3-6}\omega_d}{2\pi(6-3)\omega_n} = \frac{\ln\frac{0.16}{0.09}\sqrt{1-\zeta^2}\omega_n}{6\pi\omega_n}$
> Solving the equation above, we obtain $\zeta = 3.051\%$
> If a small damping is assumed, we have:
>
> $$\zeta = \frac{\delta_{3-6}}{2\pi(6-3)} = \frac{\ln\frac{0.16}{0.09}}{6\pi} = 3.052\%$$
>
> Virtually, the calculated damping between the two assumptions is identical from an engineering point of view

> **Example:** For a system with small damping, if the response amplitude needs to decay with a factor of W within n cycles of oscillations, what is minimum damping ratio for the system?
>
> **Solution:** $\zeta \geq \frac{\ln W}{2\pi n}$

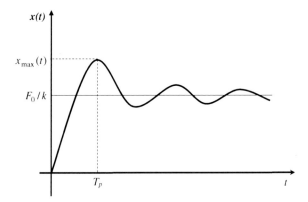

Fig. 14.12 Responses of a lightly damped system due to an initial step force excitation

As discussed in Sect. 3.3, the response decay due to viscous damping in free vibrations never ceases. The energy dissipation can be caused by many different types of damping, such as joint friction and radiation damping at supports. Therefore, other damping, typically Coulomb friction damping, must exist to stop the dynamic responses.

14.3.2 Step Response Method

In Sect. 11.3.1, it is presented that, by assuming an initial condition $x(0) = \dot{x}(0) = 0$, the responses of a lightly damped SDOF system due to a step force excitation can be written as:

$$x(t) = \frac{F_0}{k}\left[1 - \frac{e^{-\zeta\omega_n t}}{\sqrt{1-\zeta^2}}\sin\left(\sqrt{1-\zeta^2}\omega_n t + \varphi\right)\right] \quad (14.34)$$

where $\varphi = \cos^{-1}(\zeta)$.

It is noticed from the expression above that the first peak response occurs at a time T_p, which is also called peak time:

$$T_p = \frac{\pi}{\sqrt{1-\zeta^2}\omega_n} \quad (14.35)$$

And the displacement response at the peak is:

$$x_{\max}(t) = \frac{F_0}{k}\left[1 + e^{-\zeta\omega_n T_p}\right] = \frac{F_0}{k}\left[1 + e^{-\frac{\pi\zeta}{\sqrt{1-\zeta^2}}}\right] \quad (14.36)$$

14.3 Measuring Damping

By obtaining any one of the measured T_p or $x_{max}(t)$ as shown in Fig. 14.12, one can compute the damping as:

$$\zeta = \sqrt{1 - \left(\frac{\pi}{T_p \omega_n}\right)^2} \tag{14.37}$$

$$\zeta = \cfrac{1}{\sqrt{1 + \cfrac{1}{\left[\cfrac{\ln\left(\frac{x_{max}(t)}{F_0/k} - 1\right)}{\pi}\right]^2}}} \tag{14.38}$$

Readers need to bear in mind that the two equations above are valid only for SDOF systems or modal excitations in MDOF systems [165].

14.3.3 Hysteresis loop method

This method is elaborated in Sect. 14.2.6 (specific damping factor and loss factor).

14.3.4 Amplification-factor Method from Forced Vibrations

For a lightly damped system, by using the relationship between the amplification factor Q_i and the relevant damping ratio in the frequency domain, i.e., peaks of the magnification factor at each eigenfrequency and their corresponding damping as shown in Fig. 14.13, one can calculate the damping ratio as:

$$\zeta_i = \frac{1}{2Q_i} \tag{14.39}$$

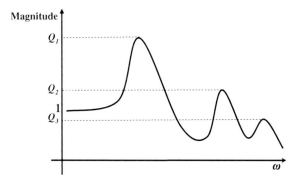

Fig. 14.13 Normalized frequency responses with respect to the maximum quasi-static deflection $X_0 = F_0/k$ of a system

14.3.5 Half-power/Bandwidth Method from Forced Vibrations

Another way of determining the equivalent damping is based on the damping that provides the same bandwidth (at resonance) in the frequency–response curve measured from a vibration test.

The equivalent damping at the ith eigenfrequency ω_{ni} (Fig. 14.14) is given by:

$$\zeta_i = \frac{\omega_{hi} - \omega_{li}}{2\omega_{ni}} = \frac{\Delta\omega_i}{2\omega_{ni}} \qquad (14.40)$$

where ω_{hi} and ω_{li} are frequencies on either side of the eigenfrequency ω_{ni}, and the amplitudes of frequency response curve at ω_{li} and ω_{hi} is defined as having $1/\sqrt{2}$ (corresponding to 3 dB) times the amplitude at the ith eigenfrequency ω_{ni}.

Therefore, the sharpness of each peak on the frequency curve indicates the value of the loss factor (η_i) at each natural frequency.

Since the half-power method is based on the frequency ratio, compared to the equivalent damping measured by the hysteretic loop, it has the merit of not requiring the measurement of static force acting on the system. In many practical cases, this force is difficult to measure.

It should also be noticed that the equation above is derived from a system that is lightly damped, which assumes that ζ_i^2 is negligible. And, strictly speaking, it is only valid for a linear system.

However, for rather low damping ratio (<1 %) system, the damping determined by frequency methods are not accurate, this is mainly due to the reason that, the frequency response curve for a rather low damping system has an extremely sharp shape near each eigenfrequency. Therefore, it is very difficult to obtain a sufficient number of points in this narrow range of frequency responses. As a result, for low damping system, it is recommended to use a time-response method to determine the damping.

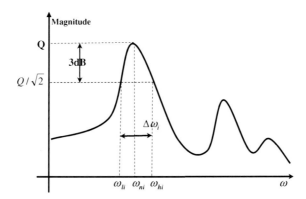

Fig. 14.14 Normalized frequency responses with respect to the maximum quasi-static deflection $X_0 = F_0/k$ of a system

14.4 Relationship Among Various Expressions of Damping

Damping can be represented by various parameters. The relationship of damping expressed by the different parameters is summarized as:

$$\eta = 2\zeta = \frac{2c}{c_n} = \frac{2c}{2\omega_n m} = \frac{1}{Q} = \tan\varphi = \frac{\delta}{\pi} = \frac{D}{2\pi U} = \frac{\Delta\omega}{\omega_n} \qquad (14.41)$$

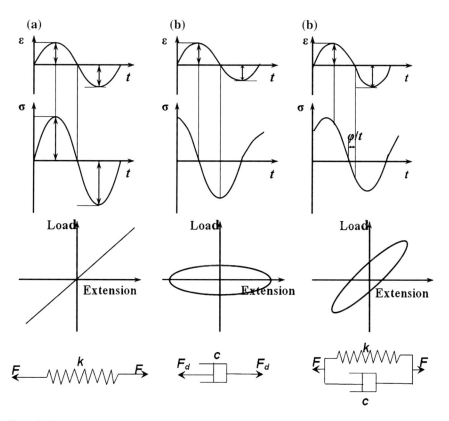

Fig. 14.15 Cyclic stress and strain time histories and load extension behavior for (**a**) linear-elastic, (**b**) viscous, and (**c**) viscoelastic materials

where η is the loss factor; ζ is the ratio of critical damping; c is the viscous damping coefficient; c_n is the critical viscous damping; ω_n is the resonant frequency; Q is the amplification factor; $\phi = \Delta t \cdot \omega$ is the phase angle between cyclic stress and strain (Fig. 14.15); δ is the logarithm decrement of transient responses (Sect. 14.3.1); D is the energy dissipation per cycle; U is the stored energy during loading; and $\Delta \omega$ is the frequency space determined from the half-power bandwidth point down from the resonance peak (Sect. 14.3.5).

14.5 Damping for Engineering Structures

In the dynamic analysis of engineering structures, damping is one of the most difficult parameters to model with sufficient accuracy. It is contributed by various sources, typically from materials and the structural joints, of which the former is typically of hysteretic type and the latter is of friction type. As discussed previously, based on the equivalent energy loss, an equivalent viscous damping can be obtained. Even if the major contribution of damping is, in most cases, not of a viscous type, for a convenient modeling and the subsequent mathematical treatment, viscous damping is still the most widely used damping measure.

14.5.1 Material Damping

Material damping occurs due to energy dissipation in a volume of macro-continuous media. Hysteretic damping modeling is normally used to represent material damping. Detailed study of material damping is a task of solid physics and thus beyond the scope of the current book.

Table 14.2 illustrates typical values of material damping, expressed as a percentage of critical damping ($2m\omega_n$), for use with the modeling of viscous damping. It is noted that the damping for an un-cracked concrete structure is lower than for a cracked one, which is due to the damping mechanism of concrete: in the un-cracked state only viscous damping presents; in the cracked state, both viscous damping in the un-cracked compression zone and friction damping between the concrete and the reinforcing steel in the cracked tension zone contribute to the damping [210].

Generally, high damping materials usually exhibit low strength accompanied with high cost, and are therefore not suitable for structures with load carrying functions. An exception is manganese copper, which has high damping at a large strain level together with high strength.

14.5 Damping for engineering structures

Table 14.2 Typical values of material damping [211]

Material	Damping of the critical (%)
Reinforced concrete	0.5 (uncracked)–3 (cracked)
Steel	0.05–0.4
Cast iron	0.15–1.5
Pure aluminum	0.001–0.1
Dural aluminum alloy	0.02–0.05
Manganese copper alloy	2.5–5
Lead	0.4–0.7
Natural rubber	5–15
Hard rubber	50
Glass	0.03–0.1
Wood	0.25–0.5

14.5.2 Structural/Slip Damping

From Table 14.2 above, it is noticed that, for most of the engineering structures made of steel or concrete, the material damping is rather small. For example, with a damping of 0.0005 for steel, the amplification factor Q is 1,000 (Eq. 14.41), which is not realistic. Therefore, there must be other damping sources to limit the dynamic responses.

The damping generated by structural joints is typically such a type of damping, which is called structural damping, or slip damping. Rather than generating damping in a volume of macro-continuous media as material damping does, slip damping arises from the boundary shear effects at joints between distinguishable parts or at mating surfaces. It can be a type of viscous in case of lubricated sliding, Coulomb friction or hysteretic. For Coulomb friction type of joint interface, efficient energy dissipation can be achieved with optimum interface pressure and geometry. However, a small deviation from this optimum condition may lead to a significant damping reduction. In addition, the optimum condition may result in serious corrosion due to wear. This leads to the development of other types of interface treatment, such as lubrication or adhesive separator at interface [212].

14.5.3 System Damping

System damping arises due to the energy dissipation from materials, joints, fasteners and interfaces. It is basically the sum of material and structural damping and in some cases also accounts for the radiation damping due to the radiation of waves in a continuous medium away from the area of excitations. This type of damping is usually what is actually used for modeling in a typical structural dynamic analysis.

14.5.4 Hydro- and Aerodynamic Damping

The relative velocity between the fluid and the encountering objects produces either hydrodynamic or aerodynamic damping, depending on the type of media. Both possess viscous characteristics and are categorized as fluid damping. This topic is discussed in Sects. 14.2.5, 12.1.1 and 12.2.1.

14.5.5 Typical Damping Levels

Table 14.3 shows the typical system damping value for land-based structures without accounting for the soil damping.

Table 14.4 illustrates typical values used for the modeling of offshore structures, which include the damping due to structures' energy dissipation, fluid–structure interactions (drag), and soil damping. It is noticed that, under extreme loading conditions such as significant seismic excitations or wave load during a major storm, a higher damping value due to structures' energy dissipation shall be used. This is because, under significant loading, the strain level is higher than that under a normal loading such as the one associated with high-cycle fatigue [243], leading to a higher damping level.

Table 14.3 Typical system damping for land-based structures and non-structural elements under seismic excitations [216, 217]

Structure type	System damping without soil damping (%)
Reinforced concrete	4 (uncracked)–7 (cracked)
Prestressed concrete	2–5
Reinforced masonry	4–7
Natural stone	5–7
Welded or bolted steel with friction connections	2–4
Bolted steel with friction connections	4–7
Large diameter (>304.8 mm) piping	2–5
Small diameter (≤304.8 mm) piping	1–3
Mechanical or electronic components	2–3
Storage tank (sloshing mode)	0.3–0.7
Storage tank (impulsive mode)	2–3
Transmission lines (aluminum or steel)	4–6

14.5 Damping for engineering structures

Table 14.4 Typical damping level for modeling of offshore structures [81, 215]

Structure type	Damping		
Jacket fatigue loading*	1–3 % for fundamental global bending mode and 2–3 % for higher-order mode		
Jacket extreme wave loading*	2–3 % for fundamental global bending mode and 3–4 % for higher-order mode		
Welded pile in soil*	0.6 % (land-based)–1.4 % (offshore)		
Welded brace*	0.3 % (land-based)–0.8 % (offshore)		
Welded mast (in air) *	0.8 %		
Bolted mast (in air) *	0.3–3 %		
Piled support structure (e.g., offshore wind turbine supporting structures) under extreme wave [215]	Radiation damping from wave creation due to structural vibrations	0.1–0.3 %	
	Hydrodynamic damping	0.1–0.2 %	
	Steel material damping (without ground connections)	0.15–0.3 %	
	Soil damping	Internal friction	2–7 %
		Geometric	damping
0.6 % (elastic)–0.8 % (plastic)			
Jacket under seismic loading	Structural damping	0.5 (elastic) %–2 % (plastic)	
	Hydrodynamic damping	0.5 (elastic) %–1.5 % (plastic)	
	Soil damping	1 (elastic structure) %–5 % (significant soil inelasticity)	
GBS under seismic loading	Structural damping	1 % (elastic)–3 % (plastic)	
	Hydrodynamic damping	0.5 %	
	Soil damping	Under calculation	

Note: *see Ref. [81], the damping includes both structural and hydrodynamic damping, but the soil damping is not accounted for

14.6 Comparison of Cyclic Responses Among Structures Made of Elastic, Viscous and Hysteretic (Viscoelastic) Materials

When structures made of elastic material vibrate, all the energy stored during loading is returned when the loading is removed. As a result, there is no time lag between the responses and loading, i.e., the displacement of the structure responds immediately (in phase), to the cyclic load. The stress and strain time histories are also completely in phase (Fig. 14.15a). Furthermore, for linear elastic materials, Hooke's law applies, where the stress is proportional to the strain.

Conversely, for a purely viscous material, no energy is conserved after the loading is removed. The input stress disappears due to "pure damping" as the vibratory energy is transferred into internal heat energy. The stress and strain time histories are out of phase as shown in Fig. 14.15b.

For all other types of damping that do not fall into one of the two category classifications above, the assumption of viscoelasticity may be used for modeling, which has both elastic and viscous properties as shown in Fig. 14.15c.

Chapter 15
Nonlinear Dynamics

15.1 From Linear to Nonlinear

Many investigations start with linear models, in which the effects are assumed to be always proportional to their causes. This assumption greatly simplifies the problem since it enables a superposition of all relevant elementary cases, which are normally well handled by physical modeling and mathematical treatment [218]. In addition, most of the more advanced problems in scientific calculations require the solution of only linear systems, having real or complex coefficients, often of very large dimensions [219]. However, strictly speaking, proportional phenomena do not exist in the real world. As a matter of fact, we often observe effects that saturate in spite of an increase of their causes, or which go in different and somehow unexpected ways. One example of this can be found in both the physical and social world, e.g., when a steel bar under axial tension is well above its yielding, the elongation of the steel bar is not proportional to the applied tension load anymore, i.e., the material nonlinearity appears due to stiffness change. Another example is an ideal rigid pendulum, with a mass at the end of a weightless rod and revolving around a horizontal axis under gravity field. Strictly speaking, the torque due to the mass's self-weight is a sine function of the revolving angle rather than the revolving angle (linear), i.e., mechanical constraints nonlinearities occur. The third example is that the income tax rate in almost all countries is not a fixed fraction of and depends on the total volume of one's income, i.e., the total tax one pays is not proportional to the total income one earns.

For treating nonlinear problems regarding dynamics, people instinctively try to go back to a problem that they know how to handle by linearizing the dynamics around it and treating small departures from it, and then reproducing this scheme as far as possible to reach other fully nonlinear states [220]. The most widely used method is perturbation theory, which has its roots in early celestial mechanics, where the theory of epicycles was used to make small corrections to the predicted paths of planets [221].

As we have discussed previously, to carry out a dynamic analysis, in a more practical sense, the finite element method, the finite difference method (Sect. 13.5), often referred to as the direct time integration method to distinguish it from that based on the modal superposition method) or modal superposition method (Sect. 13.4), and the linear iteration method (Sect. 15.5) are the three most

J. Jia, *Essentials of Applied Dynamic Analysis*, Risk Engineering,
DOI: 10.1007/978-3-642-37003-8_15, © Springer-Verlag Berlin Heidelberg 2014

commonly used numerical methods in computational solid mechanics, solving problems associated with space, time and nonlinearities, respectively, but normally in a combined manner.

Since, in linear dynamic analysis, the responses of a system/structure are proportional to the load/excitation it is subjected to, again, this enables the utilization of superposition, which brings significant convenience with regard to computation cost and convenience, and in many cases also ensures the calculation accuracy. However, when nonlinearities appear in the system/structure, the stiffness and/or load are dependent on the deformation, and the responses of a system are generally not amenable to an analytical method with exact solutions. "Exact" here means that the solution is obtained either in a closed form or in a mathematical expression that can be evaluated to any degree of accuracy by numerical means, e.g., a series of expressions [222]. The modal superposition rule is also often regarded as inapplicable [223], or at least applying only for mild and localized nonlinearity, provided that a relatively few mode shapes need to be considered. In the latter case, the same principles presented in Sect. 13.4 are also applicable to nonlinear analysis, and the eigenmodes have to be updated at each time step, so that the nonlinearities (e.g., constitutive laws) can be evaluated. However, the complete mode superposition analysis is only effective when the solution can be obtained without updating the stiffness matrix too frequently [173]. For readers interested in this topic, references [173, 224] and [223] are recommended.

It is known that a general method for obtaining the exact solution of nonlinear differential equations is not available. Most of the analytical methods that have been developed yield approximate solutions, and the available techniques vary greatly with the type of nonlinear equations [18]. In engineering practice, especially with the wide utilization of finite element method, the direct time integration for time stepping combined with decent integration methods for reaching equilibrium within each time increment is the dominant approach for obtaining nonlinear responses.

Compared to linear dynamic analysis, nonlinear dynamic analysis can detect certain critical dynamic responses due to the variation of stiffness. It can also avoid over-design of a structure through a more accurate calculation of responses. This is particularly true with respect to structural failure and collapse, which typically involve any or all types of nonlinearities. Nonlinear dynamic analysis can be used to assess structural behavior with regard to strength, stability, service configuration, reserve strength and progressive failure etc.

The essential difference between linear and nonlinear analysis is typically the treatment of stiffness, which is influenced by material properties, structural geometries and boundary conditions. For example, when the deformation of a beam structure is small, and the stress levels are well below the yielding of the structural materials, the deformation as such will not create a significant additional bending moment. Also, when the material stress–strain relationship also remains linearly elastic, the change of stiffness is insignificant. A linear elastic analysis can then be adopted for the analysis. This means that the stiffness matrix is assembled and solved only once, with no need to update it. However, when rather high loads are applied on a structure, yielding and large deformations can be expected, both

15.2 Sources of Nonlinearities

the geometry (force-deformation) and material stress–strain relationship as will be presented in Sect. 15.2 experience a significant change, leading to a change in stiffness. This requires that the stiffness matrix has to be updated during the deformation through a decent interactive solution process. This linear iteration method will be introduced in Sect. 15.5. It is essentially a procedure to find the equilibrium between the external forces and internal forces for each time step increment, which can be applied to solve both static and dynamic problems.

15.2 Sources of Nonlinearities

Although all nonlinear analyses relate to the stiffness changes, their causes can be quite different, leading to different categories of nonlinearities.

15.2.1 Material Nonlinearity

Material nonlinearity appears because the material does not exhibit a perfectly linear behavior, but its properties vary with the strain. Structural materials and components have limits on strengths coupled with different loading and unloading paths, leading to nonlinear inelastic behavior. This is illustrated in Fig. 15.1, which shows the stress–strain diagram obtained from uni-axial tensile tests for a specimen made from typical structural steel. Below the proportional point, the strain and stress has a linear (proportional) relationship, and the slope of the curve is constant and referred to as modulus of elasticity E. For typical low carbon steel, this point is in the range 200–300 MPa. With a load increase from the proportional point, the linear stress–strain relationship no longer exists. Instead, the slope of stress–strain curve becomes smaller and smaller until the yielding point is reached. After the yielding point, the curve becomes flat with a significant elongation, i.e., strain increases without noticeable change in stress. This flat part of the curve indicates a process of yielding. This is followed by strain hardening in which the stress level rises again with a declined slope (compared with the slope during the linear-elastic range) of the curve until the ultimate strength (maximum load value) is reached. The strain level corresponding to the ultimate strength is many times that at yielding point. After the ultimate strength point, the increase of strain is actually accompanied by a reduction in stress (load), and the fracture finally occurs.

Figure 15.2 shows the constitutive relationship for typical mild steel from a real measurement. It is noticed that the majority of strain increase occurs during the strain hardening, and the yielding process also contributes to a significant strain increase, while the strain at the initial yielding is comparatively small.

It should also be mentioned that, above the yielding point, with any removal of the load, the stress–strain curve will have a slope identical to that before the proportional limit point. This is illustrated as the unloading process shown in

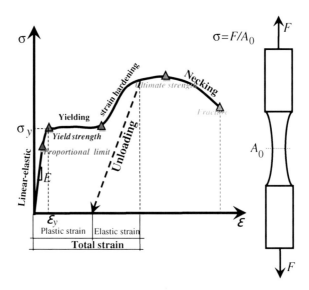

Fig. 15.1 Engineering stress (σ)-strain (ε) curve for a typical ductile steel material that has an obvious yield strength (not to scale), A_0 is undeformed cross-section area

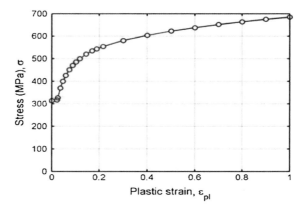

Fig. 15.2 The engineering stress-strain curve obtained from an experimental test for typical mild steel with a yield stress of 314 MPa, and an elastic modulus of 206 GPa [246]

Figs. 15.1 and 15.3. Moreover, it is noticed that, if a loading is completely removed, only the elastic strain is recovered and the plastic part of strain remains. Therefore, the plastic part of strain is permanent. This strain is often called plastic (permanent) strain, and it leads to changes of the test member's dimensions and/or shape, which is in most cases undesirable. However, there are some instances in which plastic deformation is desirable. For example, car bodies are designed in such a way that they can absorb a large amount of impact energy through plastic deformation. The second example is the utilization of work hardening to increase

15.2 Sources of Nonlinearities

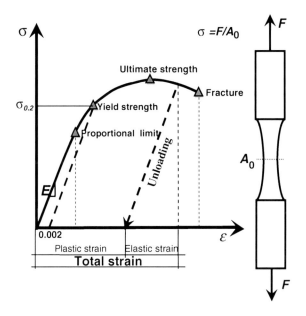

Fig. 15.3 The engineering stress–strain curve for materials that do not have an obvious yield strength (not to scale)

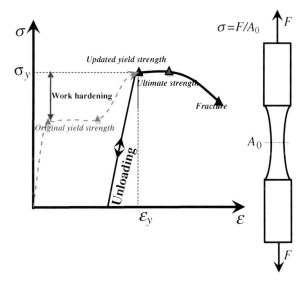

Fig. 15.4 The yield strength is increased after the work hardening

the yield strength of mild steel. Steel structural members can be preloaded beyond the original yield strength as shown in Fig. 15.4, and then unloaded, making the updated yield strength significantly higher than the original one, with this normally being accompanied by a decrease of ductility.

15.2.1.1 Yield Criterion and Hardening Rule Under Uni-Axial Stress Condition

Many ductile materials such as aluminum, heat-treated higher strength steels or cold formed steels do not have a clearly defined yield point. The yield strength of such materials is defined by projecting a line parallel to the curve at the initial linear elastic region starting at 0.2 % plastic (permanent) strain as shown in Fig. 15.3.

Because the yield strength in those cases is determined by an arbitrary rule that does not indicate an inherent physical property of the materials, it is referred to as the offset yield strength [226], or proof stress. For ductile aluminum, the yield strength point is slightly above the proportional limit, while for the higher strength or cold formed steel, the yield strength is essentially identical to the stress level at the proportional limit.

In some cases, the strain hardening may be neglected, leading to elastic-perfect plasticity as shown in Fig. 15.5 and this is generally conservative:

$$\sigma = E\varepsilon \quad \text{for} \quad \sigma < \sigma_y \qquad (15.1)$$

$$\sigma = \sigma_y \quad \text{for} \quad \varepsilon \geq \frac{\sigma_y}{E} \qquad (15.2)$$

Under this simplification and by assuming that the entire cross-section of a beam can be fully utilized when the beam reaches the plasticity, the stress distribution along the beam's cross-section can be divided into three processes: elasticity, partial plasticity and full plasticity, as shown in Fig. 15.6. There is a dilemma in this assumption such that, when yielding is initiated, the plastic deformation can be infinite. Therefore, a slight hardening is normally assigned to simulate perfect plasticity for numerical stability. For example, in finite element analysis involving plasticity, to reach numerical stability, the strain hardening is normally assumed to represent either a realistic material behavior or only a slight hardening with a slope of 1/1,000 to 1/100.

The strain hardening can be accounted for through the elastic-linear-hardening rule, elastic power-hardening relationship or by separating the plastic deformation from the total deformation and applying an exponent on the plastic hardening (Ramberg–Osgood relationship).

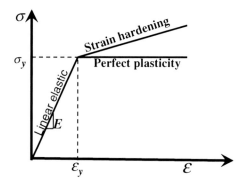

Fig. 15.5 Idealized elastic-plastic engineering stress–strain diagram

15.2 Sources of Nonlinearities

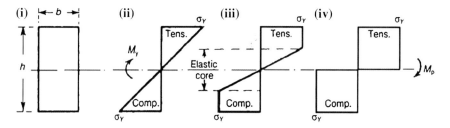

Fig. 15.6 Process of elasticity-initial yielding (*ii*), partial plasticity (*iii*) and full plasticity (*iv*) for a rectangular cross-section beam under pure bending [248]

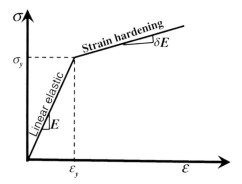

Fig. 15.7 Elastic-linear plastic hardening relationship using engineering stress-strain curve

As a rough approximation of stress–strain curves that rise appreciably after yielding, a linear hardening relationship can be used by assuming that, after yielding, the slope in the stress–strain curve is a fraction of the elastic modulus, i.e., δE (Fig. 15.7):

$$\sigma = E\varepsilon \quad \text{for} \quad \sigma < \sigma_y \tag{15.3}$$

$$\sigma = (1-\delta)\sigma_y + \delta E \quad \text{for} \quad \varepsilon \geq \frac{\sigma_y}{E} \tag{15.4}$$

where δ is a slope reduction factor after initial yielding.

Note that, in the discussions above, the engineering stress σ and strain ε are used. This is based on the assumption that the area of a beam cross-section does not change during deformation and is identical to the initial one A_0. This can be justified for small strain conditions where the reduction of cross-section area is small. When the plastic strain level is sufficiently high, the changes in the beam's cross-section are not negligible and are related to the Poisson's ratio of the material as shown in Fig. 15.8. The engineering stress and strain are then not appropriate to use for defining the stress–strain curve. Instead, the true stress should be used, in which the initial area A_0 is replaced by the updated (current) cross-section area A:

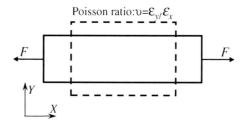

Fig. 15.8 Deformed shape (*solid line*) of a *solid bar* under uni-axial tension F

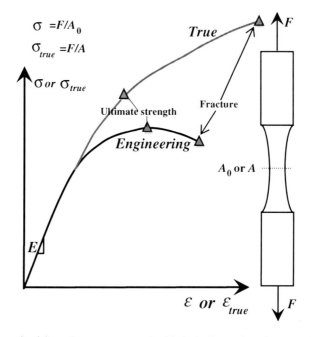

Fig. 15.9 The uni-axial tension test measured with both the engineering (σ and ε) and true stress–strain (σ_{true} and ε_{true}) curve (not to scale)

$$\sigma_{true} = \frac{F}{A} = \frac{F}{A_0} \cdot \frac{l}{l_0} = \sigma(1 + \varepsilon) \qquad (15.5)$$

where l_0 is the initial length of the specimen and l is the updated (current) length due to the deformation.

And the corresponding true strain is:

$$\varepsilon_{true} = \int_0^l \frac{dl}{l} = \ln\left(\frac{l}{l_0}\right) = \ln(1 + \varepsilon) \qquad (15.6)$$

An illustration of the relationship between engineering (nominal) and true stress–strain is shown in Fig. 15.9. The difference between the two increases with

15.2 Sources of Nonlinearities

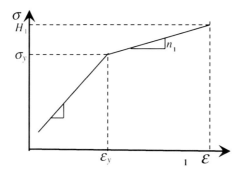

Fig. 15.10 Elastic power-plastic hardening relationship plotted on log–log coordinates using true stress–strain relationship

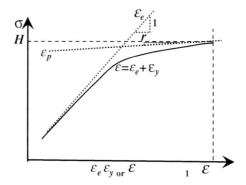

Fig. 15.11 Ramberg-Osgood relationship curve plotted on log–log coordinates using true stress–strain relationship

the increase of strain. It is also noticed that a power hardening relationship can be applied using the measure of true stress–strain curve with a log–log plot. This provides the possibility to represent a more realistic plastic deformation. In the following, we will introduce two power hardening relationships based on the measurement of true stress and strain.

Figure 15.10 shows a power hardening relationship and can be expressed as:

$$\sigma_{true} = E\varepsilon_{true} \quad \text{for} \quad \sigma_{true} < \sigma_y \tag{15.7}$$

$$\sigma_{true} = H_1 \varepsilon_{true}^{n_1} \quad \text{for} \quad \varepsilon_{true}^{n_1} \geq \frac{\sigma_y}{E} \tag{15.8}$$

where n_1 is called the strain hardening exponent, which typically ranges from 0.05 to 0.4 for metals; H_1 is the stress value when $\varepsilon = 1$.

It is noticed from Fig. 15.10 that in the log–log plot, the curve within the elastic region has a slope of unity, and it intersects with the curve in the plastic region at the yield stress σ_y.

A more popular relationship, namely Ramberg–Osgood rule, is to separate the elastic strain (ε_e) and plastic strain(ε_p), and the total strain is the sum of the two as shown in Fig. 15.11 and expressed as:

$$\varepsilon_{true} = \varepsilon_e + \varepsilon_p = \frac{\sigma_{true}}{E} + \left(\frac{\sigma_{true}}{H}\right)^{r-1} \tag{15.9}$$

where r is Ramberg–Osgood strain hardening exponent that is defined differently from the strain hardening exponent used for power-hardening relationship. H is the stress value when the plastic strain $\varepsilon_p = 1$.

It is noticed that the exponential relationship in the Ramberg–Osgood rule only applies to the plastic strain. At a low strain level, the total strain ε approaches the elastic curve with the slope of unity; while at a large strain level, the total strain approaches the plastic strain with the slope of r.

By comparing the Ramberg–Osgood and elastic power-hardening relationships, it is found that when the strains are rather large, the elastic strain is insignificant, and the two relationships are essentially equivalent. Therefore, when the total strain is rather large, the first term on the right hand side of the equation above can be neglected.

Furthermore, in many commercial finite element analysis codes for nonlinear analysis, to represent the plastic hardening relationships, the strain obtained from the material test can be directly implemented into the material modeling.

Regardless of which model is being used, numerically, the nonlinear part of the stress–strain curve is generally approximated as a series of piece-wise linear segments. Each linear segment is represented by a tangent modulus, which is the ratio between the stress and strain for that particular segment.

15.2.1.2 Yield Criterion, Flow and Hardening Rule Under Multi-Axial Stress Condition

The constitutive laws commonly adopted for structures with nonlinear inelastic material properties are characterized by a yield criterion/surface, a flow rule and the hardening rule [173, 223].

The yielding criterion and hardening rule obtained under uni-axial stress conditions must be extended to deal with multi-axial stress state. In such a condition, the yield criterion determines the onset of the yielding for multi-axial stress by accounting for the contribution from different stress components. The strain related to a stress component σ_{ij} in any location of a solid can be decomposed into strains due to change of volume and shape. It is well approved from abundant experiments that for most of the materials (especially for metal materials), the volume change (due to uniform hydrostatic pressure from all directions) is purely elastic, and does not contribute to any plastic deformation, i.e., the plastic deformation occurs under incompressibility conditions, leaving the deviator stress as the only cause for

15.2 Sources of Nonlinearities

plastic flow. For a three-dimensional stress state, the stress component due to the volume change, namely spherical stress σ_m, and the stress component due to the change in shape, namely the deviator stress σ_{ij} obey the relationship as follows:

$$[\sigma] = \begin{bmatrix} \sigma_{11} & \sigma_{12} & \sigma_{12} \\ \sigma_{21} & \sigma_{22} & \sigma_{23} \\ \sigma_{31} & \sigma_{32} & \sigma_{33} \end{bmatrix} = \begin{bmatrix} \sigma_m & 0 & 0 \\ 0 & \sigma_m & 0 \\ 0 & 0 & \sigma_m \end{bmatrix} + \begin{bmatrix} \sigma_{11} - \sigma_m & \sigma_{12} & \sigma_{12} \\ \sigma_{21} & \sigma_{22} - \sigma_m & \sigma_{23} \\ \sigma_{31} & \sigma_{32} & \sigma_{33} - \sigma_m \end{bmatrix}$$
$$(15.10)$$

where the subscript denotes the axis along three arbitrary axes that are perpendicular to each other; $\sigma_m = (\sigma_{11} + \sigma_{22} + \sigma_{33})/3$.

It is noticed that the deviator stress tensor is obtained by removing the hydrostatic pressure form:

$$s = [\sigma] - \sigma_m I = [\sigma] - (I_1/3)I \qquad (15.11)$$

In Eq. 15.11, I_1 is the first stress invariant among the three stress invariants (independent of axes as long as the three axes are perpendicular to each other):

$$I_1 = \mathrm{Tr}[\sigma] = \sigma_{11} + \sigma_{22} + \sigma_{33} \qquad (15.12)$$

$$I_2 = \frac{1}{2}\mathrm{Tr}[\sigma^2] = \frac{1}{2}\sigma_{ij}\sigma_{ji}$$
$$= -(\sigma_{11}\sigma_{22} + \sigma_{22}\sigma_{33} + \sigma_{33}\sigma_{11}) + \sigma_{12}^2 + \sigma_{23}^2 + \sigma_{31}^2 \qquad (15.13)$$

$$I_3 = \frac{1}{3}\mathrm{Tr}[\sigma^3] = \frac{1}{2}\sigma_{ij}\sigma_{jk}\sigma_{ki}$$
$$= \sigma_{11}\sigma_{22}\sigma_{33} + 2\sigma_{12}\sigma_{23}\sigma_{31} - (\sigma_{11}\sigma_{23}^2 + \sigma_{22}\sigma_{31}^2 + \sigma_{33}\sigma_{12}^2) \qquad (15.14)$$

If the three principal stress (σ_1, σ_2 and σ_3) directions are used, the three invariants can be simplified as:

$$I_1 = \sigma_1 + \sigma_2 + \sigma_3 \qquad (15.15)$$

$$I_2 = -(\sigma_1\sigma_2 + \sigma_2\sigma_3 + \sigma_3\sigma_1) \qquad (15.16)$$

$$I_3 = \sigma_1\sigma_2\sigma_3 \qquad (15.17)$$

In the same manner, the deviator stress also has three components:

$$J_1 = \mathrm{Tr}[s] = s_{11} + s_{22} + s_{33} = (\sigma_{11} + \sigma_{22} + \sigma_{33}) - 3\sigma_m \qquad (15.18)$$

$$J_2 = \frac{1}{2}\mathrm{Tr}[s^2] = \frac{1}{2}s_{ij}s_{ji}$$
$$= -(s_{11}s_{22} + s_{22}s_{33} + s_{33}s_{11}) + s_{12}^2 + s_{23}^2 + s_{31}^2 \qquad (15.19)$$

$$J_3 = \frac{1}{3}\mathrm{Tr}[s^3] = \frac{1}{2}s_{ij}s_{jk}s_{ki}$$
$$= s_{11}s_{22}s_{33} + 2s_{12}s_{23}s_{31} - (s_{11}s_{23}^2 + s_{22}s_{31}^2 + s_{33}s_{12}^2) \qquad (15.20)$$

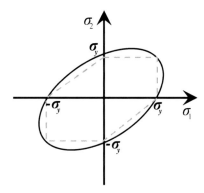

Fig. 15.12 von Mises (*solid lines*) and Tresca (*dashed line*) yield criteria in principal stress space

Typical yield criteria are Tresca theory (maximum shear stress theory), proposed by Tresca in 1864, and von Mises theory (octahedral shear), proposed by von Mises in 1913. They are both used for the application of ductile metal materials and differ slightly from each other.

Tresca theory states that the material reaches yielding when the plastic potential $Q_{Tresca}(\sigma)$ reaches zero, that is, the maximum shear ($\max|\sigma_i - \sigma_j|$) in each principal stress plane reaches yield strength as shown in Fig. 15.12 and is expressed as:

$$Q_{Tresca}(\sigma) = \max|\sigma_i - \sigma_j| - \sigma_y = 0 \qquad (15.21)$$

Different from Tresca theory, which only involves the maximum shear, the von Mises criterion (Fig. 15.12) involves the maximum shear in each principal plane $\sigma_i - \sigma_j$. It states that the materials reach yielding when the plastic potential $Q_{von\,Mises}(\sigma)$ involving the second deviator stress invariant J_2 reaches zero as expressed:

$$Q_{von\,Mises}(\sigma) = \sqrt{3J_2} - \sigma_y = 0 \qquad (15.22)$$

From the energy point of view, von Mises criterion assumes that the yielding occurs when the energy of distortion reaches the same energy at yielding in uniaxial tension. In the principal stress space, the von Mises yield criterion for two-dimensional case is shown in Fig. 15.12 and expressed as:

$$\sqrt{\sigma_1^2 + \sigma_2^2 - \sigma_1\sigma_2} - \sigma_y = 0 \qquad (15.23)$$

In three-dimensional principal stress space, both criteria above are represented by a cylinder whose axis is (1, 1, 1).

By comparing the two yield criteria, it is noticed that in the two-dimensional tension-shear plane, Tresca criterion gives:

$$\sqrt{\sigma^2 + 4\tau^2} - \sigma_y = 0 \qquad (15.24)$$

15.2 Sources of Nonlinearities

And the yield strength in shear due to Tresca criterion is:

$$\tau_y = \frac{\sigma_y}{2} \qquad (15.25)$$

While the von Mises criterion gives:

$$\sqrt{\sigma^2 + 3\tau^2} - \sigma_y = 0 \qquad (15.26)$$

And the yield strength in shear due to von Mises criterion is:

$$\tau_y = \frac{\sigma_y}{\sqrt{3}} \qquad (15.27)$$

For one dimension uni-axial tension or compression, both criteria are identical:

$$\sigma - \sigma_y = 0 \qquad (15.28)$$

Compared to the Tresca yield criterion, the von Mises criterion is less conservative and therefore can avoid an over-design when ductile materials are used.

Different from most materials, for certain materials such as soil or powders, and also for models that take damage into account, a compressive hydrostatic pressure can decrease the plastic flow [232]. The relevant yield criteria to calculate this effect include the Drucker–Prager, Mohr–Coulomb, Lade [227], Bresler–Pister [228], and Ottosen [229] theories etc.

The hardening rule prescribes the work hardening as aforementioned and describes how the yield surface changes with the progression of plastic deformation, and depends on the type of materials. Typical hardening rules includes isotropic hardening, which states that the center and shape (e.g., ellipse in the case of von Mises yield criterion) of yield surface do not change in the principal normal stress space while the size of it expands uniformly as a result of strain hardening. This is illustrated in the left figure of Fig. 15.13. The isotropic hardening is best suited for problems in which the plastic strain is significantly higher than the strain at initial yield.

From Figs. 15.13 and 15.12, it can be observed that the yield surface separates the plastic region from its elastic counterpart. Any movement of stress state toward the outside of the yield surface indicates a plastic loading, and the change of stress

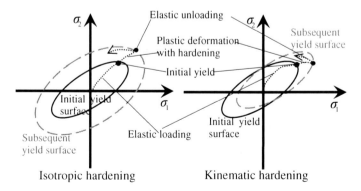

Fig. 15.13 Isotropic and kinematic hardening rule in principal stress space

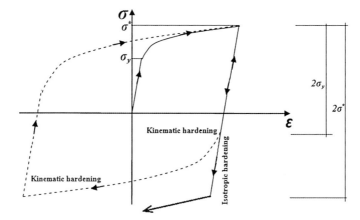

Fig. 15.14 Stress–strain diagram due to reversal loading, with isotropic hardening or kinematic hardening (due to Bauschinger effect) rule applied

state toward the interior of the yield surface indicates elastic unloading. Furthermore, the change in stress state along or within the yield surface will only cause elastic deformation.

It is also noticed that, following the isotropic hardening rule, the yield strength in tension and compression is initially the same, i.e., the yield surface is symmetric about the principal stress axes, and the yield strength remains equal as the yield surface develops with plastic strain. However, under reversal loading (the load changes its direction), yielding due to unloading normally occurs prior to the stress reaching the yield strength σ_y, as shown in Fig. 15.14, i.e., a hardening in tension will lead to a softening in subsequent compression. This early yielding behavior is called the Bauschinger effect. To account for this effect, the kinematic hardening rule has to be applied, which states that the shape and size of yield surface do not change, while the center of the yield surface changes in the stress space, this is illustrated in the right figure of Fig. 15.13.

In addition, by combining the two hardening rules shown in Fig. 15.13, one may also use the mixed hardening rule, such as Prager's rule [230], in which both the size and the center of the yield surface change, but the shape is kept unchanged, i.e., the yield surface expands uniformly.

The isotropic hardening rule is most widely used simply because of its convenience for mathematical treatment and its representativeness with regard to the hardening characteristics of a wide range of materials. However, as mentioned above, under reversal loading with repeated yielding, kinematic hardening should be used to account for the Bauschinger effect. The mixed hardening rule is rarely used due to its numerical complexity.

Readers should also bear in mind that the yield surface cannot only expand under strain hardening, but also shrink under strain softening when the necking shown in Fig. 15.1 occurs.

15.2 Sources of Nonlinearities

Besides the yield criterion and hardening rule, one also needs to describe the progression of yielding in the plastic domain, i.e., to define plastic strain rate outside the yield surface. However, different from what occurs in the elastic stage, the stress and strain in the plastic region do not generally exhibit a one to one correspondence. Therefore, a "constitutive law" has to relate the plastic strain increments to the current stress and stress increments subsequent to yielding, which is called the flow rule:

$$d\varepsilon_p = d\lambda \left(\frac{dQ(\sigma)}{d\sigma_{true}} \right) \tag{15.29}$$

where λ is a scalar factor that is determined from the yield criterion [231, 232]. Q is a plastic potential.

From the equation above, it is obvious that the plastic strain increment $d\varepsilon_p$ is perpendicular to the surface defined by the plastic potential. Therefore, the equation is also referred to as the normality rule [231]. With this geometrical explanation, one can focus on the plastic strain and rewrite the flow rule as:

$$\varepsilon_p = \lambda \mathrm{grad}(Q(\sigma)) \tag{15.30}$$

Therefore, an essential task in defining a flow rule is to define the plastic potential Q. For the majority of metal materials, Q can be assumed to be equal to the plastic potential defining the yielding surface as presented previously, such as $Q_{Tresca}(\sigma)$ or $Q_{vonMises}(\sigma)$. Such a flow rule is called an associated flow rule. If not, it is non-associated. Experimental results show that the associated rule applies well to the plastic deformation of metals, while for some porous materials such as rocks, soil and concrete, the non-associated flow rule provides a better representation of the plastic deformation [231].

The most commonly used associated rule is the Prandtl–Reuss relation (also called the Levy–Mises equation) that is suited for typical metal materials. By applying the von Mises yield criterion, one obtains:

$$d\varepsilon_p = d\lambda \left(\frac{dJ_2}{d\sigma} \right) \tag{15.31}$$

where J_2 is the second deviator stress invariants as described previously.

It should mentioned that material nonlinearity not only includes nonlinear plasticity, but is also relevant for materials with nonlinear elasticity, viscoelasticity and creep effects.

The text regarding the material plasticity covered in the current book provides a fundamental basis for performing relevant nonlinear dynamic analysis. Readers can find a complete coverage of this topic in references [231, 232].

15.2.2 Geometrical Nonlinearity

Geometrical nonlinearity is due to changes in geometry, when either the system is to support large strains and large displacements (the strains exceed the order of

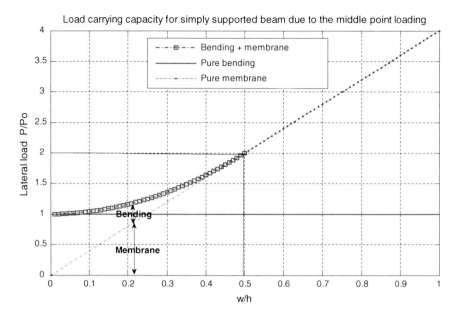

Fig. 15.15 Load carrying capacity curve for a simply supported rectangular beam due to a middle point transverse load at large deformations and with perfect plasticity involved (w deflection, h beam height, P/P_0 ratio between applied load and the load of initial yielding)

tenths) or there are large displacements (membrane stress effect) or rotations with small strains and buckling. For example, if a structural member such as a beam or plate is restrained against the in-plane deformation, the membrane forces will develop with a finite deformation. Figure 15.15 shows an idealized load-carrying capacity curve for a simply supported rectangular beam with both geometrical nonlinearities and perfect plasticity involved. It is clearly shown that with the increase of deflection, the membrane forces developed in the beam become more and more significant while the effects of bending forces behave in the opposite way. When the deflection ratio reaches 0.5, i.e., the entire cross-section becomes fully plastic, the bending contribution vanishes. Under further deformation the axial force remains constant and equal to the one at full plasticity state. Furthermore, for geometrical nonlinearity, the applied loads will either have an effect on the deformed configuration, or the configuration will have an effect on the loads, e.g., follower loads [242, 243]. The geometrical nonlinearity can be significant for cable structures and inflatable membranes, slender structures, metal and plastic forming, the stability of structures [246], and structures made of materials showing significant creep and plasticity, etc.

Figure 15.16 shows an example of the load–deflection curve corresponding to a line load applied along the center line of an HP-beam's top flange as shown in Fig. 15.17. The material's constitutive relationship for this beam is modeled according to the one shown in Fig. 15.2. It is clearly shown that the initial yielding

15.2 Sources of Nonlinearities

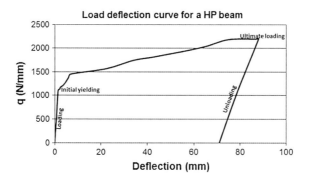

Fig. 15.16 The *line* load (applied on the top flange of an HP-beam shown in Fig. 15.17), q, versus deflection (the maximum deflection at the intersection between the web and the shell plate) with the maximum *line* loads applied along the *center line* of the top flange of the HP-beam

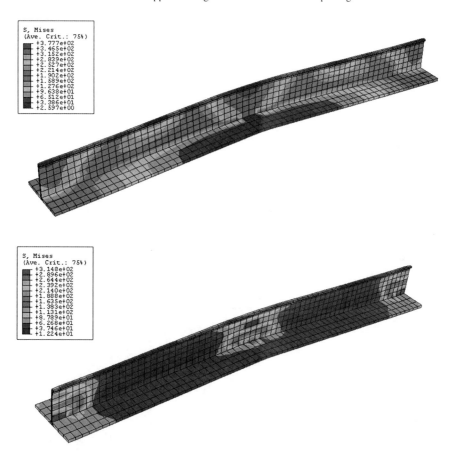

Fig. 15.17 The von Mises stress (in MPa) distribution and deformation of the HP-beam at maximum load level (*upper* figure, with the maximum *line* load of 1,400 N/mm) and after unloading (*lower* figure, with the load level decreased to zero)

occurs at a line load level of 1,134 N/mm. Then the plasticity begins to develop on the beam in that the load–deflection curve's slope dramatically decreases until it reaches a maximum line load value of 2,195 N/mm. By observing the upper figure in Fig. 15.17 that shows the von Mises stress of the beam at this maximum load level, plasticity developed at the two ends and middle of the beam can be clearly identified. After the maximum load is applied, the unloading process then takes place until the load reaches zero, at which a 71 mm of permanent deflection due to the plasticity can be identified. The von Mises stress after unloading is shown in the lower figure in Fig. 15.17. It is also observed that the slope of the curve for the unloading process is less than that of the loading process before the yielding. As discussed in Sect. 15.2.1, the slope between loading and unloading would be identical if only plasticity without geometrical nonlinearity is involved. Note that the beam's span is relatively large, this difference in slope is due to the contribution of geometrical nonlinear membrane stresses developed in the beam.

15.2.3 Buckling

Linear analysis assumes that all members work equally well in tension or compression, neglecting the tendency of slender members to buckle in compression. This alters the effective stiffness of the structure and also changes its buckling load. Buckling is a dominant failure mode for slender structures as shown in Fig. 15.18.

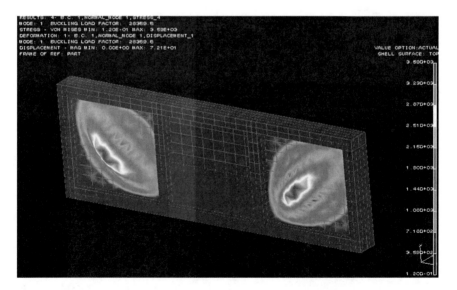

Fig. 15.18 Buckling of a beam web panel

15.2 Sources of Nonlinearities

Fig. 15.19 Conductor and conductor guide installed on an offshore platform (courtesy of Aker Solutions)

15.2.4 Displacement Boundary Nonlinearity

Displacement boundary nonlinearity arises due to a change of boundary conditions, such as a change of support conditions or loads. An example of this nonlinearity is the contact between a conductor and its guide as shown in Fig. 15.19. Under normal conditions, there is a gap between the conductor and its guide. During adverse weather, large sea wave loads along the horizontal direction can push the conductor to impact/contact its guide, and this contact immediately acts as an additional support on the conductor to restrain its deflection in the horizontal direction, leading to a significant increase of the conductor's stiffness.

15.2.5 Force Boundary Nonlinearities

Force boundary nonlinearities may be caused by the hydrostatic pressure load variation due to the variation of the wet surface, by the nonlinear drag forces induced by the passing fluid, such as the wind (Sect. 12.2.1), wave and current loads (Sect. 12.1.1) on structures, and also by the aforementioned follower loads due to geometrical nonlinearities.

15.2.6 Nonlinearities Due to Temperature Effects

Temperature dependence and degradation of mechanical material properties due to temperature change, such as elastic moduli, thermal expansion coefficients, and nonlinear stress-strain relations as a function of temperature etc., also exhibit nonlinear characteristics. An elaboration of them is omitted since they are beyond the scope of the current book.

15.3 Load Sequence Effects

During the service life of a structure, it may be subjected to a large number of loads (essential boundary conditions). The sequence/order of the loads and its application with respect to the occurrence of the plasticity, change of boundary conditions or stiffness can affect the responses of the structure, which is known as load sequence effects. In Ref. [247], Jia classified the load sequence effects encountered in structural engineering practice into four types, which are now summarized as three categories:

(1) Category 1 is associated with changes of the global and/or local stiffness. The changes of the stiffness may be due to the occurrence of plasticity, geometrical nonlinearities, cracking or damages in structures, the change of structures' support conditions, the installation of additional structural members and reinforcements.

In arctic engineering, the plasticity development of ships' hull structures or offshore structures under repeated ice floe impact [250] is an example that is relevant to load sequence effects associated with plasticity. Due to the sequence of the ice loads, it is still difficult to identify the exact cause of the structure's damage because the solutions for these types of inverse problems (knowing the consequences, causes are sought) are not unique.

The load sequence effects due to a combination of material and geometrical nonlinearities are shown by the load–deflection curve in Fig. 15.16. Obviously, the global stiffness at different stages of loads are changing.

Figure 15.20 shows the plastic utilization of a subsea tank after the permanent load is applied, with and without crack modeling. It is clearly shown that the cracks significantly decrease the structure's stiffness and induce local plasticity in the tank, i.e., the load sequence effects cannot be neglected.

The load sequence effects due to the installation of additional structural members and reinforcements may be difficult to understand. When the reinforcements or structural members are installed without pre-stressing them to the original structure, the reinforcements or the members should be considered to only carry future changed (added or deducted) loads together with the existing structural members. By performing a series of linear elastic analyses for an offshore topside frame structure shown in Fig. 15.21, which is subjected to various permanent and environmental loads, and both the support condition changes and the reinforcements are applied at different stages of the structure life, Jia [247] investigated the ultimate strength of the frame structure influenced by the sensitivities of the support condition changes and the reinforcements installations. It is found that the changes of the support conditions and the added stiffness from the reinforcements only slightly change the global stiffness of the structure, while they do influence the local stiffness and reaction forces on reinforced members and

15.3 Load Sequence Effects

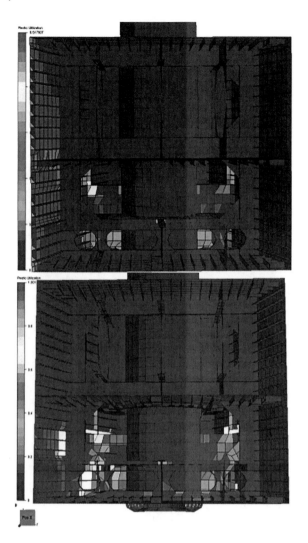

Fig. 15.20 The plastic utilization of a subsea tank without (*upper* figure) and with (*lower* figure) modeling of cracks in the tank (courtesy of Dong Energy, Denmark)

members close to the supports and reinforced members. The load and support condition sequence effects may have a significant influence on the utilizations of those members. For this particular structure, the load sequence effects are more significant from the change of the support condition than that from the installation of the reinforcements.

In order to practically account for the load sequence effects, in the same reference, Jia [247] also presented a method for specifying the consequent analysis phases that depend on factors such as whether the loads are acting at the time the structure is altered, whether only the final state of the structure is of analysis

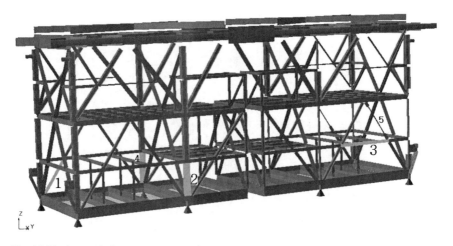

Fig. 15.21 A *topside* frame structure under investigation (courtesy of Aker Solutions)

interest or intermediate states are as well, etc. In general, the occurrence of stiffness changes should be considered as the "boundary" between analysis phases. In each phase, the pertinent loads and support conditions should be applied on the structure model corresponding to the condition in that phase. In addition, two types of loads should be applied with care: (1) the environmental loads varying with time, especially for the wave and wind loads, which always exist and vary during the entire lifetime of the offshore installations: they are recommended to be applied in the final phase/time of the analysis. (2) The operating loads (e.g., drilling machinery loads, moving crane loads) moved in different locations within structures in different analysis phase: they should be applied in the way that the relevant loads being applied in the current analysis phase should first include the operating loads with the same location and magnitude but in the opposite direction (negative sign) as the previous analysis phase; by doing this, the operating load effects from the previous analysis phase on the current analysis phase are "canceled". Then the new operating loads corresponding to the current analysis phase can be applied.

(2) Category 2 is concerned with the load sequence effects on the fatigue damage estimation under variable amplitude of fatigue loadings and the transition from the variable amplitude of fatigue loadings in reality into the constant amplitude of fatigue loadings in laboratory tests.

In engineering practice for fatigue estimations, load cycle counting by the rainflow method [243] (Sect. 17.3.5.1), the Miner rule (Sect. 17.2.5) defining the linear damage accumulation and the S–N curve (Sect. 17.2.2) representing the material performance determined from constant-amplitude fatigue tests are often used for the fatigue damage estimation of structures subjected to random loadings. Although in most cases this is the best available method, the accuracy of it remains questionable. The selection of the damage models is highly dependent on the load time history being considered. The widely adopted Miner rule may lead to a

15.3 Load Sequence Effects

conservative estimation when the load history contains an isolated overload. This is also the case when the nonlinear damage model is adopted. However, when the load history comprises large cycles separated by few small cycles, and the mean of the small cycles is highly relative to the mean of the large cycles, the fatigue estimation by using Miner rule, the interaction models or the nonlinear models will be non-conservative [305].

(3) Category 3 is the combination of the two categories of load sequence effects described above.

15.4 Eigenfrequencies Influenced by Nonlinearities

15.4.1 Material Nonlinearity

15.4.1.1 Materials with Hardening Nonlinearity

For a beam with a hardening nonlinearity (please note that this hardening non-linearity is different from a strain hardening nonlinearity where the second item in the following equation does not exist in the elasticity range and the first item does not appear in the plasticity range, see the left figure in Fig. 15.22), one may assume that the material model can be expressed as:

$$\sigma = E\varepsilon + H\varepsilon^3 \tag{15.32}$$

where σ is the normal stress, E is the elastic Youngs' modulus, ε is the strain, and H is a parameter related to the significance of hardening.

The calculation of the natural frequency with material hardening can be found in Ref. [188]. Consider a simply supported beam shown in Fig. 15.23 and assume that the beam is massless but with a concentrated mass m located at the middle of the beam where the fundamental eigenmode has its maximum amplitude of vibrations. The differential equation of beam deflection can be expressed as:

$$\frac{d^2}{dx^2}\left(EI_2\frac{d^2z}{dx^2}\right) + 6HI_4\frac{d^2z}{dx^2}\left(\frac{d^3z}{dx^3}\right)^2 + 3HI_4\left(\frac{d^2z}{dx^2}\right)^2\frac{d^4z}{dx^4} + m\frac{d^2z}{dt^2} = 0 \tag{15.33}$$

where I_2 and I_4 are the second and fourth moment of inertia of the beam's cross-section with the area of A, i.e., $I_i = \int_A z^i dA$, z is the distance from the neutral axis of the cross-section; t is the time.

By assuming an approximate vibration mode shape $X(x)$, the deflection of the beam at position x and time t is $z(x,t) = X(x)\cos(\omega_n t)$, one then has:

$$\frac{d^4X}{dx^4}\left[EI_2 + \frac{9}{4}HI_4\left(\frac{d^2X}{dx^2}\right)^2\right] + \frac{9}{2}HI_4\frac{d^2X}{dx^2}\left(\frac{d^3X}{dx^3}\right)^2 - m\omega_n^2X = 0 \tag{15.34}$$

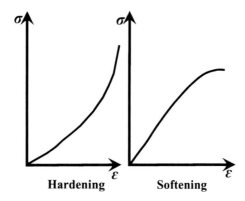

Fig. 15.22 Hardening (not strain hardening in plasticity) and softening nonlinearity

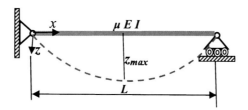

Fig. 15.23 A simply supported beam

The natural frequency can then be calculated as:

$$\omega_n = \frac{\pi^2}{L^2}\sqrt{\frac{EI_2}{m}} \frac{1}{\left(1 - \frac{27\pi^4 H I_4}{64 L^4 E I_2} z_{max}^2\right)^2} \tag{15.35}$$

15.4.1.2 Materials with Softening Nonlinearity

For a material with softening nonlinearity (right figure in Fig. 15.22), one may assume that the stress–strain relationship can be expressed as [233, 234]:

$$\sigma = E\varepsilon - \beta E^2 \varepsilon^3 \tag{15.36}$$

where β is a parameter related to significance of softening.

Considering the simply supported beam shown in Fig. 15.23. Again, by assuming an approximate vibration mode shape $X(x)$, the deflection of the beam at position x and time t is:

$$z(x, t) = X(\xi) Z(\tau) \tag{15.37}$$

where $\xi = \pi \frac{x}{L} \cdot \tau = \omega_n t$:

15.4 Eigenfrequencies Influenced by Nonlinearities

The detailed calculation of the natural frequency due to softening nonlinearity can be found in Ref. [188]. The differential equation for the time function $Z(\omega_n t)$ is:

$$\frac{d^2 Z}{d\tau^2} + \frac{v^2 a^2}{\omega_n^2} Z \left(1 - \frac{1}{3} \lambda b Z^2 \right) = 0 \tag{15.38}$$

where

$$a^2 = \frac{\pi^2 E I_2}{m L^4}$$

$$\lambda = \frac{3\pi^4 \beta E^2 I_4}{L^4 I_2}$$

$$v^2 = \frac{1}{b_0} \int_0^\pi \left(\frac{d^2 X}{d\xi^2} \right)^2 d\xi$$

$$b_0 = \int_0^\pi X^2 d\xi$$

$$b = \frac{1}{v^2 b_0} \int_0^\pi \left(\frac{d^2 X}{d\xi^2} \right)^4 d\xi$$

The natural period can then be calculated as:

$$T_n = \frac{2}{va} \left[1 + \frac{3}{8} \frac{\lambda b}{2} z_{max}^2 + \frac{57}{256} \left(\frac{\lambda b}{3} \right)^2 z_{max}^4 + \frac{315}{2048} \left(\frac{\lambda b}{3} \right)^6 z_{max}^6 + \cdots \right] \tag{15.39}$$

Table 15.1 lists the natural period of beams due to softening nonlinearity with various support conditions. In the table, z_{max} is the maximum deflection of the beam.

From the calculations above, it can be concluded that a hardening nonlinearity increases the natural frequency, while a softening nonlinearity decreases the natural frequency. In addition, it should be mentioned that both nonlinearities influence the mode shape.

15.4.2 Geometrical Nonlinearity

As explained previously, for a beam restrained against the in-plane deformation (geometrical nonlinearity), increase in stiffness can be expected due to the development of membrane forces at the finite deformation. Therefore, geometrical nonlinearity decreases the eigenperiod of a beam and also has an influence on the shape of eigenmodes.

Table 15.1 Natural frequencies of beams due to softening nonlinearity with different support conditions $\left(\lambda = \frac{3\pi^4 \beta E^2 I_4}{L^4 I_2}\right)$ [234]

Support conditions	Ratio of natural period/T_0 (T_0 is the natural period for the beam with only linear elastic material modeled)
	$1 + 0.09375\lambda z_{max}^2 + 0.013916\lambda^2 z_{max}^4$ $+ 0.002403\lambda^3 z_{max}^6 + \cdots$
	$1 + 0.47152\lambda z_{max}^2 + 0.35203\lambda^2 z_{max}^4$ $+ 0.30577\lambda^3 z_{max}^6 + \cdots$
	$1 + 0.009314\lambda z_{max}^2 + 0.000137\lambda^2 z_{max}^4$ $+ 0.0000023\lambda^3 z_{max}^6 + \cdots$
	$1 + 0.22645\lambda z_{max}^2 + 0.081193\lambda^2 z_{max}^4$ $+ 0.03387\lambda^3 z_{max}^6 + \cdots$

15.4 Eigenfrequencies Influenced by Nonlinearities

Table 15.2 Natural frequencies of beams accounting for geometrical nonlinearity with various support conditions [236]

Support conditions	Ratio of natural period/T_0 (T_0 is the natural period for the beam without geometrical nonlinearity effects)
	$1 - \frac{\pi^2}{8}\left(\frac{z_{max}}{L}\right)^2$
	$1 - \frac{1}{8}\left(\frac{z_{max}}{L}\right)^2$
	$1 - 1.55\left(\frac{z_{max}}{L}\right)^2$

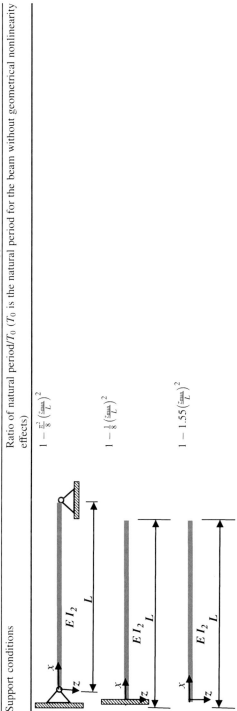

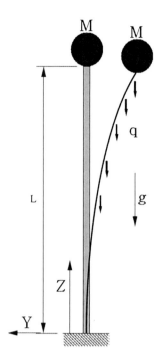

Fig. 15.24 Realization of a simplified structure modeling with P-Delta (P-Δ) effects involved: a vertical beam fixed at bottom with a concentrated top mass

Based on the differential equation involving the geometrical nonlinearity [235], the natural period of beams due to geometrical nonlinearity with various support conditions is listed in Table 15.2; the notations in Fig. 15.23 are used.

15.4.3 P-Delta (P-Δ) Effects

P-Delta (P-Δ) effects, sometimes referred to as the secondary moment effects, are due to the variation of a structure's lateral deformation in the horizontal plane. This changes the action point of the structure's resultant vertical force, and consequently induces additional actions on the structure and a change of its stiffness. This is illustrated in Fig. 15.24. Such changes in stiffness further alter the force distribution along the structure. It is obvious that the significance of P-Delta effects depends on the applied load and the structure's characteristics. For slender structures with significant weight and experiencing large lateral deflections, the P-Delta effects play an important role in increasing the effective load.

There are different procedures for including the P-Delta effects in an analysis, such as the second-order stiffness matrix, in which the stiffness matrix is the summation of the first-order stiffness matrix and the geometric stiffness (second-order

15.4 Eigenfrequencies Influenced by Nonlinearities

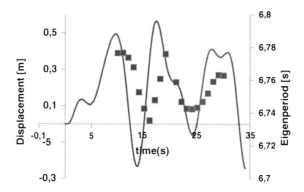

Fig. 15.25 Variation of natural period (*dot*) with the change of the displacement (*solid line*) at the top of a 100-m high offshore steel structure, which is subject to a 12-m high wave

stiffness) matrix. By applying equilibrium equation to the deformed shape of a beam-column element, the geometric stiffness matrix for a beam-column element can be written as a function of element length (L) and its axial load ($P = M \cdot g$) [237]:

$$K_P = \frac{P}{L} \begin{bmatrix} 0 & 0 & 0 & 0 & 0 & 0 \\ & \frac{6}{5} & \frac{L}{10} & 0 & -\frac{6}{5} & \frac{L}{10} \\ & & \frac{2L^2}{15} & 0 & -\frac{L}{10} & -\frac{L^2}{30} \\ & & & 0 & 0 & 0 \\ & Sym & & & \frac{6}{5} & -\frac{L}{10} \\ & & & & & \frac{2L^2}{15} \end{bmatrix} \quad (15.40)$$

For a structural system with many elements, the axial load on each element is an unknown and a decent nonlinear analysis algorithm is needed to solve the problem [238].

Except for the second-order stiffness matrix method, the negative stiffness method [239, 240] is also used to account for the P-Delta effects. In this method, The P-Delta effects are considered by either directly reducing the stiffness or indirectly introducing virtual elements in the structure. It is therefore possible to modify the stiffness matrix and include the global P-Delta effects in analysis by conducting a linear (first-order) analysis [241].

As the P-Delta effects change (normally decrease) a structure's stiffness, they also alter (normally lengthen) the eigenperiod. Figure 15.25 shows the variation of natural period with the change of the displacement at the top of a 100-m high offshore jackup structure. It is clear that the natural period increases with the increase of displacement (Delta).

However, for stiff structures with small deflection, the P-Delta effects are normally insignificant. Figure 15.26 shows the time history of the base shear, overturning moment and the resultant horizontal displacement at the center of the topside for an offshore jacket (presented in Sect. 6.3) subjected to wave loading

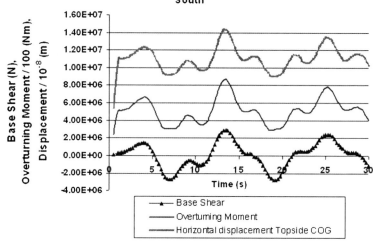

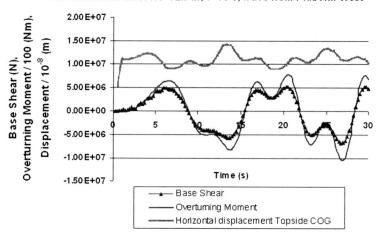

Fig. 15.26 The global base shear, overturning moment and horizontal displacement at the center of gravity of the topside due to waves from platform south (*upper* figure) and west (*lower*) with significant wave height of $H_s = 12.8$ m, and modal wave period of $T_p = 13$ s

from either platform south or west. It is noticed that the base shear, overturning moment and horizontal displacement vary almost in phase. Four time instants are selected for investigation: t = 2.0 and 4.0 s, when the global responses are small, and t = 13.75 and 19.25 s when the horizontal displacement responses approach peak values. Table 15.3 lists the calculated eigenperiods at those four time instants. It is shown that the calculated eigenperiod at different time instants are

15.4 Eigenfrequencies Influenced by Nonlinearities

Table 15.3 First three eigenperiods of the jacket-topside-flare tower structure (hydrodynamic added masses are included) for the cases with gravity and without gravity effects, waves from platform south or platform west

Eigenmode number	Wave platform direction	Eigenperiod (s)			
		2 s	4 s	13.75 s	19.25 s
1	Wave from South	4.173			
	Wave from West				
2	Wave from South	4.115			
	Wave from West				
3	Wave from South	2.453			
	Wave from West				

identical. This indicates that the P-Delta effects for this particular structure and the loadings are insignificant on tuning the structure's stiffness and eigenperiods. This is mainly due to the fact that the sea states (wave height and wave period) used for the current investigation does not induce any plastic structural deformation, i.e., no stiffness degradation due to plasticity, and the horizontal deflection of the structure also remains limited with a maximum value of 0.14 m, i.e., the stiffness changes due to geometric nonlinearities are very marginal. However, the current conclusion is based on the foundation modeling with a constant linear spring system in all six degrees-of-freedom. The P-Delta effects may have certain influence on tuning the foundation stiffness through the action of overturning moment and vertical loads. Therefore, if the topside deflection is higher under adverse sea state, and/or the structure is more slender, and/or if the nonlinear foundation stiffness is accounted for, the P-Delta effects may then be more relevant.

15.5 Numerical Solutions for Nonlinear Problem

15.5.1 Characteristics of Nonlinear Responses

The equation of motions with nonlinearities involved can often be expressed as:

$$\ddot{x}(t) + \omega_n^2 x(t) + \vartheta F(x(t), \dot{x}(t), t) = 0 \qquad (15.41)$$

where $F(x(t), \dot{x}(t), t)$ represents the nonlinear forces due to both internal and external effects; ϑ is a factor controlling the force.

It is noticed that if ϑ is small enough, the equation becomes quasi-linear, and the periodical solutions close to the ω_n exist. Several analytical methods can be adopted to find such solutions in an approximate manner, such as direct integration, free oscillation, Duffing's equation etc. [18]. Among these, the most widely used for both analytical and computer-based analysis is the direct integration method.

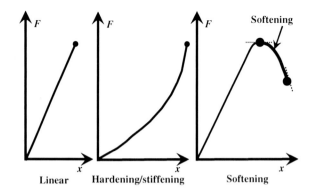

Fig. 15.27 Three basic types of nonlinear responses in the form of load-displacement paths

In order to have a reasonable understanding of solution algorithm for nonlinear responses, we will first review the characteristics of the nonlinear responses in a form of load-displacement paths. Figure 15.27 shows the three basic types of load-displacement responses, namely linear, stiffening and softening curve.

The linear curve (linear until fracture) represents the responses of structural members made of linear brittle materials, such as glass, crystals and many types of composite materials (comprising brittle fibers).

The stiffening curve shown in Fig. 15.27 illustrates the responses involving geometrical nonlinearity, such as cable structures and inflatable membranes, slender structures, metal and plastic forming, as mentioned before. This behavior is also illustrated in Fig. 15.15.

The softening curve represents typical behaviors of structural members made of ductile materials exhibiting plasticity behavior, such as aluminum, mild steel etc. It is commonly used in structural engineering.

It is noted that the tangent stiffness for both linear and stiffening curves at any state is positive, while it becomes negative for a softening curve after passing the peak load limit point, indicating an unstable equilibrium. Moreover, readers need to know that a positive stiffness is necessary but is only sufficient for stability for a single-degree-of-freedom system [249].

Based on the combination of linear, stiffening and softening curves shown in Fig. 15.27, one can reach various types of nonlinear equilibrium curves that can be realized in the physical world. The most used ones are snap through, snap back, and bifurcation curves.

Let's consider a shallow arch hinged at its two ends that is subjected to a concentrated load F in the vertical direction, shown in Fig. 15.28. As shown in Fig. 15.29, after reaching the load limit point, the path will experience a sudden drop of the load (unloading) from B to E, and these two points are two load limit points. The response portion between the two points has a negative stiffness and is therefore unstable. After passing the second load limit point E, the load increases again and the responses show a stiffening behavior (E–F—in Fig. 15.29). The snap through curve is actually a combination of softening (before load limit point E) and stiffening (after

15.5 Numerical Solutions for Nonlinear Problem   293

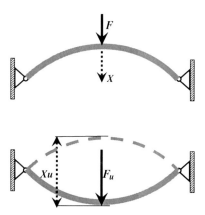

Fig. 15.28 An *arch* under transverse load F

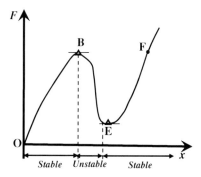

Fig. 15.29 Load-displacement path with snap through phenomenon (▲ denotes load limit points)

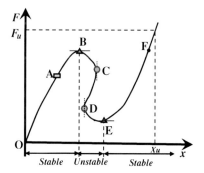

Fig. 15.30 Load-displacement curve with snap back phenomenon in nonlinear equilibrium paths (■ denotes bifurcation or stability point, ▲ denotes load limit points, and ● denotes displacement limit points/snap-back points)

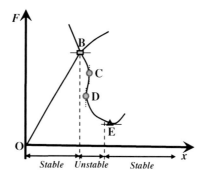

Fig. 15.31 Load-displacement path showing bifurcation (point B) combined with snap back phenomenon (■ denotes the bifurcation or stability point, ▲ denotes the load limit point and ● denotes the displacement limit points/snap-back points)

point E). The slightly curved structures such as a shallow arch under gradually increased transverse loads generally exhibit snap through responses [249].

More complicated than the snap through, the snap back (which is essentially an exaggerated snap through) curve illustrated in Fig. 15.30 has a response (displacement) turn back at the first displacement limit point (point C in Fig. 15.30). It possesses more characteristic points: load limit points (points B and E in Fig. 15.30) occur when a local maximum or minimum load is reached, with a horizontal tangent at these points. Displacement limit points, which are also referred to as snap back points or turning points (points C and D in Fig. 15.30), occur with vertical tangents on the curve. Between the two load limit points B and E, the structure is unloaded, indicating the occurrence of instability such as buckling. The structure exhibits softening between the points O and B, and stiffening after the load limit point E. Examples of structures that exhibit snap back behavior are trussed-dome, folded and thin-walled structures in which the "moving arch" effects occur [249], and cylindrical shells under compression etc.

For a member under axial compression load, in many cases, the bifurcation or buckling appears prior to the load-limit point. In addition to the load limit point, there is also a stability point (point A in Fig. 15.30) where the structure loses stability (e.g., buckling) or where bifurcation occurs (i.e., the solution switches to two or more branches) [251]. If this occurs for an axial loaded member, the strength after the bifurcation point cannot be utilized. When the bifurcation or buckling is presented, the load path may not be unique (shown in Fig. 15.31). An example of a structure that exhibits this type of behavior shown in Fig. 15.31: thin cylinder shells under axial compression [249].

As will be discussed later on, the trace of the nonlinear equilibrium path can be carried out by load control methods, which can only capture the path from O to B (the first load limit point) shown in Fig. 15.30, or displacement control method, which can normally capture the path from O to C (the first displacement limit point), or a combination of the two approaches such as the arc-length method, which can capture the entire path shown in Fig. 15.30.

15.5.2 Load Control (Newton-Type) Methods

Regardless of whether the problem is dynamic or static, as long as nonlinearity is involved, in the regime of load control methods, either an incremental procedure or an incremental-iteration is needed to find the equilibrium between the external and internal forces for each load increment. Hence, the methods are often referred to as load control. It should be noticed that, when the finite element method is used, the internal forces are actually the nodal forces that are equivalent to the element stresses.

15.5.2.1 Load Increment Procedure

Let's consider a generic load increment from t to $t + \Delta t$, and assume that a balance between external and internal forces is obtained at load level at t. As shown in Fig. 15.32, in an incremental procedure, the load is applied at relatively small increments and the structure is assumed to respond linearly within each incremental step. This method is convenient to implement and computationally efficient. However, with the increase of load, it diverges considerably from the actual equilibrium path (shown in Fig. 15.32) because the equilibrium may not be reached for every load step [251–253]. To circumvent this problem, incremental iterations have to be performed from load at t to $t + \Delta t$ until the force equilibrium is fulfilled for the load level at $t + \Delta t$. This procedure is denoted as incremental iterative [254, 255].

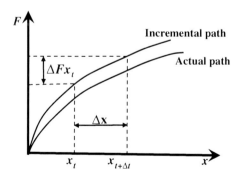

Fig. 15.32 Incremental procedure

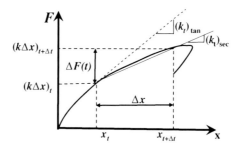

Fig. 15.33 Incremental solution of equation of motions

For obtaining the dynamic responses involving nonlinearities, one has to first formulate the difference of the governing equation of motions for a system at two adjacent time instant t and $t + \Delta t$:

$$m \, \ddot{\Delta x}(t) + c \, \dot{\Delta x}(t) + k \Delta x(t) = \Delta F(t) \tag{15.42}$$

The stiffness k in the equation above is the secant stiffness $(k_t)_{sec}$. For a nonlinear system, the stiffness depends on the displacement, and it can thus only be determined by two points at t and $t + \Delta t$ as shown in Fig. 15.33. However, the displacement at $t + \Delta t$ is an unknown. Therefore, one has to replace the secant stiffness by the tangent stiffness $(k_t)_{tan}$ at t. Due to the adoption of tangent stiffness, and the constant time interval/step also delaying the detection of transitions in the force–deformation relationship (the reversal of velocity causes the departure from the exact load-deformation path), significant errors may be introduced when solving the equation of motions [125]. To minimize the errors, load increment control using an iterative procedure can be used to obtain the solutions. This is done by rewriting the equation above and make it compatible with the Newmark method:

$$K_{tan} \Delta x(t) = \Delta F(t) \tag{15.43}$$

$$\text{where } K_{tan} = (k_t)_{tan} + \frac{\delta}{\beta \Delta t} c + \frac{\delta}{\beta (\Delta t)^2} m$$

δ and β are two constants that define the variation of acceleration over a time step, and are related to the integration accuracy and stability. c and m are damping and mass respectively. Furthermore, the equation above is based on a constant damping assumption, which in most cases can be justified from an engineering point of view.

Obviously, the nonlinear equations above can also be applied to static problems by replacing the indices of time (t) with a load step number.

Since a direct method cannot be used for solving the nonlinear equation above, within each time step or load step, iterative procedures must be adopted to solve the equation of motions. This involves a guess on an unknown at the beginning of the first iteration. Obviously, this iteration will introduce error, which in turn is essential to adjust the original guess value in the first iteration and for the second iteration. If the error is rather high, the time step (in dynamic analysis) or load step (in static analysis) can even be divided into smaller steps. This procedure continues for each time step or load step until a full load is applied on the structure model.

15.5.2.2 Newton–Raphson Method

The most popular interactive procedures are the Newton–Raphson method and modified Newton–Raphson method. For both methods, the tangent stiffness matrix is formed at the beginning of each time step and the equilibrium iterations provide convergence within specified tolerance limits at the end of each time step. While within each time step, based on a function of current internal forces and deformation state in each iteration, the formal one uses the tangent stiffness matrix in the previous iteration to calculate the current deformation using a linear solution. The calculated deformation is projected back onto the load deformation curve that is in parallel with the load axis, and the latter one uses an identical stiffness for each iteration throughout this time/load step, as illustrated in Fig. 15.34.

For the Newton–Raphson method, imagine that a load applied at time t is $F(t)$ at point A (start of each time step) in Fig. 15.34 and it is then increased to

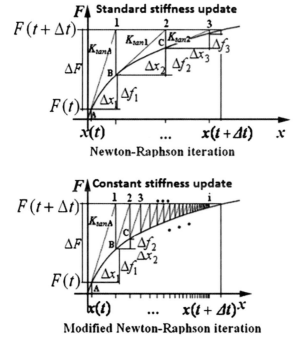

Fig. 15.34 Iteration within a time step for nonlinear systems with Newton–Raphson method (*upper*) or modified Newton–Raphson method (*lower*)

$F(t + \Delta t)$ at time $t + \Delta t$. By using the tangent stiffness $K_{\tan A}$ at point A, the first iterative step can be performed as:

$$K_{\tan A} \Delta x_1 = \Delta F \tag{15.44}$$

Associated with Δx_1 is the true force difference Δf_1 between points B and A, which is obviously less than ΔF. This then gives a residual force (load imbalance):

$$\Delta R_1 = \Delta F - \Delta f_1 \tag{15.45}$$

Again, it is worth mentioning that in the finite element analysis, the load corresponds to the element stress.

By using this residual force, one can further obtain an additional displacement Δx_2 by using tangent stiffness $K_{\tan 1}$ at point B and perform the second iterative step:

$$K_{\tan 1} \Delta x_2 = \Delta R_1 = \Delta F - \Delta f_1 \tag{15.46}$$

Again, this additional displacement Δx_2 is used to find a new value of residual force:

$$\Delta R_2 = \Delta F - \Delta f_1 - \Delta f_2 \tag{15.47}$$

The process above is repeated until the residual force is small enough (convergent) to fulfill the convergence criteria:

$$\Delta R_i = \Delta F - \sum_{i=1}^{n} \Delta f_i < \varepsilon \tag{15.48}$$

ε must be carefully chosen by considering both the accuracy and computation cost. If it is too large, the calculation may give inaccurate results. If it is too small, considerable computation efforts are required that give unnecessary accuracy.

15.5.2.3 Modified Newton–Raphson Method

It is noticed that, for systems with a large number of degrees-of-freedoms, the Newton–Raphson iteration may require a prohibitive amount of computation efforts due to the calculation/updating of the tangent stiffness in each iteration, even though this frequent updating of stiffness leads to a quadratic (i.e., fast) convergence in solution. The modified Newton–Raphson method is then introduced to overcome this difficulty. In this method, within each time step, the initial tangent stiffness $K_{\tan A}$ at point A in Fig. 15.34 is used throughout the iterations. Therefore, there is no need to recalculate the stiffness within each time/load step.

In the modified Newton–Raphson method, the choice of time/load steps (readers need to distinguish the time/load step from the load increment within each time step) when the stiffness should be updated depends on the degree of nonlinearity in the system, i.e., the more nonlinear the responses are, the more updating (i.e., smaller time/load steps) should be performed. However, as aforementioned and shown in Fig. 15.34, in general, the Newton–Raphson method converges much faster (quadratically) than the modified Newton–Raphson method.

The determination of the time/load step (ΔF in Fig. 15.34) is an essential parameter for load-control methods. As mentioned before, generally, large time/load steps are allowed where the equilibrium path is almost linear and smaller steps are normally used where the path is highly nonlinear. Furthermore, the time/load step also depends on the objective of the calculation. If one wants to study the entire equilibrium path accurately, it is recommended to use small time/load steps, while if only the responses at a limit state load level are of interest, larger time/load step can be used until the applied load is close to a load limit point. Finally, the choice of time/load step is also related to the iteration algorithm used. For example, as discussed previously, due to the utilization of an initial or non-frequently updated stiffness in the modified Newton–Raphson method, a smaller time/load step is required than that used in Newton–Raphson method, which is also shown in Fig. 15.34.

It should be noticed that, for load control methods, the hardening of structures is usually more difficult to analyze than that of softening, primarily due to the fact that the relevant iterative processes are likely to converge slowly or fail to converge [223].

Many advances in solving the nonlinear problem consist of variations of the Newton–Raphson method. For in-depth knowledge in this topic, readers may read references [257, 258].

For a more elaborated discussion of the force controlled numerical iteration method for nonlinear dynamic analysis, references [125, 173, 223] are recommended.

15.5.3 Displacement Control Methods

Note that the Newton–Raphson method or the modified Newton–Raphson method may perform poorly when the buckling of a structure is involved, in which the slope at limit points is 0. Furthermore, even though the load control method is capable of reaching the load limit point (point B in Fig. 15.36), passing it is impossible regardless of how small an increment is used.

Again, let's consider an arch hinged at its two ends that is subjected to a concentrated load F in the vertical direction, shown in Fig. 15.35. After reaching the first load limit point B, if one adopts load control methods, the path will experience a sudden "snap" through B directly to F as shown in Fig. 15.36, for which the unloading path (B–E) and the reloading path (E–F) are lost.

Therefore, to further capture more information with respect to the load path, one has to shift from load control to displacement control methods. For many cases, the displacement control methods can capture the entire equilibrium path [259, 260].

Analogous to the Newton–Raphson methods with the utilization of a load parameter, the displacement methods use a displacement component as the control parameter to trace the equilibrium path.

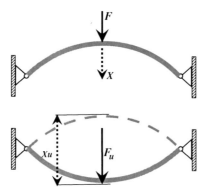

Fig. 15.35 An arch under transverse load F

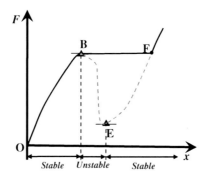

Fig. 15.36 Load-displacement path (*solid line*) when using load control methods (▲ denotes the load limit points), the path shows a snap through

However, by using a displacement method, it is difficult to obtain the equilibrium path after a displacement limit point appears, such as the ones that occur at snap back equilibrium path shown in Fig. 15.30. To solve this problem, a combined load-displacement control method has to be adopted, which will be discussed in Sect. 15.5.4.

For more details on displacement control methods, readers may read references [251, 256, 259–261].

15.5.4 Load-Displacement Control Method—Arc-Length Method (ALM)

Note that the load or displacement method keeps either the external load or displacement constant through iterations, and this introduces difficulties at load or displacement limit points, respectively. To overcome this problem, it is feasible to simultaneously change both the load and displacement levels along the

15.5 Numerical Solutions for Nonlinear Problem

incremental-iterative process [251], leading to a combined load-displacement control. It is more robust and particularly suitable for searching the collapse behavior of a structure.

In load-displacement control methods, the load is normally treated as an additional variable, so that the equilibrium configuration can be followed beyond limit points [251].

Among all the load-displacement control methods, the Arc-length method (ALM, also called Riks method [262]) is the most popular one. Unlike the Newton–Raphson method, during each time/load step, the load-factor at each iteration is treated as a variable and is modified so that the solution follows a specified path until convergence is achieved. Generally, for the first increment, the trial value of the load-factor is assumed to be 1/5 or 1/10 of the total load.

Basically, the method simultaneously controls both the load and displacement increment by imposing a constraint where a constant Euclidean norm Δs_i (arc-length) of the increment (f_i and x_i) is prescribed at the beginning of each time/load step t, and Δs_i keeps constant for every load increment iteration within that time/load step ($t \rightarrow t + \Delta t$). This procedure is illustrated in Fig. 15.37 and expressed as:

$$(\Delta s_i)^2 = x_i^2 + \psi f_i^2 \qquad (15.49)$$

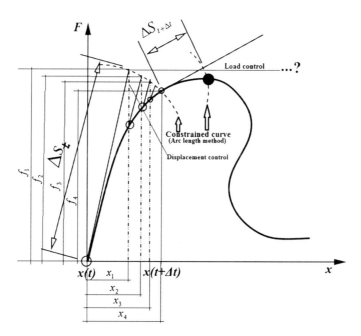

Fig. 15.37 Iteration within a time step ($t \rightarrow \Delta t$) for nonlinear systems using *Arc*-length method for a one dimensional problem, ΔS_t and $\Delta S_{t+\Delta t}$ are the constant *arc*-length for time step t and $t + \Delta t$

where ψ is a non-negative real parameter, the value of which depends on which type of arc-length method is used.

By observing Fig. 15.37, it is noticed that after an initial arc-length (ΔS_t) is determined, the subsequent load and displacement iterations (f_i and x_i) follow a constrained curve determined by ΔS_t.

All arc-length methods consist of a prediction phase to determine the arc-length and a correction phase to find a new point in the equilibrium path based on the incremental form of the equation of motions and the constraint equation. For more details of the Arc-length method, references [246, 251] merit attention.

Chapter 16
Structural Responses Due to Seismic Excitations

16.1 Seismic Ground Motions

16.1.1 Transmission of Seismic Wave from Bedrock to Ground

The determination of the earthquake ground motions is an essential part of earthquake engineering. However, as the rate and duration of energy released from earthquakes is relatively random, it is not possible to obtain the exact excitation to which the structure will be subjected. Extrapolating the effects of the energy released from sites of potential seismic activity to the location of the structure under investigation is a rather complex process. Four major characterizations enable the determination of the earthquake ground excitation (seismic input): the seismic sources (i.e. the rupture mechanism at sources); the transmission of the excitation from the sources to the sites (i.e. wave propagations); the local geotechnical effects on the motions of the soil; and soil-structure interactions (SSI) during the earthquakes. Therefore, the level of ground shaking is essentially influenced by the seismic source excitations, distance of source to site and local soil effects. Figure 16.1 shows the typical transmission from the seismic source (bedrock) to the earthquake ground (seabed) excitation.

Generally, three types of computation tools are available for calculating strong ground motions: physics-based models, stochastic models and combined (hybrid models). Physics models calculate ground motions by modeling the fault rupture, the resulting wave propagation, and the near-surface site amplification, and they require precise information about the earthquake source, wave propagation path, and SSI, leading to high computation efforts. This brings especial difficulties when calculating low period motions. Without the modeling of the fault rupture, wave propagation and SSI, the stochastic models directly calculate the ground motions, and therefore require less computation efforts and are equally applicable for both high and lower period motions. While the conventional stochastic models are based on the Gaussian process assumption, for motions with significant change with time, the time localized nonstationary process cannot be captured by the stochastic models. The drawbacks of both methods promote the combined models,

J. Jia, *Essentials of Applied Dynamic Analysis*, Risk Engineering,
DOI: 10.1007/978-3-642-37003-8_16, © Springer-Verlag Berlin Heidelberg 2014

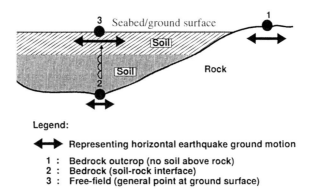

Fig. 16.1 Illustrations of transmissions from the seismic source (bedrock) to the earthquake ground (seabed) excitation [263]

which utilize the physics-based models for computing high period components of motions and stochastic models for lower period components through frequency filter [264].

16.1.2 Resonance Period of Soil—Site Period

During the seismic wave propagation from the bedrock to the ground as shown in Fig. 16.1, the soil media acts as a filter to bedrock motions and influences both the frequency and the duration of the ground motions. If the soil is stiff, e.g. the foundation is founded on rock, the ground motions will generally be short period, and vice versa. If the natural period of soil is close to the predominant period of bedrock motions, the ground motions will be amplified compared with the bedrock motions. Furthermore, such amplification may be further magnified if the natural period of the structures at the soil sites is close to the period of ground motions. To avoid double resonances (resonance of seismic wave in soil and then again resonance with the natural period of structures), in a preliminary design stage of a structure engineers should design a structure with the natural period far away from (normally above) the natural period of ground motions, which is also called the site period. By assuming the soil is elastic and the bedrock displaces dominated by shear motions, the site period can be calculated as:

$$T_{site} = 4H/v_s \qquad (16.1)$$

where H (in meters) is the depth of soil layers. v_s is the shear wave velocity of soil, which varies with different soil type and depth.

For a rough estimation, the wave shear velocity in soil media can be assumed to be 300 m/s. Therefore, for a soil with thickness of 30 m, the site period is 0.4 s ($= (4 \times 30)$ m/300 m/s). At this period, both the ground motions and structures with shallow foundations above the soil will vibrate at an amplitude much greater

16.1 Seismic Ground Motions

than the bedrock motions. This can be clearly shown by reading the response spectrum as elaborated in Sect. 16.2. However, for rather loose and soft soil, the shear wave velocity can be as low as $60 \sim 80$ m/s, then the above example gives a site period of only $1.5 \sim 2.0$ s, which is in the possible range of the natural vibration period for an offshore platform.

Site amplification can be the cause of abundant structural damages during large earthquake events. An example of this was the massive damage in downtown Mexico City during the 8.1 magnitude Mexico City earthquake 1985, which is discussed in Sect. 1.1.

For layered soils, several empirical relations [265, 266] are available for calculating the site period:

1. Based on the weighted average of shear wave velocity:

$$v_s = \left(\sum_{i=1}^{i=n} v_{si} H_i \right) / H \tag{16.2}$$

where v_{si} and H_i are shear velocity and depth of the soil layer i, n is the total layer of the soil, $H = \sum H_i$.

2. Based on the weighted average of the soil's shear modulus and density:

$$T_{site} = 4H / \left[\left(\sum_{i=1}^{i=n} \mu_i H_i \right) / \left(\sum_{i=1}^{i=n} \rho_i H_i \right) \right] \tag{16.3}$$

where $\mu_i \rho_i$ and H_i are shear modulus, density and depth of the soil layer i, n is the total layer of the soil, $H = \sum H_i$.

3. Based on the sum of site period for each layer

$$T_{site} = \sum_{i=1}^{n} (4H_i / v_{si}) \tag{16.4}$$

4. Based on the linear approximation of fundamental mode shape

$$T_{site} = 2\pi / \sqrt{3 \sum_{i=1}^{i=n} (v_i^2 H_i) / H^3} \tag{16.5}$$

For seismic design of offshore and land-based structures, based on the soil properties in the uppermost 30 m of the soil, ISO 19901-2 [267] and Eurocode 8 [268] provide the estimation value of v_s.

16.1.3 The Amplitude and Duration of Bedrock Motions

The amplitude and duration of bedrock motions are also an important factor influencing the soil responses and the subsequent ground motions. High amplitude of bedrock motions tends to cause inelasticity in the soil, i.e. the soil will absorb a

Table 16.1 The duration of ground motions associated with various levels of earthquake magnitude

Richter Magnitude	Ground motion duration (second)
8–8.9	30–180
7–7.9	20–130
6–6.9	10–30
5–5.9	2–15
4.4.9	0–5

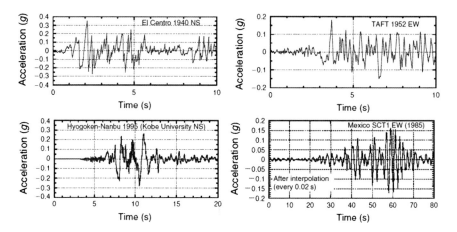

Fig. 16.2 Ground motion records from El Centro NS 1940, Taft EW 1952, Hyogoken-Nanbu, Kobe University NS 1995 and Mexico Michoacan SCT1 EW 1985 [73]

large amount of seismic wave energy, and the ground motions are then not amplified proportionally to the bedrock motions any more. Compared to a case in which the soil is elastic, this will in general decrease the ground accelerations while increase the displacements, posing higher demand on structures with medium and long natural periods. Examples of such type of structures are typical offshore platforms, long span bridges and high-rise buildings.

Long duration bedrock motions induce a large number of cycle loadings, and may cause a degradation of soil stiffness and strength and a significant increase of pore water pressure, leading to the liquefaction of saturated and partially saturated soil through the loss of cohesion [1], which is due to the loss of grain-to-grain contact in loose sandy soils media, causing the flow of material and the loss of strength and stiffness, which makes the soil behave more like a liquid than a solid. This has been occurred in many earthquakes, for example during the Niigata earthquake of 1964 as shown in Fig. 1.4.

Depending on the time required to release accumulated strain energy, the duration of ground motions is correlated to the length or area of the fault rupture. Therefore, with the increased magnitude of an earthquake, meaning an increased rupture size, the duration of the resulting ground motions also increases. Table 16.1 shows the duration of ground motion corresponding to various levels of earthquake magnitude [269]. It is also worth mentioning that during the 9.0

magnitude Sendai earthquake in 2011, strong shaking is reported to have continued for a duration of up to 5 min [270].

However, due to the variability of individual earthquake characteristics, the durations of two earthquake events can be significantly different. This is illustrated in Figs. 16.2 and 9.4, which show the ground motion records from different earthquake events.

16.1.4 Spatial Variation of Earthquake Ground Motions

For extended-in-plan structural systems (both above ground and buried ones) covering a large area such as bridges (Fig. 16.3), tunnels and dams, large gravity base structural foundations, structural system comprising several structures

Fig. 16.3 A bridge approach span collapse during the 7.2 magnitude 1995 Kobe earthquake (photo courtesy of Kobe Collection, EERC Library, University of California, Berkeley)

Fig. 16.4 Several offshore jackets and tripod structures connected by bridges at Ekofisk fields of North Sea (courtesy of Aker Solutions)

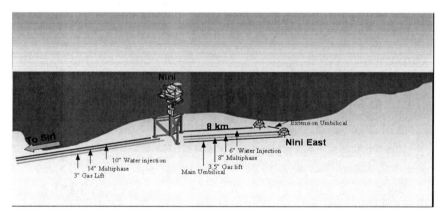

Fig. 16.5 Layout of the subsea pipelines of the Nini field on the Danish Continental Shelf (courtesy of Dong Energy, Denmark)

(Fig. 16.4), pipelines (Fig. 16.5) etc., the earthquake ground motions will arrive at different support locations at different times. During this arriving time period, the amplitude, frequency content and the phase (arriving time) of the motions are likely to change depending on the distance between the support points and the local soil conditions, local seismic motions may be enhanced with out-of-phase displacements at various locations, and different parts of the structural system may respond asynchronously with large displacement responses. This phenomenon is called spatial variation of earthquake motion (SVEGM).

16.1 Seismic Ground Motions

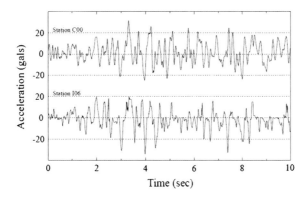

Fig. 16.6 Recorded seismic ground motion at two locations 200 m apart [271]

Figure 16.6 illustrates the recorded accelerograms at two locations 200 m apart [271, 272]. It shows that at some time instants, the accelerations at the two locations are different in both amplitude and phase. In an extreme case, the difference in motion amplitudes due to the SVEGM is comparable with the input ground motion amplitudes [273].

Numerous damages of extended land-based structures have been observed as a possible consequence of SVEGM, such as the falling off of Gavin Canyon Bridge's under-crossing and the span unseating in the 1994 Northridge earthquake [274], the unseating of deck and spans falling off (due to large longitudinal displacements between piers) at elevated highway bridges (Fig. 16.3) during the Kobe earthquake 1995 [275] etc.

It is noted that, due to the effects of SVEGM, both horizontal and vertical motions may not only induce the bending of structures, but also exert large torsional moments between the principal structural planes. If one of the relevant ground motion frequencies is close to the structures' natural frequency, resonance will occur with a magnification of structures' responses.

The simulation of SVEGM is presented in Sect. 12.4.2 by using a coherence function. For a more detailed account of SVEGM, readers may read references [139, 276] and those cited in the description above.

16.2 Seismic Response Spectrum

16.2.1 Introduction

Once the ground excitations for a structure have been determined, depending on the seismic analysis demand with respect to characteristics of excitation and structures, and also with the analysis purpose and accuracy requirement, the ground motions must be represented in certain formats using either a deterministic (time series or response spectrum) or a stochastic (energy spectrum) way. They can then be applied to a structure model representing both the superstructure together with the foundation and the effects of surrounding water (for offshore structures) and/or ice. Depending on the level of excitations accounted for, many of the properties during structural modeling may be nonlinear in nature, such as the degradation of foundation stiffness, yielding and large deformations of the structural members, fluid–structure interactions (for offshore structures). This can lead to considerable demand on structural modeling and calculation algorithm. By doing so, the structural responses can be calculated with a reasonable accuracy and a reliable demand modeling can be established.

After demand modeling is established, capacity control can then be carried out. If the structural responses are still in the elastic range, demand can be expressed as force and capacity control in terms of strength. If the structural members reach yielding, demand can normally be expressed with displacement, and the capacity control is then strain based. As strain-based control may need to reflect cyclic degradation and strain rate effects, dynamic testing is therefore required to set appropriate limits. In addition to the strength and strain capacity, the stability check should also be included in the capacity control.

Since the earthquake ground motions are by nature of short duration, non-stationary, transient and non-periodical, and also broad-band in frequency content, they can never reach steady-state vibrations. This means that even if the structure has a zero damping, the motion amplitudes are limited to finite values. Therefore, stochastic based root of mean square responses, which are utilized by the power spectrum method, are in many cases not appropriate to represent the earthquake ground motions. A deterministic time history analysis is desired to estimate the responses, but it requires considerable computational efforts for a complex structure. Moreover, for design purposes in a conservative manner, only the maximum amplitude of the response time history is needed for carrying out the seismic analysis. This also leads to a special consideration for combining each individual component of the responses.

The discussions above promote the utilization of seismic response spectrum to evaluate the seismic responses.

16.2.2 Construction of Response Spectrum

In practice, a single degree of freedom (SDOF) system with a constant damping ratio of a few percent representing the target structure is subjected to ground excitation time histories. As the ground motions typically excite a large number of vibration modes of the upper structure, by varying the natural period (T_n) of the SDOF in a range of engineering interest for the structure's eigenperiod (e.g. 0.02–10 s), and using numerical time integration methods such as Newmark's method etc., the maximum calculated relative displacement magnitude of the SDOF at each natural period is then plotted on a response spectrum graph. This

16.2 Seismic Response Spectrum

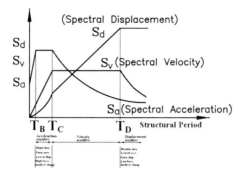

Fig. 16.7 Comparison of earthquake design spectrum shapes with acceleration, velocity and displacement measures

spectrum is often referred to as the deformation/displacement response spectrum. The value of its ordinate S_d is expressed in Eq. (16.6) and Fig. 16.7:

$$S_d = \max\left(u(t|T_n)\right) \tag{16.6}$$

By assuming harmonic vibrations at each natural period, the velocity spectrum S_v (also called pseudo relative velocity spectrum, denoted PSV) and acceleration response spectrum S_a, shown in Fig. 16.7, are expressed as:

$$S_v = \omega S_d = \left(\frac{2\pi}{T_n}\right) S_d \tag{16.7}$$

$$S_a = \omega^2 S_d = \left(\frac{2\pi}{T_n}\right)^2 S_d \tag{16.8}$$

It is noticed that the acceleration spectrum indicates the acceleration due to ground motions, without the involvement of the constant acceleration of gravity. By reviewing the concept of acceleration, velocity and displacement, it should further be emphasized that the spectra accelerations are absolute accelerations of a structure in space, since the force causing the acceleration itself is determined by the relative compression/extension of the spring with respect to the ground motions, while the spectral velocity and spectral displacement are relative values with respect to the moving ground [403].

Essentially, the seismic response spectrum is distinguished from a power spectrum in that the power spectrum is of a stochastic nature while the analysis method generating a seismic spectrum is deterministic. This is because the response spectra are generated from deterministic time history responses [23]. The unit of the ordinate for a response spectrum should exhibit the same unit as the name of the spectrum stands for, i.e., acceleration spectrum has a unit of acceleration, velocity spectrum has a unit of velocity, etc.

16.2.3 Modal Combination Techniques for Response Spectrum Analysis

In order to obtain the peak modal responses for a multi-modal period structural system and calculate the responses such as forces and displacements, an appropriate method for combining the modal response at each individual period must be adopted. Three popular methods used by engineers to perform modal combination are:

(1) Sum of absolute value: by assuming that the maximum modal values for all modes occur at the same time and their algebraic sign can be neglected, the straightforward but most conservative method is to sum the absolute modal response value together, expressed as:

$$R = R_1 + R_2 + \cdots + R_n \qquad (16.9)$$

where $R_1\ R_2 \cdots R_n$ are peak modal response, i.e., peak response in each mode. It can be a displacement, force, stress etc.

It is obvious that this method is over-conservative, and is not popular in engineering applications.

(2) Square Root of the Sum Squares (SRSS) method: by assuming that the maximum modal values in each individual mode are statistically independent and randomly phased, the peak response can then be calculated as SRSS:

$$R = \sqrt{R_1^2 + R_2^2 + \cdots + R_n^2} \qquad (16.10)$$

This combination rule provides excellent response estimates for systems with well separated natural period. However, for a structure with large degrees of freedom or high redundancy, such as complex offshore topside modules, complex piping systems and multistory buildings with asymmetric plans etc., in which a large number of modes appear at close or even identical frequencies, the use of SRSS method is then limited [124].

The SRSS method can be non-conservative when the eigenfrequencies of two modes are closely spaced, i.e. the two modes are coupled and the responses due to the two modes are dependent. In this case the method can be modified by using absolute summation of all modes whose eigenperiods are less than 10 % apart and the SRSS of those and all other modes [81], which is expressed as:

$$R = \sqrt{\left(R_1 + R_2\right)^2 + \ldots + R_n^2} \qquad (16.11)$$

where R_1 and R_2 are peak modal responses of closely spaced modes.

(3) Complete Quadratic Combination (CQC) method: in order to account for potential phase correlation when modal frequencies are close, as a replacement

16.2 Seismic Response Spectrum

for the SRSS method, the most popular method used nowadays is the Complete Quadratic Combination (CQC) method [404], which overcomes the limitations of SRSS methods by including cross modal contributions:

$$R_k = \sqrt{\sum_{i=1}^{N}\sum_{n=1}^{N} \rho_{in} R_{ik} R_{nk}} \qquad (16.12)$$

where ρ_{in} is the correlation coefficient of mode i and mode n, it varies between 0 and 1 for $i = n$. Therefore, the equation above can be rewritten as:

$$R_k = \sqrt{\sum_{n=1}^{N} R_{nk}^2 + \sum_{i=1}^{N}\sum_{n=1}^{N} \rho_{in} R_{ik} R_{nk}} \quad \text{for } i \neq n \qquad (16.13)$$

If the eigenfrequencies of structures are well spaced, i.e., the modal responses are uncorrelated, the off-diagonal terms of ρ_{in} tend to be zero ($\rho_{in} = 0$ for $i \neq n$), and the CQC method approaches SRSS method [81].

It should be noticed that the response parameters must be always calculated directly by summing the response parameter of interest, and cannot be calculated from other response parameters [81]. For example, the peak response at the bottom of a jacket structure must be calculated by combining peak modal response at the jacket bottom. It cannot be calculated by summing up the peak forces at each elevation of the jacket. This is because the modal combination does not contain the information of signs.

The SRSS and CQC methods would be most accurate for ground motions with wide-banded frequencies and long phases of strong ground motions, which are several times longer than the natural period of structures having not too light damping (>0.5 %). If the ground motions are short duration impulsive or contain many cycles of essentially harmonic excitations, the two modal combination methods will become less accurate [124].

Furthermore, based on the random vibration assumption, the peak response R can be interpreted as the mean of the peak values of responses to an ensemble of ground motions. Therefore, the two modal combination methods are well applicable by using a smoothed design response spectrum, which is calculated from the mean or median or even more conservative spectra (e.g. mean plus one standard deviation spectrum) obtained from many individual ground motion histories. The calculated peak response may be either conservative or non-conservative, but is generally within a few percent. However, if the two combination methods are used for calculating peak response due to a single ground motion characterized by a jagged response spectrum, the errors are larger, in the range of possibly 10–30 %, depending on the natural period of a target structure [124].

Other modal combination rules are also available, such as the A. Der Kiureghian [404, 405] or Rosenblueth-Elorduy [406] methods. Interested readers may further read relevant literatures.

16.3 Characteristics of Seismic Responses Varying with Frequencies

Different from dynamic loads due to wind, wave and ice impact on structures, the loads generated on a structure during an earthquake are in a sense purely due to the inertia of the structure, which is caused by the acceleration of the structural masses. The acceleration is the sum of the ground acceleration and the acceleration of the structural masses relative to the ground. Chapter 11 elaborates the response characteristics under harmonic excitations. Similarly, the dynamic structural responses measured with different units (acceleration, velocity or displacement) under seismic excitations are here quantitatively discussed by categorizing the dynamic characteristics of both structures and seismic ground motions.

When the structure is stiff (such as a low rise building) with a natural period below 0.5 s, it is more sensitive to acceleration than displacement, and the structure tends to move in the same acceleration amplitude as the ground, the acceleration of the structural masses relative to the ground is negligible, and the resultant earthquake loading is then purely proportional to the structure's mass.

However, when a structure is more flexible with a natural period above 2 s (such as high-rise buildings or many fixed offshore structures, some fluid tanks and base-isolated buildings), by observing the earthquake design spectrum shown in Fig. 16.7, it is found that the structure is more sensitive to ground displacement than acceleration, i.e., the structure undergoes large relative horizontal displacement, which may result in damage to non-structural elements, equipment etc. The acceleration of the structural masses tends to oppose the ground acceleration motions, and the sum of the acceleration is therefore low. For a preliminary design for those more flexible structures, the resultant acceleration responses are approximately proportional to the square root of the sum of structure masses. Figure 16.8 shows the comparison between the excitations at the bottom and the responses at a location of a topside for a fixed gravity-based offshore structure with a water depth of 100 m. The natural period of the structure corresponding to the first global bending vibration mode is around 2 s. It shows that the acceleration responses at the topside in all three directions are much less than the excitations, indicating that this structure may work as a filter to decrease accelerations due to the earthquake excitations transferred from the foundation.

For a very flexible structure, such as a compliant offshore tower structure with a natural period above 10 s, the structural masses tend to remain motionless, which do not generate any significant loading, and the structure subjected to seismic ground motions is free from damage.

Furthermore, if a structure's natural period is close to the period of earthquake ground motions, the energy is effectively fed into the structure and the structural responses are amplified. In this case, the damping in the structure will absorb the energy from the structural response and dissipate it, then slow down the build-up of the resonant responses and reduce the response amplitudes.

16.3 Characteristics of Seismic Responses Varying with Frequencies 315

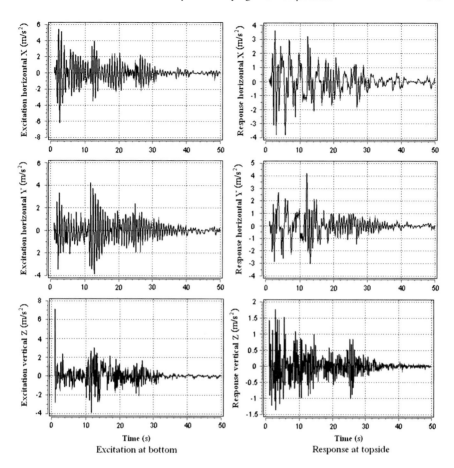

Fig. 16.8 Comparison between the seabed ground excitations and responses at topside of a fixed offshore structure with a natural period of 2 s

Because the peak velocities and peak accelerations are typically associated with motions at different frequencies, in order to determine the significance of seismic responses and the potential damage, the seismic excitations can be categorized based on the ratio between the peak ground accelerations and peak ground velocities, namely *a/v* ratio [262]. This ratio is interpreted as the angular frequency of the equivalent harmonic motions. It provides a rough indication of which frequency contents of ground motions are most significant, and reflects the characteristics of sources, travel path, site conditions, and structural responses. A low *a/v* ratio smaller than 0.8 g/ms^{-1} (g is the acceleration of Earth's gravity) indicates that significant responses are contained in a few long duration ground motion acceleration pulses, which are likely to occur at a soft soil site. Such ground motions can amplify responses of flexible structures with high natural period. A high *a/v* ratio greater than 1.2 g/ms^{-1} indicates that the ground motions contain many high frequency oscillations with large amplitude, which are likely to occur at a rock and rather stiff soil site, a stiff structure is sensitive to oscillations within this frequency range. Ground motions with an a/v ratio in between 0.8 and 1.2 g/ms^{-1} have significant energy content for a wide range of frequencies [278].

Except for inertia forces due to ground accelerations transferred to structures, for offshore structures, the relative motions between the submerged structural members and their surrounding fluids also create hydrodynamic damping forces. Furthermore, the surrounding fluids will also enhance the inertia effects of the submerged structural members, which are normally referred to as the effects of added mass. The hydrodynamic damping in a seismic analysis can normally be neglected. However, when a strong earthquake occurs together with a significant storm (i.e., large wave height), the hydrodynamic damping forces can dramatically increase. Note that the joint probability of occurrence of both events is practically extremely low, the simultaneous occurrence of both events is therefore not considered by typical offshore structural design codes.

16.4 Influences from Structures' Orientations and Ice Covering

It is obvious that the orientation of the structure also influences seismic responses. For horizontally-oriented structures such as a horizontal cantilever, its seismic responses can be more sensitive to the vertical excitations at its base. However, for vertically-oriented structures such as a tower structure or a building, the horizontal acceleration is more dominating. For oblique structures such as a flareboom

Fig. 16.9 A flare boom (marked with a circle) installed on a gravity based offshore structure in the North Sea (courtesy of Aker Solutions)

16.4 Influences from Structures' Orientations and Ice Covering   317

(shown in Fig. 16.9), the responses are sensitive to both vertical and horizontal accelerations at its base.

For ice-covered offshore installations, interaction between the ice and structures during earthquakes may significantly affect the responses of offshore or coastal structures under seismic loading. The ice can affect a structure's dynamic responses with respect to both motions and their frequency content [279]. Ice sheets can also stiffen a structure by providing lateral support to it. This beneficial effect to resist miscellaneous horizontal seismic loads was witnessed during the 9.2 magnitude earthquake that occurred in Anchorage in March 1964, which caused massive destruction in the Anchorage area, but the marine structures in the Port of Anchorage survived with only minor damage [280].

16.5 Whipping Effects

Note that if a structure has a rather slender tip, seismic responses at this tip may be significantly amplified. This phenomenon is called the whipping effect, and is mainly due to the decrease of lateral stiffness at the tip compared to the rest of the structure [23]. Subject to ground motions with acceleration spectra shown in Fig. 16.11, Table 16.2 illustrates the maximum accelerations and the relative displacements at 14 representative locations on the platform shown in Fig. 16.10. The first three eigenperiods are marked in the acceleration spectra in Fig. 16.11. It is shown that the accelerations at these three periods are not high and well below the peak value of spectra acceleration. Furthermore, whipping effects can be clearly identified in that the responses at the tip of the derrick (location 10) and the

Table 16.2 Maximum seismic accelerations and relative displacements in all three directions (X, Y and Z, shown in Fig. 16.10) on 13 locations of the topside and 1 location at the top of the jacket structure

Location	a_x (m/s^2)	a_Y (m/s^2)	a_z (m/s^2)	d_x (cm)	d_Y (cm)	d_z (cm)
1	1.0	1.5	2.1	9.7	8.9	4.0
2	1.3	1.2	1.7	11.5	10.3	3.0
3	1.4	2.3	1.4	12.4	12.0	1.9
4	1.9	2.1	2.2	13.3	12.7	3.2
5	1.6	2.1	1.5	12.7	11.5	1.7
6	2.0	2.6	1.6	13.7	12.0	2.4
7	1.9	2.7	2.0	13.8	11.5	3.1
8	1.0	1.0	1.4	10.8	11.5	1.8
9	1.8	2.6	1.7	14.0	13.2	1.4
10	5.2	9.1	1.7	20.5	16.6	1.7
11	1.1	1.5	1.5	9.8	9.1	3.1
12	3.7	2.9	3.2	16.3	13.5	6.5
13	5.3	3.7	4.0	18.9	15.2	7.9
Jacket top	1.4	1.7	1.1	8.2	7.1	1.7

Fig. 16.10 14 representative locations on the structure under investigation

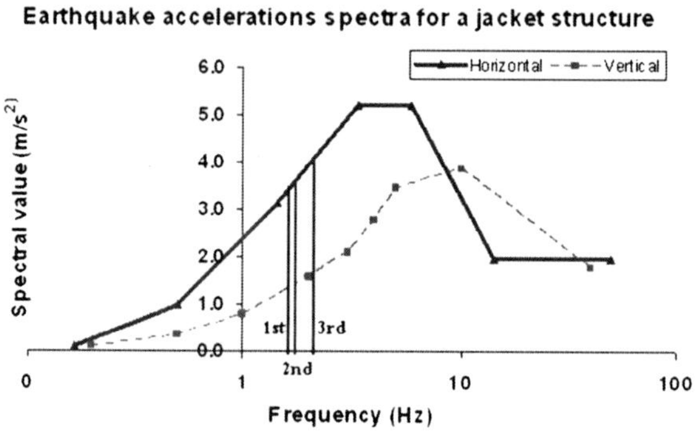

Fig. 16.11 First three eigenperiods marked on the accelerations spectra graph with 10,000 years of return period for the target offshore jacket structure

tip of the flare boom (location 13) have significantly higher accelerations and displacements than the rest of the topside and jacket.

Readers who are interested in more detailed knowledge on seismic response calculation and earthquake engineering may read Ref. [23].

16.6 Seismic Analysis Methods

The major objective of seismic analysis is to develop a quantitative measure or a transfer function that can convert the strong ground motions at a structure's foundation to loading and displacement demands of the structure, which can finally lead to a reliable assessment of structural capacity.

Traditional methods calculate various aspects of structural effects due to non-linearity and dynamics. With respect to ground motion characteristics, different methods can also account for the effects due to spatial variation, non-Gaussian and non-stationary properties. Five traditional seismic analysis methods are as follows:

- Simplified Static Coefficient Method
- Response Spectrum Analysis (Sect. 16.2)
- Nonlinear Static Pushover Analysis
- Random Vibration Analysis
- Nonlinear Dynamic Time Domain Analysis

With the advent of Performance Based Design (PBD) [281] for land-based structures, which is a design philosophy for engineers to manage the cost of construction as well as maintaining the safety of structures, various seismic analysis methods are emerging. Note that the traditional seismic analysis methods aim to pursue accuracy of the calculated responses, while the recently developed methods focus more on the compatibility between the structural response calculation and the evaluation of detailed performance demand, on revealing a structure's intrinsic seismic response and essential performance characteristics, and on improving the robustness of analysis results. The most widely presented or researched methods are:

- Incremental Dynamic Analysis (IDA), also named Dynamic Pushover Analysis [282]
- Endurance Time Analysis (ETA) [283]
- Hybrid method
- Probability-Based Seismic Design (PBSD)
- Critical Excitation Analysis
- Wavelet Analysis

However, it is noted that almost no industry sectors have kept pace with the newly developed methods. This is mainly due to the difficulties of implementing PBD into structural design. It is therefore a strong intention of the present author to promote them for facilitating the modern seismic design of both offshore and land-based structures.

Furthermore, readers should note that regardless of the sophistication of the numerical method, it is not exact. Many uncertainties still exist. Therefore, in the development of new methods for future seismic analysis, more attention should be paid to their robustness.

For more details of various seismic analysis methods, Ref. [23] is recommended.

Chapter 17
Fatigue Assessment

17.1 Failure of Structural Components

The failure of a component can be related to extensive deformation, which is the loss or impairment of functions for components due to changes in their physical dimensions or shape. This type of failure can be either time dependent or independent. An example of the former is so-called creep, which is the accumulation of deformation with time. This occurs for almost all types of materials, but it is more pronounced for plastics and low-melting-temperature metal materials or at high temperatures. An example of time-independent failure is yielding and subsequent plasticity development, which can cause the collapse of entire structures or the significant deformation normally related to the loss of serviceability. In addition, buckling, as previously presented in Sect. 15.2.3, can also be regarded as a deformation failure.

The failure of a component can also be due to fracture, which includes brittle fracture under impact loading, ductile fracture, and creep rupture. For more details, readers may refer to the abundant textbooks on the mechanics of materials, such as Ref. [243].

An even more typical failure type for many structures exposed to dynamic loadings is due to fatigue. Under cyclic (repeated) loading, a component can reach premature failure or damage well below the yielding stress of the component material, known as fatigue. This is because small crack-like defects exist and when they are subject to a sufficiently large cyclic tension stress, they will grow in size and eventually cause the member to reach fatigue failure. The cracks are developed in four stages:

1. Crack initiation (usually starts from the surface and can be detected by common technical means, e.g., 1 mm in length and 0.5 mm in depth)
2. Stable crack-growth
3. Unstable crack-growth (rupture)
4. Ultimate ductile failure.

J. Jia, *Essentials of Applied Dynamic Analysis*, Risk Engineering,
DOI: 10.1007/978-3-642-37003-8_17, © Springer-Verlag Berlin Heidelberg 2014

Fig. 17.1 A crack through the entire height of a stiffener's web inside a subsea steel tank at a depth 165 m below the sea surface

What distinguishes fatigue failure from plastic deformation failure is that when the former occurs, the atomic bonds of the material break at a right angle to the applied tensile stress, while during plastic deformation, even if the atomic bonds break due to the shear stress, they form new atomic bonds with their neighbors, thus returning to a stable configuration with the new neighbors after the dislocation has passed. From a continuum mechanics point of view, the changes in the material's atomic bonds at the microscopic level have inherent changes in stress, strain and energy density at the macroscopic level, which are used for studying various stages of crack development in fatigue assessment or plasticity development for strength evaluation.

Fatigue crack initiation is rather unpredictable, and crack defects are present immediately after fabrication, such as the ones in welds. Crack initiation can be quite rapid. Once the cracks are initiated, they propagate perpendicular to the direction of tensile stress.

Figure 17.1 is a photo taken by a remote control vehicle that shows a crack going through the entire height of a stiffener's web inside a subsea steel tank. This crack has the potential to cause the global failure of the entire subsea tank and the repair of it can be rather costly.

In terms of the number of loading cycles, a small number of cycles (not more than a few thousands) with significant amplitude (approaching ultimate tensile strength with plasticity development) can be relevant to low cycle fatigue, which is generally accompanied by a significant amount of plastic deformation (Sect. 15.2.1). High cycle fatigue, in contrast, is a more typical failure mode than low cycle fatigue, and is normally associated with a large number of cycles (more than 10^4 cycles) and low amplitude of loading inducing only elastic deformation. For example, for offshore structures, the main contribution to fatigue damage is caused by frequently occurring wave load effects that are of the order of 10–20 % of the extreme load effects in the service life [24].

17.1 Failure of Structural Components

In addition, corrosion (loss of material due to chemical reaction, and/or creation of surface notches, can finally lead to fatigue crack initiation), wear (removal of surface due to abrasion or sticking between solid surfaces) and erosion (wear caused by a fluid) can also cause failures of structural component.

17.2 Fatigue Damage Assessment

17.2.1 Classification of Fatigue Assessment Approaches

The assessment of fatigue damage can generally be performed in three different approaches: stress (S–N curve)-based approach related to high cycle fatigue, strain-based approach related to low cycle fatigue, and fracture mechanics. These three approaches will be presented in Sects. 17.2.2, 17.2.3, and 17.2.4. Among them, the S–N curve-based fatigue assessment approach is most commonly used due to its convenience and accuracy for the assessment of high cycle fatigue. The dynamic analysis methods presented later in Sect. 17.3 are mainly associated with the stress-based approach.

Note that, except for the fracture mechanics approach, other fatigue assessment methods are based on the concept that the fatigue damage increases with applied cycles in a cumulative manner. The existing fatigue assessment approaches can also be divided into cumulative fatigue damage theory and fatigue crack propagation theory [284].

On the other hand, the fatigue assessment methods can be divided into global and local approaches [285]. In a global approach, external forces and moments or the nominal stresses (Sect. 17.2.2.2.3) in the critical cross-section are measured to assess the fatigue damage. Developed from the global approach, in a local approach, local stress, local strain or other local parameters are used. Such measures are related to the notch stress method described in Sect. 17.2.2.2.5 or strain-based method introduced in Sect. 17.2.3, and the fracture mechanics approach introduced in Sect. 17.2.4. By this categorization, crack initiation, propagation and final fracture are all termed local approaches [286]. Yet another method acting as a link between the global and local approach is the hot-spot or structural stress method described in Sect. 17.2.2.2.4, in which the stress increases due to the joint geometry (size and type of welded joints), but not local toe geometry (notch effects), which is implicitly accounted for by degrading the S–N curve. For simplification, we here still classify the hot-spot stress method as a local approach.

It should be noted that the local parameters such geometry, material and loading have a predominant influence on the structural performance with regard to fatigue, apart from the dynamic response calculation. The treatment of these parameters is an essential part of fatigue assessment.

Fatigue assessment for complex structures is often performed with the aid of FE analysis. This requires a reasonably good understanding of FEM, the influence of calculated fatigue damage by various dynamic analysis methods, and the philosophy behind each type of fatigue assessment methods. This entire chapter highlights these three essentials, especially the latter two.

Apart from the stress- and strain-based approaches and fracture mechanics, which are the most popular ones, other methods are available for fatigue assessment, such as the energy-based approach [287], which uses an energy-based damage parameter to assess the damage caused by various types of loadings such as creep, fatigue and thermal and damage mechanics [288]. This uses state variables to represent the damage on the stiffness, and remaining life of the material that is damaging as a result of thermomechanical load and aging.

17.2.2 Stress-Based Approach

17.2.2.1 General Method

As mentioned in the previous section, failure due to fatigue is not rare for structures exposed to dynamic loading. With respect to high cycle fatigue, the material performance is typically characterized by the S–N curve (Wöhler curve), which defines the log-linear dependence between predicted number of cycles to failure N and a stress range S as shown in Fig. 17.2. For practical fatigue design of welded structures, welded joints are divided into several classes according to the geometrical arrangement of detail, the direction of the fluctuating stress relative to the detail, and the method of fabrication and inspection of the detail, each with a corresponding design S–N curve.

It should be noted that, for a harmonic stress history, the stress range S is twice that of the stress amplitude σ_a as shown in Fig. 17.3:

$$S = 2\sigma_a = \sigma_{max} - \sigma_{min} \tag{17.1}$$

Based on sufficient data obtained from test samples, known as coupon testing, where a regular harmonic stress is applied by a testing machine that also counts the number of cycles to failure, the regressed S–N curve (mean curve in Fig. 17.4) is formulated as:

$$\log_{10}^{N} = \log_{10}^{A} - m\log_{10}^{S} \tag{17.2}$$

where N is the predicted number of cycles to failure under stress range S; A is a constant relating to the mean S–N curve; m is the inverse slope in the S–N curve; both A and m are obtained from test data; $\log_{10}^{A}$ is actually the intercept of $\log_{10}^{N}$ axis by mean S–N curve;

Because the relationship between S and N cherishes a high uncertainty, the parameters A and m in the equation above can be regarded as random variables [289].

For structure components made of low-alloy steels or plain carbon [311], a threshold stress (also called fatigue limit, endurance limit or non-damaging stress)

17.2 Fatigue Damage Assessment 325

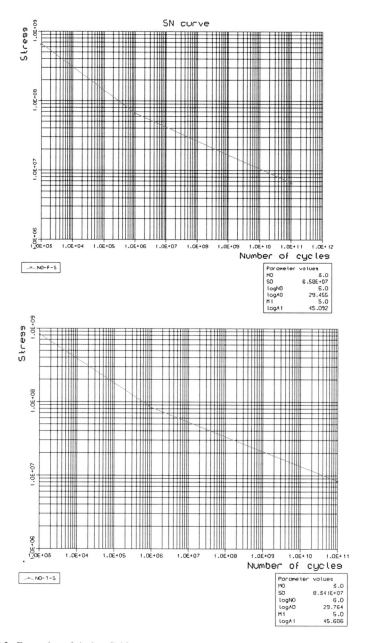

Fig. 17.2 Examples of design S–N curves

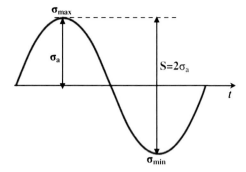

Fig. 17.3 Definition of stress amplitude σ_a and stress range S

S_0 is introduced below, for which the number of cycles leading to fatigue damage will be assumed to be infinite, i.e., no fatigue damage or infinite fatigue life:

$$N = \begin{cases} A \cdot S^{-m}, & \text{for } S > S_0 \\ \infty, & \text{for } S \leq S_0 \end{cases} \quad (17.3)$$

From the equation above, it is noticed that the fatigue damage is proportional to the stress amplitude with the power (m) typically in the range from 3.0 to 5.0. A slight variation of the stress amplitude may induce significant change of the calculated fatigue damage. This requires a more dedicated dynamic analysis to reduce uncertainties and to increase the accuracy regarding the calculated responses.

Threshold stress can also be explained from a fracture mechanics point of view, as will be discussed briefly in Sect. 17.2.4.1.

It is always desirable to design a structure with an operating stress level below the threshold stress level. An example of this is the power trains that operate at high speed. However, for many types of structures such as offshore structures, the fatigue limit has lost much of its significance because it is typically in the range of 20–80 MPa depending on type of joints and test conditions, and it would be rather uneconomical to design such types of structures with design load below the initial threshold level. Furthermore, the fatigue limit obtained from small scale tests may not be representative of fatigue limits in large scale structures, e.g., the probability of exceeding a certain size of defect for a detail of an offshore structure will be significantly higher than that for a small test specimen. In addition, localized corrosion can also introduce pits and crevices on welds, causing a time-dependent growth of initial defects.

Note that the mean S–N curve implies a 50 % failure probability, which is definitely not acceptable from an engineering point of view. Based on statistical analysis of fatigue data, the S–N curves with various probabilities of failure can be obtained, which are sometimes referred to as S–N–P curves. For design purposes, the design S–N curve is normally based on the mean-minus-two- (e.g., for civil engineering applications) or mean-minus-three- (e.g., for the application of aircraft industry) standard-deviation of the mean curve from relevant experimental test

17.2 Fatigue Damage Assessment

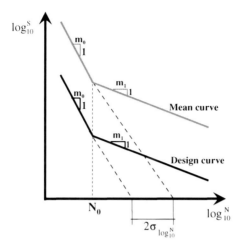

Fig. 17.4 Mean and design S–N curves (m_0 and m_1 are the inverse slope for the first and second segment of the S–N curves)

data. For the case with the mean-minus-two-standard-deviation, the design S–N curve is associated with a 97.6 % probability of survival:

$$\log_{10}^N = \log_{10}^A - 2\sigma_{\log_{10}^N} - m\log_{10}^S \tag{17.4}$$

where $\sigma_{\log_{10}^N}$ is the standard deviation of $\log_{10}^N$; $\log_{10}^A - 2\sigma_{\log_{10}^N}$ is actually the intercept of $\log_{10}^N$ axis by design S–N curve.

Figure 17.4 illustrates a comparison between a mean and the corresponding design S–N curve based on the mean-minus-two-standard-deviation.

17.2.2.2 Stress Measures

Stress Measure for Welded Plated Structures

As welding is extensively used as an effective and economical way of connecting metal plates, various codes describe methods for fatigue assessment of welded joints based on the stress-based approach, such as DnV RP-203 [330], Eurocode 3 [331], ASME Boiler and Pressure Vessel Code [332], and British Standard BS7608 [333], etc. In those codes, the welds are categorized by the type of joint, the geometry, the loading direction and the potential failure direction. Different design S–N curves are assigned to each weld category (class) for fatigue damage calculation. The stress level is typically based on the principal stress with the largest range.

In this process, the location and nature of stress for fatigue calculation is somewhat difficult to determine. The commonly used stress definitions associated with the stress-based approach are nominal stress, structural hot-spot stress and local notch stress. Fatigue assessment using all three stress definitions are widely applied in various industry sectors, which will be elaborated in the subsequent sections.

Stress Measure for Welded Tubular Joints

For tubular joints, i.e., brace to chord connection, the utilization of tubular members gives rise to significantly high stress concentrations in the joints. Fatigue life is one of the major concerns for tubular structures. Apart from the other elaborated methods in this section, the stress to be used for design purpose is the range of idealized hot-spot stress, which is defined by the greatest value of the extrapolation of the maximum principal stress distribution, immediately outside the region affected by the geometry of the weld. A relationship between the hot-spot stress (σ_{ht}) at weld toe and the nominal stress (σ_{nom}) is:

$$\sigma_{ht} = SCF \cdot \sigma_{nom} \tag{17.5}$$

where SCF is the stress concentration factor of tubular joints depending on the geometrical parameters of tubular members, which can be obtained through experiments or FE analysis. Various parametric formulas are available for a simple calculation of SCFs, with Efthymiou's [326], Kuang's [328], and Gibstein's [329] formulas being the most popular. Many design codes and standards present similar parametric formulas for calculating the SCFs for tubular joints.

Unless otherwise specified, the stresses in this section are defined as principal stresses.

Global Stress Approach: Nominal Stress

Nominal stresses are those stresses derived from simple beam models or from FE calculations based on coarse mesh models. Even though the stress concentrations resulting from the gross shape of the structure are included in the nominal stress measure, it does not include any stress increase/concentration due to the structural detail or the welds. Therefore, the nominal stress is only valid in the parts of structure at some distance from the welded joints.

As shown in Fig. 17.5, typically, the stress increase (concentration) at a weld toe is contributed by both the joint geometry (size and type of welded joint) and local toe geometry (toe angle and toe radius). It is clearly shown that the stress decreases with the increase of distance from the weld. Only above a certain distance that mainly depends on the stressed plate thickness t does the stress approach the constant nominal stress.

However, for a given weld and structural geometry, the nominal stress can be used with an S–N curve associated with the "notch class," "detail class" or "fatigue class" (FAT), which is typically determined experimentally for that specific weld and structure geometry, i.e. this S–N curve already implicitly includes the influence from material, geometry (joint and toe geometry) and/or surface (residual stress introduced in Sect. 17.2.2.3) with statistical scatter due to parameter variations. Therefore, the nominal stress is not suitable for fatigue damage evaluation of various new welds and joint geometries.

17.2 Fatigue Damage Assessment

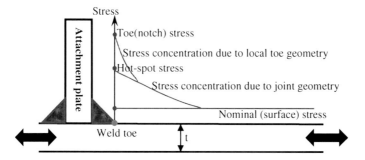

Fig. 17.5 Stress concentration at a weld toe

In engineering practice, for an initial screening of fatigue critical locations on a structure, the nominal stress can be used with a rather conservative S–N curve. After the fatigue critical locations are identified, a more dedicated stress measure such as hot-spot stress can be used with an appropriate S–N curve to calculate the fatigue damage.

Local Stress Approach: Hot-Spot Stress Method

Introduction to the Hot-Spot Method. Despite of simplicity of the method itself, in many cases the nominal stress is difficult to estimate due to geometric and/or loading complexities of welded structures.

Therefore, local stress approaches have to be adopted, in which the stress in the vicinity of the crack initiation is obtained by either numerical analysis (mostly by FE modeling) or experimental measurements. This approach includes the hot-spot stress and notch stress method, both of which decrease the scatter in the fatigue life predictions associated with the stress concentration.

The hot-spot is the critical point in a structure from where a crack can be assumed to propagate. In a welded structure this point is usually located close to the weld toe or weld root. This method is only applicable for fatigue assessment with fatigue failures starting from the weld toe [290].

Unlike nominal stress, hot-spot stresses, also known as structural stresses or geometric stresses, include nominal stresses and stresses due to the structural discontinuities and the presence of attachments, i.e., it contains the stress increase due to the structure geometry, but not the nonlinear stress peak due to the local weld geometry (which usually refers to the weld toe), which is implicitly taken into account in the hot-spot stress method by degrading the relevant S–N curve. This is because singularities at the weld toe are difficult to predict with a reasonable accuracy even by using FE modeling [291, 292]. The relationship between the hot-spot stress (σ_{ht}) and the nominal stress (σ_{nom}) is:

$$\sigma_{ht} = K_g \sigma_{nom} \tag{17.6}$$

where K_g is the structural/geometric stress concentration factor.

Therefore, unlike the nominal stress method, in the hot-spot stress method, a unique structural class for each joint type is not required, leading to a decrease in the number of joint classes and a reduced number of S–N curves needed. The relevant S–N curves that are linked to the hot-spot methods can be found from various design codes and recommendations such as DnV-RP-203 [330] for offshore structures, DnV CN 30.7 [339] for ship structures, and IIW [338] for air environments. In addition, misalignment of plates at welded connections and weld imperfections can also be roughly taken into account by suitable stress magnification factors from various design codes and recommended practices.

Without a sound theoretical basis, the hot-spot stress method is rather an engineering means to perform more efficient fatigue assessment. A failure criterion is no longer well defined when the fatigue test data based on nominal stress is transferred into a hot-spot stress associated S–N curve [293]. Developed in the 1970s in a combined effort by classification societies, offshore platform operators as well as research institutes, the general method for calculating the hot-spot stress is to determine the largest value at the weld toe by using FE modeling. This has been widely used in engineering design since the 1990s. FE models are normally created by assuming an ideal geometry of structures. As the hot-spot stress cannot be read directly from the FE calculation, it is often obtained by linearly extrapolating stress at two reference locations at some distance from the weld toe, which is illustrated in Fig. 17.6. Sometimes, three reference locations are used to make a quadratic stress extrapolation for calculating the hot-spot stress. Research [294] has shown that no significant difference can be found between the results of the linear and quadratic extrapolations. Furthermore, Fricke [295] suggested a simple calculation in which the hot-spot stress can be obtained by multiplying the stress at a reference point of 0.5 t from the weld toe with a factor of 1.12, where t is the thickness of the stressed plate.

FE modeling for the hot-spot method. The stresses calculated by FE analysis are highly mesh sensitive, as the structural hot-spot stresses are often in an area of high strain gradients. Depending on the type and size of elements and the procedure used to extract the values of the hot-spot stresses, the resulting stresses may differ substantially. This is regarded as a drawback that leads to the increased effort for FE modeling.

Many design codes and recommended practices provide a method to determine the hot-spot stress. For example, DnV [330] recommends that, by using either 20 node solid elements with a size of $t/2 \times t/2$ or 8 node shell element with a size of $t \times t$, linear extrapolations of the component stresses from points at $t/2$ and $3t/2$ can be used to derive the hot-spot stress at the weld toe, which is illustrated in Fig. 17.6. Slightly different from the DnV recommendation, ABS [334] recommends that the solid or shell element can be applied with a size of $t \times t$. Eurocode 9 [335] defines hot-spot stress as the largest value of component stress extrapolated in the normal direction to the weld. Figure 17.7 illustrates typical meshes and evaluation paths for stress extrapolation. A benchmark study by Fayard et al. [336]

17.2 Fatigue Damage Assessment

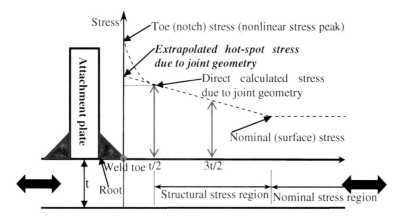

Fig. 17.6 Derivation of hot-spot stress and its relation with nominal and notch stress

suggests that, compared to that of the shell element, the modeling with solid elements does not increase the accuracy of calculated hot-spot stress.

In engineering practice, if 4 node shell elements are used for modeling, the recommended mesh sizes at the hot-spot region are from t/2 × t/2 to 2t × 2t; in this case, larger mesh size may result in non-conservative results. In cases with steep stress gradients, 8 node shell elements are recommended for modeling, the recommended mesh sizes at the hot-spot region are then from t × t to 2t × 2t; in this case, both larger and smaller mesh size may result in non-conservative results.

It is also possible to obtain hot-spot stress with coarse meshing, such as in that conducted by Hobbacher [316], who presented a review of the procedure to determine the FE meshing with coarse mesh.

If possible, it is strongly recommended to read the stress values at the element integration points rather than the element nodes, as the nodal stress is normally taken as the average of two elements located on both sides of the weld toe normal. If the element size is t × t, for shell or plate elements, the surface stress may also be evaluated at the corresponding mid-side points; for solid elements, the stress may first be extrapolated from the integration points to the surface. Then these stresses can be interpolated linearly to the surface center or be extrapolated to the edge of the elements [330].

Figure 17.8 [337] presents an example of meshing of two FE models that represent a panel joint. In order to calculate the stress concentration factor due to joint geometry, which is defined as the ratio between the hot-spot stresses and the nominal stresses, two mesh modelings are used: the coarse mesh shown in the left figure is used for nominal stress calculation, and the fine mesh in the right figure is for hot-spot stress calculation.

Readers need to bear in mind that the S–N curves are generally based on test results from uni-axial loading, and the hot-spot stresses are normally derived from stresses normal to the weld toe. However, in a real structure, the hot-spot stresses are in many cases bi-axial, and the principal stress range direction can deviate

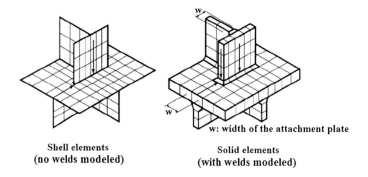

Fig. 17.7 Typical meshes with shell and solid elements to calculate the hot-spot stress. Extrapolations are normally performed along the stress evaluation paths (marked with *arrows*)

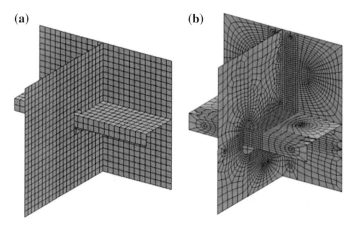

Fig. 17.8 FE models of a panel joint: **a** coarse mesh for nominal stress calculation and **b** fine mesh for hot-spot stress calculation with t × t mesh [337]

significantly from the weld toe normal. It then becomes conservative to use the principal stress range together with a classification of the connection for stress range normal to the weld toe [339, 286]. There are various recommendations on how to determine hot-spot stresses under multi-axial stress state. One of the treatments is to define a range of the deviation angle from the weld toe normal direction, so that if the principal stress range direction lies within this angle range, it can be used to calculate the hot-spot stresses. This angle range is normally within ±60°, but varies from one guideline document to another. For example, for ship structures, DnV [339] recommends an angle range of ±45°, while IIW [338] recommends that this range be ±60°. If the direction of principal stress range is outside this angle range, the stress normal to the weld toe may be taken as the hot-spot stress. Normally the stress range in both the two principal directions should be assessed with respect to fatigue. On the other hand, the effects of deviation of the

17.2 Fatigue Damage Assessment

Table 17.1 Selection of S–N curves using nominal stress and hot-stress method by DnV DnV-RP-C203 [330] for fatigue assessment of offshore structures

The angle between principal stress range direction and the weld toe normal direction	Detail classified as F for stress direction normal to the weld	Detail classified as E for stress direction normal to the weld	S–N curve when using the hot-spot stress methodology
0°–30°	F	E	D
30°–45°	E	D	C2
45°–60°	D	C2	C2
60°–75°	C2	C2	C2[a]
75°–90°	C2[a]	C2[a]	C2[a]

[a] A higher S–N curve may be used in special cases

principal stress range direction from the weld toe normal can also be accounted for by assigning an appropriate class of S–N curve. Table 17.1 shows the selection of S–N curves using the nominal stress and hot-stress methods, which is recommended by DnV DnV-RP-C203 [330] for offshore structural design. Obviously, when the angle between the principal stress range direction and the weld toe normal direction becomes larger, a better class of S–N curve is assigned to reduce the conservativeness.

For fatigue check of a complex structure with a large number of degrees-of-freedoms, or for a structure in which only a part of it is of interest for fatigue assessment, it is not economical or even possible to model a complete structure down to the last detail. In these cases, the hot-spot stress is often calculated by using the sub-modeling technique.

In the sub-modeling technique, without excessively increasing the calculation efforts, an analysis based on the global model with coarse mesh is first performed. Based on this global model, a dynamic analysis can then be performed to obtain the nominal stresses. A fatigue screening check can then be carried out to identify the critical parts with regard to fatigue damage. By using the sub-modeling technique, the fatigue critical parts are then modeled with finer mesh and analyzed separately by transferring the prescribed displacement (computed from the previous analysis of global model) along the boundary/interface where it has been cut free from the rest of the global structure. If the refinement of the mesh adds more nodes along the boundary that is not presented in the global model, a decent interpolation method should be adopted to obtain the prescribed displacement on the added nodes.

Figure 17.9 illustrates the sub-modeling process for a transition ring on the top of an offshore gravity based structure. In the global model, the transition ring is modeled mainly by 4 node shell elements with coarse mesh, and some parts of it having sharp edges are modeled with 3 node triangle elements. A dynamic analysis subjected to wave loadings is performed followed by the fatigue screening check based on the calculated nominal stress and the associated (conservative) S–N curves. The welds on the transition ring are identified as the possible critical area with regard to fatigue damage. In its sub-model, the areas close to the transition welds are modeled by 8 node shell elements with a finer mesh size comparable to

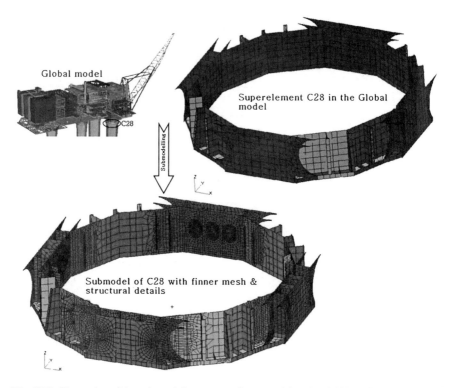

Fig. 17.9 Illustration of the sub-modeling process for a transition ring C28 installed on the top of a gravity-based structure (courtesy of Aker Solutions)

the thickness of the plate ("t × t" mesh). Figure 17.10 shows an example of the sub-model with "t × t" mesh, and the total number of nodes for the global and sub-models are 3898 and 142952, respectively. "Node to element" coupling is used between the global model and the sub-model boundaries. This means that the spatial (element shape function) interpolation of the global model's solution is used to calculate the prescribed displacement values for both the global model and the sub-model. Different from the screening process, more relevant S–N curves are assigned for fatigue check for each identified critical area.

Figure 17.11 illustrates an example of the fatigue life distribution for welds based on the hot-spot stress calculation. It is found that part of the welds between the MSF cellar deck and transition rings are likely to be fatigue critical, with calculated fatigue lives of less than 44 years.

Various studies have been carried out to find a hot-spot method that is not mesh sensitive. Xiao and Yamada [340] suggested computing the hot-spot stress at a depth 1 mm below the surface at the weld toe in the direction of the expected crack path, which is assumed to represent the stress gradient over the plate thickness. This approach has been demonstrated to be valid by the authors [340] for non-load-carrying fillet welds and by Noh et al. [341] for load-carrying fillet welds. Dong and

17.2 Fatigue Damage Assessment

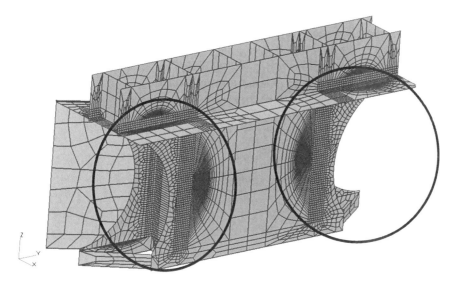

Fig. 17.10 "t × t" mesh on the fatigue critical weld areas of an extended geometry from the original transition ring boundary (courtesy of Aker Solutions)

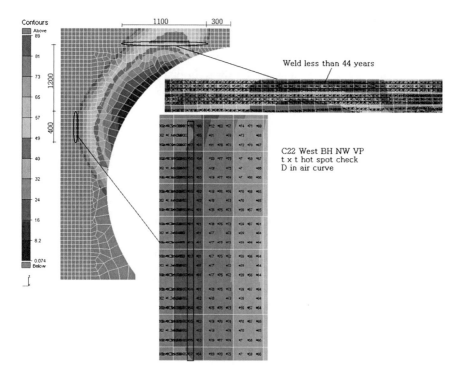

Fig. 17.11 Hot-spot check of fatigue life (based on D curve in air from DnV DnV-RP-C203 [330]) at the critical welding area (fatigue lives are between 40 and 88 years) of the transition ring (courtesy of Aker Solutions)

336   17   Fatigue Assessment

his co-workers [342–344] proposed that the hot-spot stress can be calculated as the sum of membrane and bending components at a small distance from a weld toe. However, this method is demonstrated to be valid mostly in simple two-dimensional structure details such as fillet weld lap joint [292, 340, 345].

It should be noted that, since the hot-spot stresses are calculated on the basis of the stresses read at plate surfaces, this can lead to an erroneous assessment when stresses vary significantly over the plate thickness. An example of this is that in which the plate stresses due to bending are dominant. In such a condition, other methods such as the fracture mechanics method are more suitable for assessing the fatigue life.

Finally, it is worth mentioning that the hot-spot stress method is so far only applied to thick plate structures in engineering applications. However, developments for thin plate structures are receiving more and more research efforts, especially in the automotive industry [354, 355].

Local Stress Approach: Effective Notch Stress Method

Introduction to the effective notch stress method. Even though the hot-spot stress method is a powerful tool, its application is restricted to weld toe cracking [296, 297]. For fatigue damage assessment with root cracking, the notch stress method is applicable, which can handle not only root crack, but also the toe cracking.

When notches emanate from geometrical discontinuities such as holes, joints, or defects from welds, they contribute to higher stress concentrations.

The effective notch stress consists of the sum of both geometrical (hot-spot) stress and nonlinear stress peak at the root taking into account the stress concentration caused by the local notch, which is shown in Fig. 17.6. It can be calculated by multiplying the hot-spot stress with a stress concentration factor K_s, which is defined as the ratio of the maximal notch stress (σ_{max}) and the nominal stress (σ_{nom}):

$$K_s = \sigma_{max}/\sigma_{nom} \qquad (17.7)$$

The notch stress is strongly affected by several weld details, most of which are very difficult to measure even by using non-destructive tests. Therefore, by assuming the microstructural support effect [346, 347], the effective radius ρ_e is used to simulate the notch, which is represented by increasing the actual radius ρ with an additional material-dependent microstructural length ρ^* multiplied by the support factor s, which is illustrated in Fig. 17.12 and expressed by:

$$\rho_e = \rho + s\rho^* \qquad (17.8)$$

17.2 Fatigue Damage Assessment

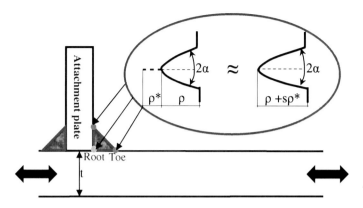

Fig. 17.12 The effective radius ρ_e is represented by increasing the actual radius ρ with an additional material dependent microstructural length ρ^* multiplied by the support factor s

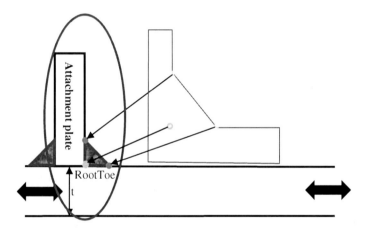

Fig. 17.13 Modeling of notch for both weld root and toes

As an conservative estimate, for welded steel joints, Radaj [346] recommended using $\rho = 0$ mm, s = 2.5 and $\rho^* = 0.4$ mm, i.e., a effective radius of 1 mm.

FE modeling for the effective notch stress method. Obviously, in FE modeling aiming for obtaining notch stresses, welds need to be modeled. The notch for the weld root is modeled with a radius, and similarly, the edges in the weld toes are replaced by a hole cut-out with a radius typically of 1 mm. This is shown in Fig. 17.13. Very fine mesh should be applied to the parts close to the weld toes and root. It is recommended that at least six elements be used around the root hole or toes per 90° [348]. In addition, subdivision should be sufficiently fine normal to the surface in order to model the steep stress gradient in thickness direction [349]. Alternatively, the weld root can also be modeled as an oval shaped cut-out as shown in Fig. 17.14 [349]. Since the stiffness with an oval shaped cut-out for

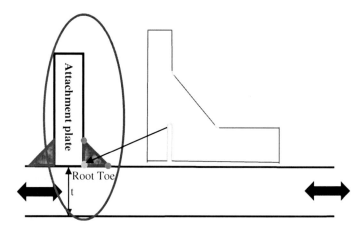

Fig. 17.14 The weld root can also be modeled as an oval shaped cut-out

Table 17.2 Pros and cons of different stress measures for fatigue assessment [298]

Stress measures	Pros	Cons
Nominal stress method	Simple calculations Well-defined Widely used Available experimental data Available parametric formula Available fatigue classes in design codes Suitable for weld root and toe cracking	Fatigue detail category dependency Limitation for misalignment and macrogeometric changes Less accuracy in complex structures Thickness effect not included
Hot-spot stress method	Reduced number of S–N curves needed The use of existing stress analysis Acceptable accuracy Less FE modeling effort Macro geometric effect included Utilized for tubular structures for many years	Dependent on element size Dependent on element arrangement Different stress determination procedures Thickness effect not included Limited to weld toe cracking
Effective notch stress method	Thickness effect included in calculations Not affected by the stress direction Suitable for weld roots and toes cracking A single S–N curve	Applicable only with FE analysis Dependent on mesh density Dependent on radius size Larger FE models Time consuming for modeling

17.2 Fatigue Damage Assessment

modeling the root notch is generally lower than that of the hole cut-out, this leads to an increased stresses at the weld toe. Therefore the stress evaluation with an oval shape cut-out is generally conservative [349].

It is noticed that, in the calculation of effective notch stress, if a structure is modeled by solid elements, the geometry and stiffness of the welds can be conveniently modeled. However, if shell elements are used for modeling, additional efforts need to be made in order to correctly model the stiffness of the welds [350, 351]. Available methods are to represent the welds by using shell elements [352], by increasing the thickness in the intersection region of welded joints [353], or by using rigid links [336].

Finally, it should be mentioned that, similar to the hot-spot stress method, in engineering applications, the effective hot-spot method is so far mainly applied to thick plate structures with the thickness of more than 5 mm.

Pros and Cons of Different Stress Measures

Based on the literature study by Aygül [298], Table 17.2 summarizes the advantages and disadvantages of three stress measures for fatigue damage assessment.

17.2.2.3 Influencing Parameters

Increasing the thickness of the adjoining plates in relation to the weld toes would reduce the fatigue life. The design S–N curve for a weld influenced by the adjoining plate thickness can be expressed as:

$$\log_{10}^N = \log_{10}^A - 2\sigma_{\log_{10}^N} - m \log_{10}^{S\left(\frac{t}{t_{ref}}\right)^k} \qquad (17.9)$$

where t_{ref} is a reference thickness. In DnV RP-203 [330] for offshore structural design, for welded connections other than tubular joints or bolts, t_{ref} is normally taken to be 25 mm, for tubular joint it is 32 mm, and for bolts it is 25 mm; t is the plate thickness through which a crack will most likely to grow; when t is less than t_{ref}, $\frac{t}{t_{ref}} = 1$; k is the thickness exponent on the fatigue strength, e.g. $k = 0.10$ for tubular butt welds made from one side, and 0.25 for threaded bolts subject to stress variation in the axial direction [330].

The internal stress in a material, known as residual stress, also has an effect on the fatigue damage. Compressive residual stress has a positive effect to decrease fatigue damage, which can be introduced through permanently stretching a thin surface layer by yielding it in tension; this is realized in reality through shot peening [243], even though it may have negative effects on corrosions of base materials. High tensile residual stress can induce crack propagation even if the applied external stress is compressive. For welded structures, residual stress normally exists in areas close to the welded joints. It is therefore necessary to use the full stress range in the fatigue calculation for welded details unless they have been subjected to stress relief [81]. In practical fatigue assessments of welded joints, for simplicity, it is assumed that all stress cycles effectively drive the crack due to the presence of the tensile residual stresses. However, it should be noticed that residual stresses due to welding and construction are reduced over time due to the external loading, known as residual stress relief.

For fatigue analysis at regions of the base material not significantly affected by residual stress due to welding, mean stress needs to be accounted for. Mean stress can be caused by slowly varying stress due to payloads, which has a similar effect as residual stress. With the presence of compressive stress, the stress range can be reduced. Goodman, Gerber or Söderberg relations [60, 243] are typically used to account for the mean stress effects. Rather than modifying the S–N curve, the effects of mean stress can be accounted for by multiplying a reduction factor f_m of the calculated stress range before entering the S–N curve, such as the one proposed by Joint Tanker Project [244, 245] and adopted by A number of fatigue design codes such as DnV RP-203[330]:

$$f_m = \frac{\sigma_t + 0.6\sigma_c}{\sigma_t + \sigma_c} \tag{17.10}$$

where σ_t and σ_c are the maximum tension and maximum compression stress of the stress range, respectively. The numerator in the above equation $\sigma_t + 0.6\sigma_c$ is normally referred to as the effective stress range.

This effect is more significant in case of through-thickness cracks [24].

In addition, as elaborated in Sect. 17.2.2.2, the local stress concentration caused by the shape of welds will also reduce the fatigue life. In order to calculate the hot-spot stress used in the fatigue damage calculation, a stress concentration factor (SCF) is adopted as the factor by which the nominal stress due to pure axial force or pure in-plane/out-of-plane bending (at the stress point in question) needs to be multiplied by. The majority of the design codes account for the stress concentration with designated design S–N curves for plate and stiffener details. For tubular joints, the SCFs need to be calculated through various parametric equations [60, 330] as briefly mentioned in Sect. 17.2.2.2.

The adverse environment with regard to fatigue, e.g., freely corroded condition at sea, will also reduce the fatigue life. The influence of the environmental conditions is accounted for by most of the design codes (e.g. "Tubular in Air" or "Tubular at Sea" in DnV Recommended Practice [330]).

For welds, their non-homogeneous material properties, welding defects and imperfections are not usually accounted for in a local approach [356], and the material characteristic value of the welds are normally assumed to be the same as that of the base material.

As mentioned above, even though the stress-based method is widely used, it does not deal with any of the physical phenomena within the material, i.e., it does not separate the crack initiation from the propagation stage, and only the total life to fracture is considered [289].

17.2.3 Strain-Based Approach

For low cycle fatigue associated with plastic deformation, the account in terms of stress is less useful, and the strain in the material offers a simpler description. This can be justified because, in reality, the critical location is often a notch in which plastic strains are imposed by surrounding elastic materials, and the situation will then be strain-controlled [284]. Therefore, the strain-based approach is normally adopted. The low cycle fatigue is often characterized by the Coffin-Manson relationship, the name of which arises from the separate development of related equations in the late 1950s by LF Coffin and SS Manson [243, 357]:

$$\frac{\Delta \varepsilon_p}{2} = \varepsilon_f'(2N)^c \qquad (17.11)$$

where $\frac{\Delta \varepsilon_p}{2}$ is the plastic strain amplitude; ε_f' is an empirical constant known as the fatigue ductility coefficient, which is the failure strain for a single reversal; 2 N is the number of reversals to failure (N cycles); and c is an empirical constant known as the fatigue ductility exponent, commonly ranging from -0.5 to -0.7 for metals, applied for time-independent fatigue. Slopes can be considerably steeper in the presence of creep or environmental interactions.

A similar relationship for the material zirconium is also used in the nuclear industry [366].

In contrast to high cycle fatigue, for which the weld geometry and initial defects are the most important parameters, in the assessment of low cycle fatigue, tensile strength and ductility of materials are important parameters for determining structural performance [60].

17.2.4 Fracture Mechanics Approach

17.2.4.1 Introduction to Fracture Mechanics Method

Note that both the stress- and strain-based approaches are empirical kinds. Though they allow life prediction and design assurance, life improvement or design optimization can be enhanced by using fracture mechanics, which can account for various stages of crack development, and are essential for inspections and repairs. In the fracture mechanics method, the steady (region II described in Sect. 17.2.4.2) and the subsequent unstable crack growth (fracture, region III described in Sect. 17.2.4.2) are explicitly modeled, while the crack initiation is only based on empirical data. It can also model crack growth beyond what is regarded as fatigue failure (e.g., through-thickness crack) in the S–N-based approach [24].

A cracked body can be loaded in any one or combination of the three displacement modes shown in Fig. 17.15. In mode I (opening or tensile mode), the crack faces move apart due to the tension loading, and this mode gives rise to a

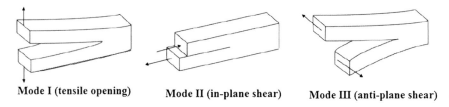

Fig. 17.15 Three basic modes of crack surface displacement

significant stress intensity factor, which characterizes the magnitude (intensity) of the stress in the vicinity of an ideally sharp crack tip in a linear-elastic and isotropic material [243]. The stress intensity factor is hereby defined as:

$$K_I = \sigma_{nom} \sqrt{\pi \cdot a} \cdot f \tag{17.12}$$

where σ_{nom} is the nominal stress, a is the crack length, and f is a dimensionless function depending on the crack geometry and loading, and usually also on the ratio of crack length to another geometric dimension, such as the member width or half width, b.

The stress intensity factor can be checked in many handbooks [307–309]. It can also be calculated by FE analysis. For an elaborated discussion of this concept, readers may read Ref. [243].

Both mode II and mode III are due to shear. In mode II (sliding or in-plane shear), the surfaces of the crack slide over each other. In mode III (tearing or antiplane shear), the surfaces of the crack move parallel to the leading edge of the crack and relative to each other.

Mode I is the predominant loading mode for most of the engineering problems associated with fracture failure or potential fracture failure. Therefore, the majority of assessment using linear elastic fracture mechanics is based purely on mode I. Similar treatment can be readily be extended to mode II and mode III.

The crack growth rate can be described by the sigmoidal shape relationship between cyclic crack growth rate $\frac{da}{dN}$ and stress intensity range ΔK as shown in Fig. 17.16. Three regions in the plot are of engineering interest. At low stress intensity (region I), the crack growth rate is rather slow and goes asymptotically to zero before ΔK reaches a crack growth threshold ΔK_{th}, which is regarded as the lower limiting value of ΔK below which the crack does not grow or grows at too slow a rate (2.5×10^{-7} mm/cycle, which corresponds to the spacing between atoms in metals) to measure, i.e. this corresponds to the endurance or fatigue limit mentioned in Sect. 17.2.2.1. The crack growth in this region has been formulated by several researchers such as Donahue et al. [310]. The crack growth threshold ΔK_{th} mainly depends on the stress ratio (the ratio between the minimum and maximum stress), frequency of loading and environment. Engineering design with ΔK below ΔK_{th} would be very desirable, but this is in many cases not practical due to the low stress range required, which leads to an uneconomical design. As an alternative, limiting defect size so that the ΔK is below ΔK_{th} would also be

17.2 Fatigue Damage Assessment

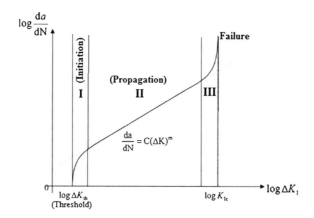

Fig. 17.16 Mode I sigmoidal crack growth rate curve with three regions of behavior assumed by linear elastic fracture mechanics

desirable, but this is even more difficult than limiting the stress range, and practically unattainable [311].

In the intermediate region (region II), the crack shows a stable growth and the curve is essentially linear when using a log–log plot. Many engineering structures with fatigue failure potential operate in this region.

At region III, the crack shows a unstable rapid growth prior to final failure (the growth rate is approaching infinity) of the cracked body and little fatigue life is involved, i.e., in engineering practice, this region may be neglected for the fatigue life estimation. The boundary between region II and region III is dependent on the yield strength of the material, stress intensity factor K_I and stress ratio. The curve approaches the fracture toughness K_c for the material and thickness of interest, which has been formulated by Forman et al. [324] to present the nonlinear and fast growing portion of the curve. This will be presented at the end of Sect. 17.2.4.2. This rapid unstable growth at high ΔK sometimes also involves fully plastic yielding (the size of the associated plastic zone is comparable to or even larger than the size of the crack), i.e., ductile tearing and/or brittle fracture, which should be analyzed using the elastic–plastic fracture mechanics approach such as J-integral or the crack-tip opening displacement (COD) method [311, 312].

17.2.4.2 Formulation of Crack Growth at Regions II and III

Most of the current applications of linear elastic fracture mechanics (LEFM) concepts to describe crack growth are associated with the intermediate region (region II) [311], and Paris' law [313, 314] is widely used to describe the crack growth curve in this region:

$$\frac{da}{dN} = C(\Delta K)^m \tag{17.13}$$

Table 17.3 Representative values of m and C

m	C (mean value) (10^{-12})	Material type
3.0	5.38	Weld metal [321]
3.0	3.9	Base and weld metal [322]
3.0	5–6	Low and medium strength steel [323]

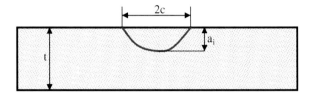

Fig. 17.17 A semi-elliptic crack at a weld toe with an initial crack size of a_i

where a is the crack length; C is the crack growth rate constant found by extending a straight line to, for example, $\Delta K = 1$ MPa $\sqrt{m}$; and m is the slope on the log–log plot of $\frac{da}{dN}$–ΔK relationship; for welded steel joints, m is typically in the range between 3.0 and 5.0, with m = 3 being more typical [315–318]. Both C and m are dependent on the environmental conditions, and can be found in many handbooks and other studies, such as the ones suggested by NORSOK N-004 [326]. Some representative values of them are shown in Table 17.3.

The stress intensity range ΔK can be expressed as:

$$\Delta K = f(a) \cdot S\sqrt{\pi \cdot a} = (F_{mb} \cdot S_{mb} + F_b \cdot S_b)\sqrt{\pi \cdot a} \quad (17.14)$$

where f(a) is a one-dimensional compliance function, depending on the geometry and the relative crack length; subscripts mb and b refer to membrane (axial) and bending stress effects, respectively.

When ΔK is less than the crack growth threshold of stress intensity factor ΔK_{th}, the crack does not grow.

By using Paris' law, the number of the cycles for a crack to propagate from the initial crack size a_i (Fig. 17.17) to the final critical crack size a_f is calculated as:

$$N = \int_{a_i}^{a_f} \frac{da}{C(\Delta K)^m} \quad (17.15)$$

The initial crack size is a rather difficult parameter to define. A typical initial crack size (depth a_i shown in Fig. 17.17) is between 0.05 mm and 0.2 mm. Many researchers use an initial crack size (depth) of 0.1 mm [306, 315]. In BS 7608 [333], an initial crack size (depth) of 0.15 mm is specified. For offshore structures, a_i can be taken to be 0.5 mm [319] or lower [320]. The final critical crack size a_f is mainly dependent on the type of structure and the level of safety requirement. For example, for ships, the utilization of residual strength allow the criteria of "leak before break" [24], even though a formal assessment of crack sizes of the order of the plate thickness is specified by codes such as BS 7910 [327].

17.2 Fatigue Damage Assessment

It has been found that the fracture mechanics approach using Paris' law gives results comparable to the S–N approach when certain material parameters C and m are used and a reasonable initial crack size is assumed.

Since ΔK is a function of the crack length and one dimensional compliance function f(a) expressed in Eq. (17.14), the equation above cannot be integrated analytically and is normally solved numerically. To obtain the solution in a closed form, F(a) can normally be determined at the initial crack length.

It is worth mentioning that, when the initial crack size a_i is much smaller than final critical crack size a_f, the number of cycles to fatigue failure is strongly dependent on a_i, but not sensitive to the changes of a_f, i.e. the fatigue life is not sensitive to the fracture toughness. If the initial crack size a_i is close to the final critical crack size a_f, such as in a case in which a rather hard material is subjected to large stresses, such as the working condition for gears [311], the fatigue life is also sensitive to the fracture toughness.

For cracks with rather small sizes, the application of Paris' law must be calibrated based on the S–N data. This is because the uncertainties involved in the initiation of cracks are difficult to quantify. For more details, see references by Moan et al. [358] and Ayala-Uraga and Moan [359].

Note that Paris' law only models the region II (linear portion of the crack growth curve), as the stress intensity factor range increases to region III with a much faster growth rate than that of region II. Paris' law was modified by Forman [324] to not only account for stress intensity range ΔK, as Paris' law does, but also to explicitly take account of the influence of the mean stress, by introducing a factor depending on $(1-R)$ in the denominator, where R is the stress ratio for cyclic loading (= minimum stress/maximum stress). For a given ΔK, the crack growth rate increases with the increase of R. This is known as Forman's equation:

$$\frac{da}{dN} = \frac{C(\Delta K)^m}{(1 - R)K_C - \Delta K} \tag{17.16}$$

The stress intensity factor ΔK_{th} and fracture toughness K_c in Paris' law were accounted for by Vasudevan and Sadananda [360] who presented a two-parameter unified approach. Cui et al. [361] extended this two-parameter model so that all three regions of crack growth shown in Fig. 17.16 can be calculated.

As mentioned before, if the crack is loaded in a combination of the two or three displacement modes shown in Fig. 17.15, the stress intensity factor in Paris' law can be modified by an effective stress intensity factor, K_{eff}. Readers may read Ref. [315] for a detailed formulation of ΔK_{eff}.

It should also be noticed that, in reality, straight cracks seldom exist. Therefore, the straight crack assumption involved in the conventional fracture mechanics method introduces certain uncertainties.

Finally, readers also need to bear in mind that the curve illustrated in Fig. 17.16 is for a cracked body with little or no plasticity. For a condition of full plasticity, the definition of stress intensity is no longer valid.

17.2.4.3 FE Modeling for Fracture Mechanics Analysis

In linear fracture mechanics, by assuming the presence of a small crack, the propagation of the crack through the material can be simulated. It is normally assumed that a crack is initiated perpendicular to the maximal principal stress. However, unlike the notch stress method, the fracture mechanics method does not normally account for the radius of the transition between the welds and the base materials. Therefore, the transition is typically modeled sharp as shown in Fig. 17.18.

In a number of commercial FE analysis codes to perform the fatigue check using fracture mechanics, quadratic elements are used to simulate the singularity effects at the crack tip so that the crack tip can be simulated by skewing the midside nodes to a shorter (typically 20 to 30 %) distance from the crack tip as shown in Fig. 17.19. Since the crack tip has much higher stress and strain gradients than the remaining part of the structure (the stresses and strains are proportional to

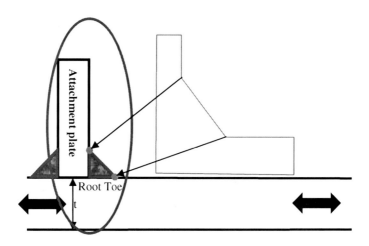

Fig. 17.18 Modeling of the transition between the weld and the base plate material

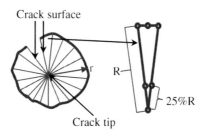

Fig. 17.19 Modeling of crack surface and crack tip in fracture mechanics

17.2.5 Cumulative Damage

In reality, the stress amplitudes experienced by a structural member often vary under dynamic loading. The direct use of S–N curve, which is derived from experiments under constant force amplitude operation, is not possible. Therefore, a proper method to calculate the fatigue damage under variable amplitude of loading has to be adopted. The method most typically used in civil and mechanical engineering fields is the Miner summation [362], also called the Palmgren [363]-Miner rule, which assumes that the fatigue damage on a structure produced by an individual cycle i is constant:

$$D_i = \frac{1}{N_i} \tag{17.17}$$

where N_i is the number of cycles to fatigue failure at a given stress range S_i.

The Miner summation also implies that, in a stress history with various stress ranges S_i, each with a number of cycles n_i, the fatigue damage is a result of a linear accumulation of partial fatigue damage produced by individual cycles, and cycles of different stress ranges do not interact to retard or accelerate crack growth:

$$D = \sum \frac{n_i}{N_i} \tag{17.18}$$

Experimentally, fatigue failure occurs when D ranges from 0.5 to 2.0 [304]. However, for design purposes, it is assumed that fatigue failure occurs when the sum of partial damage is unity:

$$D = \sum \frac{n_i}{N_i} = 1.0 \tag{17.19}$$

A number of researchers have reported that damage produced by individual cycles in variable amplitude load time histories is strongly affected by prior loading history, i.e., the damage in one cycle is not a function of that cycle, but also depends on the preceding cycles. Note that the Miner rule does not account for these sequence effects (stress interaction) [60, 243, 289], i.e., it assumes that the damage caused by a stress cycle is independent of where it occurs in the load history. Therefore, Miner's rule cannot be justified under certain choice of stress sequence, such as applying several high stress cycles followed by several low stress cycles, and vice versa [46]. This may introduce a significant bias. What makes things even more complicated is that no general rule exists to predict if Miner's rule is conservative or non-conservative, primarily because people still lack knowledge of exactly which parameters are related to Miner's rule that influence the fatigue damage results. If Miner's rule is applied in a time history

Table 17.4 Fatigue design factors for offshore structures [325]

Classification of structural components based on damage consequence	Not accessible for inspection and repair or in the splash zone	Accessible for inspection, change or repair and where inspection or change is assumed
Substantial consequences	10	2
No substantial consequences	3	1

with isolated overload, the cumulated fatigue damage will normally be conservative. On the other hand, if the model is applied to a time history containing large cycles separated by few small cycles, and the mean of the small cycles is high relative to the mean of the large cycles, the calculated fatigue life will generally be non-conservative [305].

To account for various types of uncertainties mainly related to Miner's rule, in most of the design codes, an additional safety factor is introduced as fatigue design factor (FDF) ranging from 1.0 to 20, which depends on the damage consequence in terms of fatalities, economic losses, pollution, and inspection accessibility. FDF generally results in a much lower Miner's sum than unity:

$$D \leq \frac{1.0}{\text{FDF}} \tag{17.20}$$

Table 17.4 shows required FDFs for offshore structures in accordance with Norwegian standard NORSOK N-001 [325].

For ships and floating production storage and offloading vessels (FPSOs), the FDF is normally taken to be 1.0. This is because cracks in the majority of those marine vessels do not cause an immediate risk of global failure.

For the fatigue assessment of TLPs' (tension leg platforms) tendon, due to the high consequence, this FDF is typically taken to be 10.

However, one can argue that, in addition to the general condition of inspection and consequence, the selection of FDF is also a function of a more precise measure of residual strength and explicit measure effects from inspection, including the quality of the inspection [24].

For vehicle design, there is no unified FDF specification and it is normally company specific, i.e., it differs from one company to another, and this factor is in many cases confidential to protect the interests of relevant companies. It varies at different parts of a vehicle.

Fortunately, the uncertainty associated with the relative Miner sum seems to be smaller than the inherent uncertainty in the S–N curve [60], and the test series also show a reduced scatter in Miner sums as compared to the S–N curve data [299].

Miner's summation can also be applied to the fracture mechanics method to assess fatigue. Interested readers may refer to references [60, 243].

Despite the drawbacks of using Miner's rule, in engineering practice of fatigue estimations, load cycle counting by the rain-flow method (Sect. 17.3.5.1), Miner's rule defining the linear damage accumulation rule, and the S–N curve (Sect. 17.2.2) representing the material performance determined from constant-amplitude

17.2 Fatigue Damage Assessment

fatigue tests represent the state of practice, and are often used for fatigue damage estimation of structures subjected to random loadings.

17.3 Dynamic Analysis Methods for Calculating Fatigue Damage

There are several analysis methods for calculating fatigue based on the S–N-based method: deterministic approach, simplified fatigue analysis approach (based on a proper assumption of long-term probability distribution of stress responses), stochastic approach for narrow band and broad band stress ranges, and semi-stochastic time domain approach. In the following, we will discuss each method by presenting a procedure for calculating wave- or wind-induced fatigue. However, readers can generalize each method to calculate the fatigue life due to other types of dynamic loadings.

17.3.1 Deterministic Fatigue Analysis Method

Deterministic fatigue analysis is based on the assumption that each load and corresponding responses possess one individual frequency as well as a stress range with zero mean. A schematic procedure for performing a deterministic fatigue analysis is illustrated in Fig. 17.20.

In this type of analysis, a deterministic load (hydrodynamic loading in the case of wave or wind load) analysis is performed followed by a dynamic or static structural analysis. For example, to calculate sea wave-induced fatigue, regular waves each with a design wave height (H) and wave period (T) combination need to pass through a target structure, generating a cyclic wave loading. It is noticed that in order to adopt a regular wave in the deterministic analysis, design wave height and period instead of significant wave height (H_s) and the corresponding zero crossing period (T_z) should be used. At required locations on the structure, the stress time history responses through a structural analysis, covering the steps through the entire wave cycle, are calculated at sufficiently small time intervals. Thereafter, the stress range at those locations for each individual wave (sea state q and direction j) can be obtained and then fitted into the design S–N curve to find the corresponding fatigue damage ($D_{1cycle(j,q)}$) for one cycle.

For each given load state (sea state or wind state) from a defined direction specified in the hydrodynamic analysis, it is necessary to specify the total number (n) of loads passing through the structure within a certain period τ (in seconds), for example, 1 year:

$$n = \frac{\tau}{T} \tag{17.21}$$

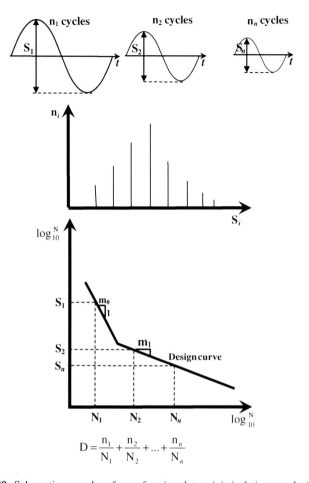

Fig. 17.20 Schematic procedure for performing deterministic fatigue analysis

where $\tau = 365$ days $\times$ 24 h $\times$ 60 min $\times$ 60 s $= 31{,}536{,}000$ s; and T is the period of the loading.

Hence, the individual (partial) fatigue damage ($D_{j,q}$) by each load state q with the corresponding load direction j is calculated as the fatigue damage ($D_{1cycle(j,q)}$) due to one cycle of stress range under that load state multiplied by the number of cycles n for 1 year.

$$D_{j,q} = n \cdot D_{1cycle(j,q)} \tag{17.22}$$

where $D_{1cycle(j,q)} = \frac{1}{N_{(j,q)}}$; $N_{(j,q)}$ is the number of cycles to fatigue failure under load state q with the corresponding load direction j.

From load (wave or wind) statistics, one can obtain that a given load state q (sea state or wind state) from a defined direction j occurs a fraction $\Psi_{j,q}$ (probability of occurrence) of the entire life time for the target structure.

17.3 Dynamic Analysis Methods for Calculating Fatigue Damage

Based on Miner's rule, partial damages are then weighted over this probability of occurrence $\Psi_{j,q}$ with various load states and directions, the total damage (D_{total}) due to all load states from all directions can then be summed up as:

$$D_{total} = \sum_j \sum_q \left(\Psi_{j,q} \cdot D_{j,q} \right) \qquad (17.23)$$

Since the fatigue damage is based on one year's duration for all load cases together, the fatigue life in years can simply be calculated as the inverse of the 1 year fatigue damage:

$$L_{total} = \frac{1}{D_{total}} \qquad (17.24)$$

For sea wave-induced fatigue analysis, deterministic fatigue analysis applies especially to members within splash zone (areas of the structure that are periodically wetted due to waves and tidal variations), where the local wave loading on members and wave load nonlinearities (due to the variation of wave surface and drag-induced nonlinearities) can be significant, neither of which the stochastic fatigue analysis presented in the Sect. 17.3.3 can handle properly.

17.3.2 Simplified Fatigue Analysis Approach

For establishing the general acceptability of fatigue resistance, or as a screening process to identify the most critical details [300] to be considered in a stochastic fatigue analysis, as will be elaborated in Sect. 17.3.3, an even more simplified deterministic fatigue analysis can be performed through a proper assumption of long-term probability distribution of stress responses. This is similar to the case in which environmental data measured for only a few years are extrapolated in order to estimate the long-term environmental loads. This enables a significant reduction of analysis efforts, i.e., one only needs to carry out hydrodynamic and subsequent structural analysis using one wave from each wave direction, and the entire life time distribution of the stress can be obtained and used for fatigue assessment. The method is a type of "simplified fatigue analysis," which is efficient but conservative as a tool to extrapolate results of detailed fatigue analyzes among similar offshore structures [81].

For example, based on a Weibull distribution of stress exceedence, Marshall and Luyties [301] proposed that the number of stress cycles exceeding stress σ in n_0 cycles is:

$$n(stress > \sigma) = n_0 e^{\left[-\left(\frac{\sigma}{\sigma_0} \right)^h \ln(n_0) \right]} \qquad (17.25)$$

Table 17.5 Values of $m\sqrt{\Gamma\left(1+\frac{m}{h}\right)}$

h m	3.0	3.1	3.3	3.5	3.7	4.2
0.4	24.12	25.78	29.28	33.04	37.08	48.38
0.5	8.96	9.43	10.41	11.42	12.49	15.35
0.7	3.35	3.46	3.69	3.93	4.17	4.79
0.9	2.10	2.15	2.26	2.37	2.47	2.47
1.0	1.82	1.86	1.94	2.02	2.10	2.29
1.1	1.63	1.66	1.72	1.78	1.85	2.00
1.3	1.39	1.42	1.46	1.50	1.54	1.64
1.5	1.26	1.28	1.31	1.34	1.37	1.45

where σ_0 is the stress range that is exceeded once in n_0 cycles; and h is the Weibull shape parameter determined according to the characteristics of load and structural responses, typically ranging from 0.5 to 1.5.

From the equation above, it is noted that the determination of the Weibull shape parameter h is an essential task in the simplified fatigue analysis method.

When the long-term stress range distribution is defined using two parameter Weibull distribution for various load conditions, the fatigue damage in n cycles is then:

$$D_n = \frac{\left[\Gamma\left(1+\frac{m}{h}\right)\right] \cdot n_0 \sigma_0^m}{A[\ln(n_0)]^{\frac{m}{h}}} \tag{17.26}$$

where $\Gamma(x)$ is the gamma function (in cases where x is a positive integer, $\Gamma(x) = (x-1)!$).

By setting the allowable cumulative fatigue damage as D, one obtains the maximum allowable stress range as [60]:

$$A_{allow} = m\sqrt{\frac{DA}{n_0}} \cdot \frac{[\ln(n_0)]^{\frac{1}{h}}}{m\sqrt{\Gamma\left(1+\frac{m}{h}\right)}} \tag{17.27}$$

It is noted that the two equations above are based on a single slope S–N curve, which generally yields conservative assessment with regard to fatigue damage.

Table 17.5 gives some values of $m\sqrt{\Gamma\left(1+\frac{m}{h}\right)}$.

Using the allowable stress calculation above, and by giving the estimated maximum number of cycles (typically 10^8 for offshore structures), Weibull shape parameter h, the allowable cumulative fatigue damage D, and the type of S–N curve (e.g., W3, W2, ... B1 in DnV RP 203 [330]), one can calculate the allowable local stress range that serves for pre-engineering design. In some codes, a design chart with various combinations among allowable stress range A_{allow}, Weibull shape parameter h and type of S–N curves is given for a convenient check.

As mentioned above, the determination of Weibull shape parameter h is essential for performing simplified fatigue analysis. However, there is no explicit answer for determining its value. The value is normally selected either based on experience from fatigue analysis of similar structures or by carrying out relevant numerical analyses. And in practice, the parameter is influenced by the structure type, dynamic amplification, water depth, wave climate (long term distribution of wave height) and position of the joint in the structure [60].

Alternatively, by assuming a log-linear long-term wave height distribution, a similar method is proposed by Williams and Rinnie [302].

17.3.3 Stochastic Fatigue Analysis Method with Narrow-Banded Responses

Since the deterministic fatigue analysis approach does not contain the information regarding the energy content at various frequencies, it is not suitable to be directly used to calculate the dynamic responses. Specifically, it cannot successfully handle dynamic sensitive structures for which the dynamic loading has a band of periods close to the structure's important eigenperiods. This is due to the fact that a slight variation of assumed period in the loading can significantly alter the responses. In addition, dynamically, it is possible that the number of response cycles is not identical to the number of loading cycles.

Moreover, for performing deterministic fatigue analysis, a significant element of engineering judgment is needed to properly select the collection of discrete deterministic waves, which need to be sufficient to establish the fatigue demand that a structure would experience. Therefore, when an explicit fatigue assessment is to be pursued for offshore structures that are designed on a site-specific basis, preference is given to stochastic fatigue assessments over a deterministic approach [303].

However, mobile offshore units are not based on site-specific sea state data, and self-elevating units could experience significant variations with regard to water depth and large variations of wave-induced responses can be expected. For example, the fatigue assessment on the most important locations on the legs of a mobile jack-up structure are site specific [303]. In this case, preference is given to deterministic fatigue assessments over a stochastic approach.

Furthermore, the deterministic approach requires a significant amount of data storage for each combination of wave height and period, stress responses at a number of (typically more than nine) steps need to be stored. With a large amount of waves, the required data storage is massive. For example, considering an off-shore structure subject to waves from 12 directions with a 30° direction interval, with 100 combinations of wave height and period per direction and responses at 15 steps per wave, the stresses at only 1 location require 18,000 ($12 \times 100 \times 15$) response calculations to determine the 1,200 stress ranges.

354   17   Fatigue Assessment

In addition, as mentioned previously, deterministic fatigue analysis implies that loading is produced by a regular wave, which does not take into account the stochastic nature of the sea waves.

All the drawbacks and difficulties of deterministic fatigue analysis above can be solved by using stochastic analysis, as will be elaborated in the following.

A stochastic fatigue analysis requires a linearized frequency domain hydrodynamic analysis followed by a quasi-static or dynamic structural analysis. It is applicable for structures subjected to dynamic loading that have statistically stationary properties for a large number of stress cycles [81]. The essential feature of stochastic analysis is the fact that the stress response ranges and periods (i.e., number of cycles) can be determined from the stress response spectrum, which is calculated by multiplying load spectrum with the square of the modulus of stress transfer function. For each calculated stress response spectrum, by assuming that the short-term stress range responses follow Rayleigh probability distribution (whereby it is assumed that the variation of stress is a narrow-banded random Gaussian process as discussed in Sect. 10.3), the number of cycles in each stress range category can be simply obtained and used to calculate the partial fatigue damage contributed by each stress range. When a narrow band assumption is not valid for the stress process, a correction factor can be applied in the calculation of short-term fatigue damage as will be presented in Sect. 17.3.5.2. Having calculated the short-term damage, by using Miner's rule the fatigue damage from all stress ranges can be calculated for the defined stress response spectrum. The entire procedure is illustrated in Fig. 17.21 and elaborated as follows.

Load (stress) transfer functions (for waves, this is with respect to the wave height and period) are obtained by passing a harmonic loading function (for waves, this is for example the airy wave [364]) with predefined load conditions (for waves, this is the wave height at different periods and directions; for wind, this is the wind speed at different directions) through the target structure. Figure 17.22 shows a set of the modulus of the stress transfer functions ($|H(f)|$, see Sect. 11.1.1) to unit wave height (from various directions) at a selected structure location obtained through a structural analysis. The stress transfer function must be determined on the basis of analyses of a sufficient number of wave/wind directions. For each direction of stress transfer function, tens of load (wave or wind) periods are incorporated at a period range of engineering interests (accounting for the influence from both loading and responses).

Thereafter, the load spectrum is multiplied by the square of the modulus of hotspot stress transfer function for each of the load (wave or wind) directions in order to provide the hotspot stress response spectrum $S_{\sigma\sigma}(f)$:

$$S_{\sigma\sigma}(f) = |H(f)|^2 S(f) \qquad (17.28)$$

where $S(f)$ is the load (wave or wind) energy spectrum at various frequencies f. For waves, this is the wave spectrum corresponding to each sea state with predefined significant wave height and zero crossing period (not design wave height and period) combinations from each direction; for wind, this is the wind

17.3 Dynamic Analysis Methods for Calculating Fatigue Damage

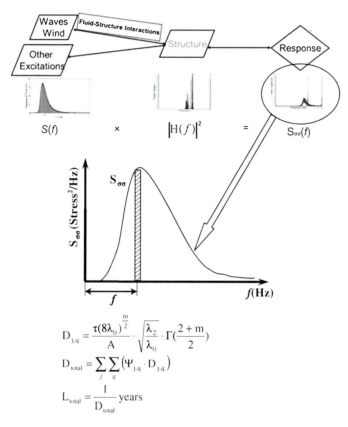

Fig. 17.21 Schematic procedure for a stochastic fatigue calculation (τ is 1 year measured in seconds, $\Psi_{j,q}$ is the probability of occurrence for a given load state q (sea state or wind state) from the load direction j)

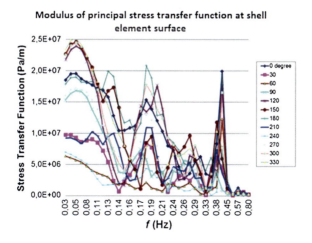

Fig. 17.22 Modulus of transfer functions ($|H(f)|$) of principal stress at a shell element surface at a location on the top of a GBS shaft (12 wave directions with an interval of 30°, wave periods range from 1.25 s ($f = 0.80$ Hz) to 30.0 s ($f = 0.033$ Hz))

spectrum at each wind speed from each wind direction; f is the frequency in Hz; $H(f)$ is the stress transfer function at various frequencies f.

It should be noted that the abscissa of the spectra used here have a unit of frequency Hz instead of an angular frequency, which is used in Sect. 10.2.

The zero moment (variance) λ_0, which is equal to the area under the stress spectrum curve about the vertical axis (at $f = 0$), is also equal to the square of the root of mean square of the stress process:

$$\lambda_0 = \sigma_{RMS}^2 = \int_0^\infty S_{\sigma\sigma}(f)df = \int_0^\infty |H(f)|^2 S(f)df \qquad (17.29)$$

For every load state from a defined load direction, there is an associated average load (wave or wind) period T_z:

$$T_Z = \sqrt{\frac{\lambda_0}{\lambda_2}} \qquad (17.30)$$

where $\lambda_2 = \int_0^\infty S_{\sigma\sigma}(f)f^2 df$ is the second moment with respect to the vertical axis.

Similar to the deterministic analysis method, one can assume that the given load state (sea state or wind state from a defined direction) occurs with a predefined duration τ in seconds, for example 1 year as $\tau = 365$ days $\times$ 24 h $\times$ 60 min $\times$ 60 s $= 31{,}536{,}000$ s.

The corresponding number of cycles that occur for this load state within a year is then:

$$n = \frac{\tau}{T_Z} = \tau \sqrt{\frac{\lambda_2}{\lambda_0}} \qquad (17.31)$$

By assuming that the short-term stress range follows Rayleigh distribution (Sect. 10.3) in a given stress response spectrum, the probability density of stress range banded around σ_s is given as:

$$p(\sigma_s) = \frac{\sigma_s}{4\lambda_0} e^{-\left(\frac{\sigma_s^2}{8\lambda_0}\right)} \qquad (17.32)$$

Within τ seconds, the number of stress cycles banded around σ_s is:

$$dn = n \cdot p(\sigma_s)d\sigma_s \qquad (17.33)$$

The fatigue damage dD associated with this band of stress cycles using the S–N curve data is:

$$dD = \frac{dn}{N} = \frac{dn}{A\sigma_s^{-m}} = \frac{n \cdot p(\sigma_s)d\sigma_s}{A\sigma_s^{-m}} = \frac{n \cdot \frac{\sigma_s}{4\lambda_0} e^{-\left(\frac{\sigma_s^2}{8\lambda_0}\right)} d\sigma_s}{A\sigma_s^{-m}} \qquad (17.34)$$

17.3 Dynamic Analysis Methods for Calculating Fatigue Damage

The 1 year partial fatigue damage for an entire stress response spectrum corresponding to a load state q and a load direction j is simply obtained by integrating the expression above:

$$D_{j,q} = \int_0^\infty dD d\sigma_s = \int_0^\infty \frac{n \cdot \frac{\sigma_s}{4\lambda_0} e^{-\left(\frac{\sigma_s^2}{8\lambda_0}\right)}}{A\sigma_s^{-m}} d\sigma_s = \frac{n}{4A\lambda_0} \int_0^\infty \sigma_s^{(1+m)} e^{-\left(\frac{\sigma_s^2}{8\lambda_0}\right)} d\sigma_s \quad (17.35)$$

For a constant m, the integral above has a gamma function solution [81]:

$$D_{j,q} = \frac{(8\lambda_0)^{\frac{m}{2}}}{A} \cdot \tau \sqrt{\frac{\lambda_2}{\lambda_0}} \cdot \Gamma\left(\frac{2+m}{2}\right) \quad (17.36)$$

By equating the fatigue damage calculation above to the fatigue damage under constant stress amplitude, one has

$$\frac{(8\lambda_0)^{\frac{m}{2}}}{A} \cdot \tau \sqrt{\frac{\lambda_2}{\lambda_0}} \cdot \Gamma\left(\frac{2+m}{2}\right) = \frac{nS^m}{A} \quad (17.37)$$

One obtains the equivalent fatigue stress range σ_e (also called effective stress range) that would produce the same amount of fatigue damage:

$$\sigma_e = (8\lambda_0)^{\frac{1}{2}} \left[\Gamma\left(\frac{2+m}{2}\right)\right]^{\frac{1}{m}} \quad (17.38)$$

Note that the root of the mean square of stress range is:

$$\sigma_{RMS} = \sqrt{8\lambda_0} \quad (17.39)$$

And the significant stress range is:

$$\sigma_s = \sqrt{4\lambda_0} \quad (17.40)$$

One can also obtain the relationship among the three stress terms as:

$$\sigma_e = \sigma_{RMS} \left[\Gamma\left(\frac{2+m}{2}\right)\right]^{\frac{1}{m}} = \sigma_s \frac{\left[\Gamma\left(\frac{2+m}{2}\right)\right]^{\frac{1}{m}}}{\sqrt{2}} \quad (17.41)$$

From load (wave or wind) statistics, a given load state q (sea state or wind state) from a defined direction j occurs with the probability of occurrence $\Psi_{j,q}$. Partial damages are then weighted over this probability of occurrence with various load states and directions in order to assess the total damage:

$$D_{total} = \sum_j \sum_q \left(\Psi_{j,q} \cdot D_{j,q}\right) \quad (17.42)$$

And the total fatigue life L in years is calculated as:

$$L_{total} = \frac{1}{D_{total}} \tag{17.43}$$

It should be noted that almost all responses due to environmental loadings are not narrow banded, or at least not strictly narrow banded. The narrow band assumption always leads to a conservative evaluation with regard to fatigue, even though this in many cases fulfills the requirement of engineering accuracy. The fatigue assessment due to wide-band load effects can be addressed in the frequency domain, by either modifying the fatigue damage calculated from a narrow band assumption or by manipulating the probability density of stress ranges calculated from rain-flow counting, which will be discussed in Sect. 17.3.5.2.

Example: Assume that the wave-induced responses for fixed offshore structures are narrow banded and dominated by a single vibration mode corresponding to the fundamental eigenfrequency. Establish the relationship between fatigue lives and the variation of fundamental eigenfrequencies.

Solution: Note that the fatigue damage accumulation experienced by a fixed offshore structure, such as a jacket or a jack-up structure, is mainly caused by the inertia force-dominated (rather than drag force-dominated) wave loading due to small and medium sized sea waves. Therefore, by neglecting the wave-induced drag forces, and assuming that the fundamental eigenmode dominates the dynamic responses of the structure, i.e. the quasi-static contribution to the mean square stress is assumed to be small, the ratio of fatigue damage at two different natural frequencies can be expressed by:

$$\frac{FL'}{FL} = \left(\frac{f_1'}{f_1}\right)\left(\frac{\sigma_{dl}'}{\sigma_{dl}}\right)^m \tag{17.44}$$

where σ_{dl} and σ_{dl}' are the mean square root stress from the dynamic responses of the vibration mode at the two different natural frequencies, and m is the negative inverse slope of the S–N curve for the material.

Depending on whether the cause of natural frequency change is due to stiffness or mass change, the mean square root stress is influenced differently [213].

For stiffness change-dominated dynamic responses, σ_{dl} can be expressed as:

$$\sigma_{dl} = \sqrt{\frac{A^2}{M_1} \frac{S(\omega_1)}{\omega_1^5}} \tag{17.45}$$

For mass change-dominated dynamic responses, σ_{dl} can be expressed by:

$$\sigma_{dl} = \sqrt{\frac{A^2}{K_1} \frac{S(\omega_1)}{\omega_1^3}} \qquad (17.46)$$

where A is a constant depending on structural and wave characteristics such as damping, wave spreading, etc. M_1 and K_1 are modal stiffness and modal mass, and are related by:

$$K_1 = M_1 \cdot \omega_1^2 \qquad (17.47)$$

For wind-driven sea states, the upper bound of the wave spectrum value can be modeled at frequencies higher than the frequency of the peak [214]. The wave spectrum can then be written as:

$$S_{\max}(\omega) = 2.5783 \cdot 10^{-4} \left(\frac{\omega}{2\pi}\right)^{-4.6} \qquad (17.48)$$

By assuming that, for small variation in natural frequency ω_1, the ratio between mean square modal deflection and mean square stress remains constant, i.e. the mode shape does not change with the variation of natural frequency [213], and by further assuming that all of the constants of the spectra in Eq.(17.48) are absorbed into a constant A, Eqs. (17.45) and (17.46) can then be rewritten as:

For the stiffness change-dominated case:

$$\sigma_{dl} = \sqrt{\frac{A^2}{M_1} \frac{1}{\omega_1^{9.6}}} \qquad (17.49)$$

For the mass change-dominated case:

$$\sigma_{dl} = \sqrt{\frac{A^2}{K_1} \frac{1}{\omega_1^{7.6}}} \qquad (17.50)$$

By substituting the two equations above into Eq. (17.44), one finally obtains the fatigue damage ratio as expressed in Eq. (17.51) for the stiffness change-dominated case, and Eq. (17.52) for the mass change-dominated case:

$$\frac{FL'}{FL} = \left(\frac{f_1'}{f_1}\right)^{(-4.8m+1)} \qquad (17.51)$$

$$\frac{FL'}{FL} = \left(\frac{f_1'}{f_1}\right)^{(-3.8m+1)} \qquad (17.52)$$

where f_1' and f_1 are the natural frequencies that are assumed to dominate the dynamic structural responses.

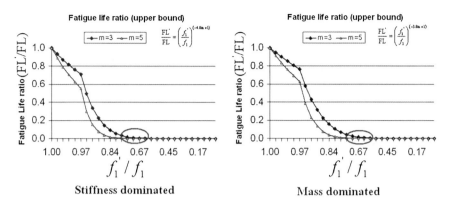

Fig. 17.23 The fatigue life ratio due to the variation of natural frequency, m is the negative inverse slope of the S–N curve

Based on the two equations above, Fig. 17.23 shows the fatigue life ratio varying with the ratio of natural frequency. It is noticed that a 1 % variation of natural frequency can result in a fatigue life change of up to 13 %, 17 %, and 21 % for m = 3.0, m = 4.1 and m = 5.0, respectively. If the natural frequency decreases to less than 80 % of the original one, the fatigue life is close to zero, as marked by the circles in Fig. 17.23.

It should be noticed that, in a practical case, both quasi-static and dynamic responses contribute to the mean square stress, while the derivation of the two equations above omits the contribution from the quasi-static part of the responses. Therefore, the results from these equations are regarded as the upper-bound solutions. In reality, the sensitivity of fatigue lives due to the variation of natural frequencies may not be so high.

17.3.4 Deterministic Versus Stochastic Fatigue Analysis for Structures Subjected to Wave Loads

As discussed in Sects. 17.3.1 and 17.3.3, for sea wave-induced fatigue analysis, deterministic fatigue analysis applies especially to members within the splash zone because the load nonlinearities involved in the splash zone are mainly due to variations of the water surface, while stochastic analysis applies to dynamic sensitive structures. Experience shows that the fatigue life calculated by deterministic fatigue life is often lower than that calculated by stochastic fatigue analysis.

17.3 Dynamic Analysis Methods for Calculating Fatigue Damage

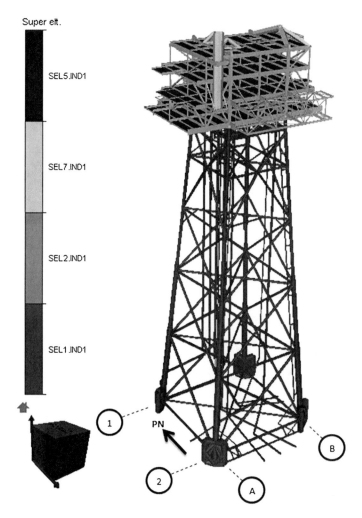

Fig. 17.24 Geometry model of a jacket structure for fatigue analysis (*PN* stands for platform north direction) (courtesy of Aker Solutions)

The fatigue lives of an offshore jacket structure shown in Fig. 17.24 are calculated by both stochastic and deterministic fatigue analysis. The fatigue lives on different locations of legs are selected for checking as shown in Table 17.6. It is clearly shown that the stochastic fatigue calculations generally result in a higher calculated fatigue life.

Table 17.6 Fatigue lives for leg nodes along row A and row B shown in Fig. 17.24

Joints	Fatigue life (years)	
	Stoch.	Determ.
Joints on various	242	92
locations along	367	273
row A and B	358	181
	379	229
	990	158
	1190	252
	1480	154
	1500	359
	1740	468
	2200	1240
	1880	310
	1820	336
	1980	318
	1810	399
	2050	237
	2250	663
	2330	570
	2700	507
	2750	833
	3590	3400
	3470	688
	2980	321
	3520	716
	3530	356
	3930	898

17.3.5 Fatigue Analysis Methods Accounting for Bandwidth, Multi-modal Frequency and Nonlinearities

It is noticed that the assumed Rayleigh distribution of short-term stress range (stress responses follow a Gaussian distribution) in the stochastic fatigue calculation is only valid if the responses are approximately narrow banded. When the loading frequencies are far from a structure's important eigenfrequencies, the structural responses will have a noticeable variety of frequencies, and jagged and irregular time history, which are obviously a broad-banded process.

Furthermore, when structures' oscillation frequencies or important eigenfrequencies are well separated from that of the loading, the resulting responses may exhibit multi-modal frequency components. The fatigue calculation of the relevant responses then requires special assumptions and mathematical treatment.

In addition, if significant nonlinearities are involved, the stochastic analysis is no longer valid, and the responses are non-Gaussian. As discussed in Sect. 10.2, examples of these nonlinearities for offshore structures are the variation of the water surface causing the intermittency of the wave loading, the variation of

17.3 Dynamic Analysis Methods for Calculating Fatigue Damage

buoyancy forces on members in the splash zone [53–57], large structural deformation and nonlinear plasticity, and the nonlinearities induced from drag forces (Morison's equations) [52, 53, 201]. The spectrum fatigue analysis method cannot efficiently handle non-Gaussian responses in an accurate manner. For instance, the fatigue damage induced by the wave-induced drag force can be as much as several times that under the Gaussian assumption [24].

The problems with regard to the bandwidth can be solved by a simple modification to fatigue damage calculated from a narrow-banded assumption (Sect. 17.3.3), which will be illustrated in Sect. 17.3.5.2. The problem regarding both bandwidth and nonlinearities can be solved by the semi-stochastic analysis in the time domain, as will be elaborated in the Sect. 17.3.5.4, even if the method in the time domain requires a significant increase of computer memory storage.

17.3.5.1 Cycle Counting of Stress Response Time History

Figure 17.25 shows two stress time history samples that are wide and narrow banded. For the narrow-band stress history, the stress ranges can be easily identified between the two adjacent upcrossing points. However, with the increase of bandwidth, numerous small peaks occur and it is not immediately obvious how to count stress cycles to be used in Miner's accumulation rule, as is shown in the lower figure of Fig. 17.25. The stress history is normally reduced to a sequence of events that can be regarded as compatible with constant stress range in S–N curves, such a reduction is cycle counting. It is also known that pairing large peaks and troughs gives an upper-bound solution with regard to fatigue damage, while paring sequential peaks and troughs gives a lower bound.

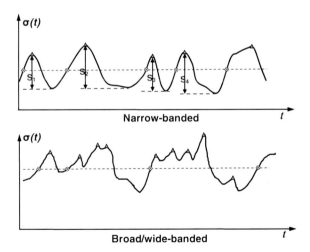

Fig. 17.25 Samples of random stress response time series showing narrow- (*upper*) and wide-banded (*lower*) characteristics, respectively (▲ indicates indicates a local maxima (positive and negative peak), ⊗ indicates an upcrossing)

A number of algorithms [367, 368] have been proposed for stress counting for the wide-band process, such as peak counting, range counting, level crossing counting, and rain-flow counting. All of these construct effective stress ranges based on series of peaks (local maxima) and valleys (local minima). The output from each counting method is the stress range with the associated number of occurrences, which can be further used in combination with relevant S–N curves and cumulative laws such as Miner's summation, so that the final fatigue damage can be obtained. The most widely recognized method is the rain-flow counting method [243] originally proposed by Matsuishi and Endo [369]. The method received its name because it resembles rain flowing off a pagoda roof [369]. It uses a specific cycle counting scheme to account for effective stress ranges and identify stress cycles related to closed hysteresis loops in the material stress–strain diagram [81, 304].

Over time, several versions of rain-flow counting methods have been proposed, such as the Range-Pairs count, Wetzel's method, and pagoda roof method, all of which are fairly complicated. To make this description simpler, one needs to perform the counting with a time history that starts and ends at either the highest peak or lowest trough [81]. Each local maximum is paired with one particular local minimum such that the minimum is the lowest drop before reaching the value of the local maximum again both forward and backward in time. See references [60, 243, 370] for elaborations of the background for rain-flow counting methods.

By using a Gaussian load process, Rychlik [376] has formulated the rain-flow counting method in an explicit mathematical manner and provided the basis for deriving the long-term distribution of rain-flow cycle amplitudes. As illustrated in Fig. 17.26, suppose that the stress $\sigma(t)$ ($-T < t < T$) has a local stress maximum $\sigma(\delta)$ at time δ. One needs to search for the lowest values in both forward and backward direction (t^+ for forward direction and t^- for backward direction) in time between the time point of the local maximum and the nearest at which the stress exceeds the value of the local maximum $\sigma(\delta)$, i.e., t^+ is the first time at which the first up crossing point occurs after time δ, and t^- is the first time at which the last down crossing occurs before time δ. If one cannot find the first up crossing point, $t^+ = T$; similarly, if one cannot find the last down crossing point, $t^- = -T$. Thereafter, one can define the difference between the local stress maximum $\sigma(\delta)$ and the stress value at t^+ and t^-:

$$\sigma_a^+(\delta) = \max\{\sigma(\delta) - \sigma(\tau)\} \text{ for } \delta < \tau < t^+ \tag{17.53}$$

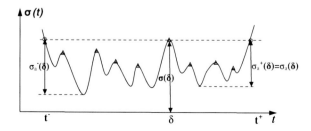

Fig. 17.26 Illustration of rain-flow counting method proposed by Rychlik [376] (▲ indicates a local maxima (positive and negative peak), ⊗ indicates a crossing point)

17.3 Dynamic Analysis Methods for Calculating Fatigue Damage

$$\sigma_a^-(\delta) = \max\{ \sigma(\delta) - \sigma(\tau)\} \text{ for } t^- < \tau < \delta \tag{17.54}$$

The rain-flow cycle amplitude $\sigma_a(\delta)$ is the minimum of the two values above:

$$\sigma_a(\delta) = \min\{ \sigma_a^+(\delta), \sigma_a^-(\delta)\} \tag{17.55}$$

For a more elaborated description of the rain-flow counting method, readers may read references [46, 243, 368].

Rychlik [377], and Frendhal and Rychlik [378] have proved the following rigorous relation with regard to fatigue damage by various counting rules:

$$D_{RC} \leq D_{RFC} \leq D_{LCC}(= D_{NB}) \leq D_{PC} \tag{17.56}$$

where $D_{RC}, D_{RFC}, D_{LCC}, D_{NB}, D_{PC}$ represent the fatigue damage calculated by range counting, rain-flow counting, level-crossing counting, narrow band approximation, and peak counting.

For an elaboration of range counting, level-crossing counting, and peak counting, ASTM Designation E 1049-85 [367] is recommended for reading.

From the equation above, it is important to know that fatigue damage based on rain-flow counting has an upper-bound limit that is equal to the damage estimated from the narrow-band approximation. Furthermore, range counting is less conservative than either rain-flow counting or narrow band approximations.

The rain-flow counting method has its limitation in that it is difficult to determine the probability distribution of rain-flow stress amplitudes.

Since rain-flow stress counting does not have a closed form solution in the frequency domain for a bi- or tri-modal or generally wide-band Gaussian process [24], in a spectrum fatigue analysis, it can be implicitly accounted for either by amending the narrow band fatigue damage with a correction factor or by a direct calculation of the rain-flow counting [243] from the response spectrum of stress range. Both approaches provide closed form expression, and they result in a lower fatigue damage than their narrow band counterpart. For example, Gao and Moan [379] presented that for bandwidth parameter (defined in Sect. 10.2) ranging between 0.3 and 0.5, the narrow-band assumption gives an overestimation of fatigue damage by less than 10 % and 30 %, respectively.

17.3.5.2 Spectrum Fatigue Analysis Methods Accounting for Bandwidth

Modifying Narrow-Banded Fatigue Damage

Bandwidth can conveniently be accounted for by modifying the fatigue damage calculated from a narrow-band assumption (elaborated in Sect. 17.3.3).

In Sect. 10.2, the average bandwidth parameter of energy density spectrum [67] ψ is introduced as:

$$\psi = \sqrt{1 - \left(\frac{\overline{T}_p}{\overline{T}_z}\right)^2} = \sqrt{1 - \frac{\lambda_2^2}{\lambda_0 \cdot \lambda_4}} \tag{17.57}$$

$\psi = 1.0$ indicates a broad/wide-band process and 0.0 a narrow-band process.

To calculate the fatigue damage with responses being wide-banded, by analyzing the time series samples from 17 uni-modal and 17 bi-modal spectra, with the inverse slope in the S–N curve m varying at 3.0, 4.0, 5.0, 6.0 and 10.0 and with the bandwidth parameter ranging from 0.45 to 1.0, Wirsching and Light [371] proposed an empirical rain-flow correction factor $\Phi(m, \Psi)$ to the narrow band fatigue damage D as:

$$D_{WB} = D \cdot \Phi(m, \Psi) \tag{17.58}$$

where D_{WB} and D are the wide- and narrow-band fatigue damage, respectively; The correction factor $\Phi(m, \Psi)$ is expressed as:

$$\Phi(m, \Psi) = 0.926 - 0.033m + \left[(0.074 + 0.033m) \cdot (1 - \Psi)^{(1.587m - 2.323)}\right] \tag{17.59}$$

A number of researchers [372, 373] have presented approaches similar to the effective stress range (weighted average stress) discussed in Sect. 17.3.3. One such is that proposed by Kam and Dover [372]:

$$\sigma_{eWB} = (8\lambda_0)^{\frac{1}{2}} \left[\Phi(m, \Psi) \cdot \Gamma\left(\frac{2+m}{2}\right)\right]^{\frac{1}{m}} \tag{17.60}$$

Direct calculation of Rain-Flow Counting From the Stress Response Spectrum

Note that the methods introduced above are based on the narrow-band fatigue damage. A completely different approach is based on the probability density of stress ranges calculated from rain-flow counting.

Here we first introduce another set of parameters to define bandwidth as functions of spectral moments:

$$\alpha_X = \frac{\lambda_X}{\sqrt{\lambda_0 \lambda_{2X}}} \text{ for } X = 1, 2, 3\ldots \tag{17.61}$$

Thus we have:

$$\alpha_1 = \frac{\lambda_1}{\sqrt{\lambda_0 \lambda_2}} \geq 0.0 \tag{17.62}$$

17.3 Dynamic Analysis Methods for Calculating Fatigue Damage

$$\alpha_2 = \frac{\overline{T}_p}{\overline{T}_z} = \frac{\lambda_2}{\sqrt{\lambda_0 \lambda_4}} \le 1.0 \tag{17.63}$$

α_2 is sometimes also named the irregularity factor.

where λ_i is the spectral moment defined in Eq. (10.6).

Both α_1 and α_2 approach unity when the random process is narrow banded and zero when the process is wide banded, and the two fatigue estimation methods introduced as follows are explicit functions of α_1 and α_2.

- Dirlik's method

Based on the Monte Carlo simulation to generate a large amount of signals that are fitted into 17 power spectrum density functions varying in shape, Dirlik [374] presented an empirical probability density function $p(\sigma_r)$ of rain-flow counting stress ranges σ_r as:

$$p(\sigma_r) = \frac{\frac{D_1}{Q} e^{-\frac{Z}{Q}} + \frac{D_2 Z}{R^2} e^{-\frac{Z^2}{2R^2}} + D_3 Z e^{-\frac{Z^2}{2}}}{\sqrt{4\lambda_0}} \tag{17.64}$$

where $D_1 = \frac{2\left(x_m - \alpha_2^2\right)}{1+\alpha_2^2}$; $x_m = \frac{\overline{T}_p}{\overline{T}} = \sqrt{\frac{\lambda_2}{\lambda_4}/\frac{\lambda_0}{\lambda_1}} = \alpha_1 \alpha_2$; $\overline{T} = \frac{\lambda_0}{\lambda_1}$ is the mean period of motions described in Sect. 10.2; $R = \frac{\alpha_2 - x_m - D_1^2}{1-\alpha_2 - D_1 + D_1^2}$; $D_2 = \frac{1-\alpha_2-D_1+}{D_1^2 1 - R}$; $D_3 = 1 - D_1 - D_2$; $Q = \frac{1.25\left[\alpha_2 - D_3 - (D_2 R)\right]}{D_1}$; $Z = \frac{\sigma_r}{2\sqrt{\lambda_0}}$;

Based the probability density function described above, the effective stress range can be obtained as:

$$\sigma_{\text{er}} = \int_0^\infty \sigma_r^m p(\sigma_r) d\sigma_r \tag{17.65}$$

Assuming that the number of cycles n occurs with a predefined duration τ in seconds, when the duration is 1 year, n is calculated as:

$$n = \frac{\tau}{T_p} \tag{17.66}$$

One finally obtains the fatigue damage within duration τ [376, 380]:

$$D_{\text{WB}} = n\frac{\sigma_{\text{er}}}{A} = \frac{\tau}{T_p} \frac{\int_0^\infty \sigma_r^m p(\sigma_r) d\sigma_r}{A} \tag{17.67}$$

where A is a constant related to the mean S–N curve.

Though obviously lacking of theoretical justification, Dirlik's method has been proved to be far superior to other existing solutions for rain-flow damage prediction [381, 382].

368   17 Fatigue Assessment

- Benasciutti and Tovo's method

A more recent method proposed by Benasciutti and Tovo [383] states that
fatigue damage of wide-banded responses can be approximated by a linear combination of narrow-banded and range counting results, both of which have a closed
form expression in the frequency domain:

$$D_{WB} = \rho D_{NB} + (1 - \rho)D_{RC} = \left[\rho + (1 - \rho)\alpha_2^{m-1}\right]D \tag{17.68}$$

where D_{NB} and D_{RC} are fatigue damage under narrow-band assumption and using
range counting method, respectively; and ρ is a weighting parameter dependent on
power spectral density, which again is obtained from extensive numerical
simulations:

$$\rho = \frac{(\alpha_1 - \alpha_2)\left\{1.112[1 + (\alpha_1\alpha_2) - (\alpha_1 + \alpha_2)]e^{2.11\alpha_2} + (\alpha_1 - \alpha_2)\right\}}{(\alpha_2 - 1)^2} \tag{17.69}$$

The equation above is calibrated on the results from 286 numerical simulations
considering Gaussian random processes with different spectral densities having
various combinations of bandwidth parameters α_1 and α_2.

An independent investigation performed by Gao and Moan [384, 385] shows
that the empirical formulas proposed by both Dirlik and Benasciutti and Tovo are
accurate enough for all types of wide- and narrow-banded Gaussian processes.

17.3.5.3 Fatigue Damage Accounting for Multi-modal Frequency Components

For structures with eigenfrequencies or flexible oscillation frequencies that are
well separated from that of the loading, the responses appear with multi-modal
characteristics, typically of bi-modal or tri-modal type. Structures relevant to
bi-modal responses are typical land-based structures under dynamic wind loadings,
responses of large ships and the marine mooring system, wave-induced responses
or springing responses of tension leg platforms. Tri-modal process could also be
relevant for structural responses involving vortex-induced vibrations [385]. In
order to calculate fatigue damage, the spectrum is normally divided into several
well separated frequency regions.

Let's take the bi-modal spectrum as an example. As shown in Fig. 17.27, the
bi-modal spectrum can be divided into two regions: a low frequency (LF) region
and high frequency (HF) region, where within each region the spectrum is narrow
banded. Applying rain-flow counting, essentially, two types of cycles are extrac-
ted: large cycles accounting for interaction between the LF and HF regions, and
small HF cycles only [24].

A number of methods are proposed for calculating fatigue due to responses
possessing a bi-modal spectrum, the majority of which are based on the
assumption that the process follows a Gaussian distribution.

17.3 Dynamic Analysis Methods for Calculating Fatigue Damage

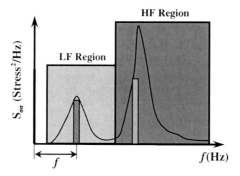

Fig. 17.27 A bi-modal stress spectrum with low frequency (*LF*) and high frequency (*HF*) region

Sakai and Okamura [388] showed that fatigue damage under a bi-modal spectrum can simply be calculated as the sum of each narrow-band component without interaction between the two components. Therefore, the fatigue damage calculated from this method is often underestimated.

Fu and Cebon [387] discussed a method that involves assuming that the number of large cycles is directly associated with the LF cycles, while that of small cycles is equal to the difference of the HF number and the LF number by keeping the total number of counted cycles the same as that of the HF cycles.

Based on Fu and Cebon's work, Benasciutti and Tovo [386] proposed that the number of long-period cycles can be determined by using the mean zero up-crossing rate of the equivalent process.

By combining an envelope HF process and an LF process, Jiao and Moan [389] presented that the fatigue with bi-modal spectrum can be calculated by summing up the HF damage caused by small cycles and the damage due to an equivalent process of the HF envelope plus the LF process. They proposed a bi-modal fatigue damage as a correction of the narrow band approximation:

$$D_{Bi-modal} = \Phi D \quad (17.70)$$

where D is the fatigue damage under narrow band assumption (Sect. 17.3.3); and Φ is a correction factor due to the bi-modal responses, which is formulated as:

$$\Phi = \frac{v_{HF}}{v_Y} \eta_{HF}^{\frac{m}{2}} + \frac{v_P}{v_Y} \left[\eta_{LF}^{\left(\frac{m}{2}+2\right)} \left(1 - \sqrt{\frac{\eta_{HF}}{\eta_{LF}}}\right) + \left(\sqrt{\pi \eta_{HF} \eta_{LF}} \frac{m\Gamma \frac{m+1}{2}}{\Gamma \frac{m+2}{2}}\right) \right] \quad (17.71)$$

where v_{HF} is the mean zero up-crossing rate (inverse of zero upcrossing period) for spectrum within HF region; $\eta_{HF} = \frac{\sigma_{HF}^2}{\sigma_{HF}^2 + \sigma_{LF}^2}$ and $\eta_{LF} = \frac{\sigma_{LF}^2}{\sigma_{HF}^2 + \sigma_{LF}^2}$ are the normalized variance of process in HF and LF regions, respectively; σ_{HF} and σ_{LF} are the standard deviation of spectrum in HF and LF region, respectively. $v_Y = \sqrt{\eta_{HF} v_{HF}^2 + \eta_{LF} v_{LF}^2}$; $v_P = \sqrt{\eta_{HF} \eta_{LF}; v_{HF}^2 \Psi_{HF}^2 + \eta_{LF}^2 v_{LF}^2}$; $\psi_{HF} = \sqrt{1 - \frac{\lambda_2^2}{\lambda_0 \cdot \lambda_4}}$ is the bandwidth parameter for spectrum within HF region, and can be set as 0.1; ν_{LF} is the mean zero up-crossing rate for LF components; and m is the inverse slope in the S–N curve.

From the equation above, it is noticed that the first item on the right hand side of the equation represents the number of small cycles, determined by the mean zero up-crossing rate of the HF component. The number of large cycles is given by that of the equivalent process in the second item.

The method by Jiao and Moan presented above has been implemented in a few offshore standards including ISO 19901-7 [390], API RP 2SK [391] and DnV OS-E301 [392], to estimate the fatigue damage in mooring lines.

All the methods for calculating fatigue damage mentioned above with a bi-modal spectrum provide accurate assessment [385].

17.3.5.4 Fatigue Assessment Accounting for Both Bandwidth and Nonlinearities: Semi-stochastic Fatigue Analysis Method in the Time Domain

It is noted that all the methods above cannot completely handle the problems induced from a combination of bandwidth and nonlinearities.

If such situation occurs, a semi-stochastic fatigue analysis (sometimes also called a direct calculation method) is recommended. In this method, the stochastic load histories in the time domain are generated from various load spectra for different load conditions with a scattered probability of occurrence. After applying each load time history on a structure and performing a dynamic structural analysis, the stress responses on various locations of the structure can be obtained, and are for example rain-flow counted to calculate the partial fatigue damage. All the partial fatigue damages are then summed up to obtain the total fatigue damage or life.

Let's take the wave-induced fatigue damage as an example. The procedure is illustrated as follows:

1. Establish the FE modeling of a target structure.
2. Divide scatter diagram of waves into representative blocks.

- q sea state blocks with relevant probability.
- j wave directions with relevant probability.
3. Generate the load history of wave loading from a relevant wave energy spectrum (Sect. 12.1.2).
4. Include hydrodynamic coefficients, accounting for the buoyancy effects.
5. Dynamic analysis.

- Start with the first sea state and the first wave direction and proceed through the $q \times j$ analysis.
- Time histories of the forces/stresses for selected locations are documented.
6. Calculate the stress concentration factors.

17.3 Dynamic Analysis Methods for Calculating Fatigue Damage

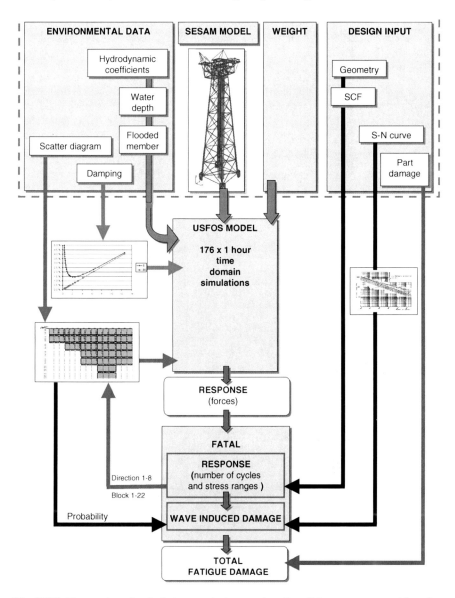

Fig. 17.28 The semi-stochastic fatigue analysis procedure for offshore structures subjected to dynamic wave loading (courtesy of Aker Solutions) [365]

7. Choose a suitable S–N curve. For welded joints, the mean stress effects may not be taken into account due to its insignificant influence compared to the presence of residual stress. However, if a structure is constructed with mean stress sensitive material such as composites, its effects should be considered.
8. Calculate the fatigue damage for each sea state:

- For tubular members, calculate the hot-spot stresses for joints at the eight spots in the crown and eight spots in the saddle side from the force time histories.
- The number of cycles and stress ranges are calculated using the rain-flow counting rule.

9. Calculate the 1 year fatigue damage at each hotspot of a joint.

10. The fatigue damage at each hotspot of a joint for $q \times j$ load cases are multiplied by the probability for each sea state and accumulated to the 1 year fatigue damage. Then select the maximum damage among those hotspots as the 1 year fatigue damage.

The procedure shown above is also illustrated in Fig. 17.28.

For a detailed description of the semi-stochastic fatigue analysis procedure, readers may read Ref. [365].

Chapter 18
Human Body Vibrations

18.1 General

Even if most vibrations do not cause damage to structures and equipment, they may introduce a sudden change of forces on human bodies, causing discomfort and local injury known as whiplash.

The balance organs located in the inner ear can detect changes in the magnitude and the direction of gravitational and angular accelerations. For most individuals, excessive stimulation of these organs will cause motion sickness. Motion sickness can be exacerbated if the individual is inside a room and cannot see the horizon. Motion sickness can also be magnified by anxiety, fatigue, hunger, odors (e.g. cooking and fuel oil, greasy food, reading, and carbonated alcoholic drinks) [91]. Some studies indicate that human beings are more sensitive to horizontal translational motions than vertical ones because horizontal motions can directly "throw people off their feet." People are more sensitive to rotations about their vertical axis than that other axes because the rotations about their vertical axis produce a large sight displacement of distant objects. Experiments [393] also indicate that young children and women are more susceptible to motion sickness than men. Generally, middle-aged people are more affected by motion than elderly people.

It is noted that the human body does not directly perceive displacement or velocity, but perceives accelerations. This can be easily explained by an example of a passenger on board an airplane. He or she can feel comfortable at a normal speed ranging from 400 to 1,000 km/hour. But when the plane suddenly meets turbulence and strongly shakes, the change of speed, i.e. accelerations, will be perceived by the passenger with discomfort or even fear, simply because the accelerations induce forces on the human body and balance organs in the inner ear. For this reason, the effects of motions on human beings are normally measured with accelerations.

Motion effects are also dependent on a number of parameters [394]:

- The duration of motions: motions acceptable for a short duration may not be acceptable for long durations.

J. Jia, *Essentials of Applied Dynamic Analysis*, Risk Engineering,
DOI: 10.1007/978-3-642-37003-8_18, © Springer-Verlag Berlin Heidelberg 2014

- The activities in question: general office work, operating work, or precision tasks.
- The form of excitations: e.g. earthquake with a Richter scale 5.0, average sway amplitude of a ship lasting for minutes due to a storm, traffic at rush hour etc.
- The kind of vessels or structures in which the human beings are located: e.g. office buildings, residential houses, chemical plants, bridges, ships, and offshore platforms.
- The time of day: day time or night time.

As mentioned above, the kinematics of the human body make people more sensitive to accelerations in the horizontal direction (along the axis from the spine to breast bone X and from the right shoulder to the left Y) than the vertical plane (the axis from feet to head Z). Therefore, an effective amplitude a_e contributed from accelerations from all three directions (a_X, a_Y, and a_Z) is formulated as:

$$a_e = \sqrt{2(a_X^2 + a_Y^2) + a_Z^2} \qquad (18.1)$$

The formula above is widely used for evaluating the limiting amplitudes for various activities and situations.

There are no unified limits of acceptable and unacceptable levels of motions. The sensitivity of a human being to motions strongly depends on his/her individual's characteristics, the activities in which he/she is engaged, the environmental noise, and even his/her emotions [394].

As mentioned in Sect. 2.1, jerk, which is the time-derivative of acceleration, is also an important parameter when evaluating the discomfort caused to passengers onboard vehicles. For passenger comfort, a train in operation will typically be required to keep the jerk less than 2 m/s^3. In the aerospace industry, a type of instrument called a jerk-meter is used for measuring jerk.

18.2 Criteria Related to Human Body Vibrations

When a person is subjected to motions, each part of the body has its own resonance frequency. Vibrations transmitted to human bodies can be amplified at or close to the resonance frequencies of each part of the body, giving stretching or compression of tissues and limbs to a variable degree depending on intensity, frequency and directivity of vibrations. Figure 18.1 illustrates a mechanical model of a human body with resonance frequencies for each part of the body.

The most important parts of the human body with respect to vibrations are the abdomen, head and neck area. Posture is another example of human body function that can be influenced by external vibrations. For vibrations with frequencies ranging from 1 to 30 Hz, people experience difficulty in maintaining correct posture and balance [2]. Specifically, in the longitudinal direction of the human body (feet to head), the human body is most sensitive to vibrations in the

18.2 Criteria Related to Human Body Vibrations

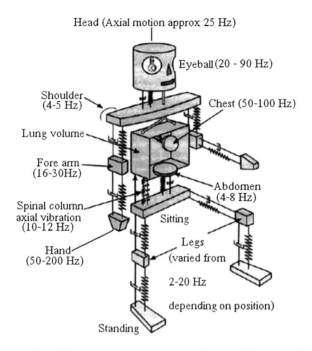

Fig. 18.1 Mechanical model and resonance frequencies of a typical human body [396, 397]

frequency range between 4 and 8 Hz, while in the transverse direction, the body is most sensitive to vibrations in the frequency range 1–2 Hz. Some vibration effects on the human body may be difficult to relate to specific frequencies, but are related to the cumulative effects of exposure over a range of frequencies. In general, they cause imbalance, disorientation, lack of coordination, and motion sickness [395].

ISO 2631-1 [2] provides guidance on the evaluation of vibration perception and comfort of humans in sitting, standing and reclining positions.

Based on the abundant measurements and experiences on how vibrations affect human beings, ISO 2631 provides probable subjective reactions of people subject to vibrations at various acceleration levels, as shown in Table 18.1.

For offshore structures, based on boundaries presented in ISO 2631, NORSOK S-002 [399] specifies the maximum limits for continuous whole body vibrations due to the vibrations of machinery and equipment. The limits are derived from the acceptability of the exposure of human beings to vibrations based on a 12 h working day. The vibration limits are specified graphically as combined levels for vertical and horizontal movements as shown in Figs. 18.2 and 18.3. It should be noted that extrapolation beyond this range is not allowed. In the two figures, the vibration limits are categorized as follows:

- Category 1—Limits for central control room and living quarter areas.
- Category 2—Limits for workshops, laboratories, control rooms, offices and equipment rooms outside living quarters.

Table 18.1 Human perception of motions [2, 403]

Acceleration level m/s^2	Degree of discomfort	Acceleration level m/s^2	Degree of discomfort
<0.05	Human beings cannot perceive motions	<0.315	Not uncomfortable
0.05–0.1	Sensitive people can perceive motions; hanging objects may move slightly		
0.1–0.25	Most people can perceive motions; the level of the motions may affect desk work; long-term motions may produce motion sickness		
0.25–0.315	Desk work become difficult		
0.315–0.5	People strongly perceive motions; it is difficult to walk naturally; standing people may lose their balance	0.315–0.63	A little uncomfortable
0.5–0.6	Most people cannot tolerate the motions and are unable to walk naturally		
0.6–0.7	People cannot walk or tolerate the motions	0.5–1	Fairly uncomfortable
0.8–1.6	Uncomfortable; objects begin to fall and people may be injured	1.25–2.5	Very uncomfortable
		>2	Extremely uncomfortable

(continued)

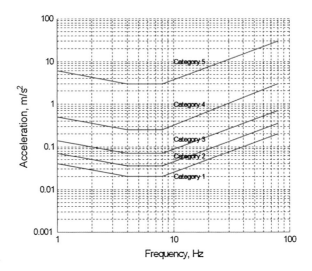

Fig. 18.2 Vertical vibration limits defined by NORSOK S-002 [399]

- Category 3—Limits for process, utilities and drilling areas.
- Category 4—Limits for vibrations locally to equipment.
- Category 5—Maximum limits (normally unmanned areas).

18.2 Criteria Related to Human Body Vibrations

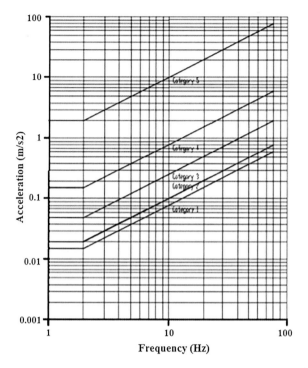

Fig. 18.3 Horizontal vibration limits defined by NORSOK S-002 [399]

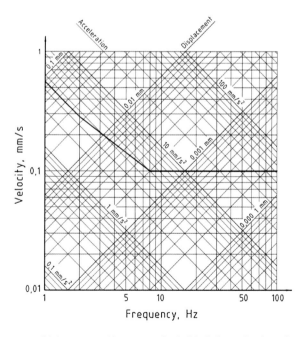

Fig. 18.4 Human sensitivity curve (the curve in bold defines the boundary between the unacceptable vibration level in the *upper part* and acceptable level in the *lower part*) [400]

Table 18.2 Operability criteria by NATO Standard 4154

Responses	Criteria
Vertical acceleration (RMS)	1.96 m/s^2
Lateral acceleration (RMS)	0.98 m/s^2
Roll amplitude (RMS)	4°
Pitch amplitude (RMS)	1.5°
Motion sickness incidence	20 % in 4 h
Motion-induced interruption	1 per minutes

It should be noted that higher levels than those given in category 4 may be tolerated for exposure shorter than 12 h. Categories 1, 2 and 5 shall also apply for intermittent operation.

For ships, by considering the influence of work, recreation and rest for the crew as well as the sensitivity of advanced electronic equipment on board, ISO 6954 [400] establishes the limits of maximum desirable vibration levels varying with frequency, which are known as the human sensitivity curve (shown in Fig. 18.4). The curve in bold defines the boundary between the unacceptable vibration level in the upper part and acceptable level in the lower part. It should be noted that the value in the graph is the highest peak value (maximum repetitive value) of each harmonic component over 1 min duration [401], which is the measurement in a wider frequency band or root of mean square values multiplied by a factor of 2.5. NATO 4154 [402] has also specified the operating criteria for sea vessels, as shown in Table 18.2

More detailed elaborations of human body vibrations can be found in studies by Driffin [398], Harris and Piersol [212] and ISO 2631 [3].

Chapter 19
Vehicle-Structure Interactions

The vehicle-structure interactions elaborated in this chapter are a summary of research by Jia [179, 410], Jia and Ringsberg [411], and Jia and Ulfvarson [68, 177, 178, 412–414].

19.1 Introduction to the Topic

Vehicle-structure interaction is an interdisciplinary topic that relates to several fields of engineering such as bridge and road engineering, railway engineering, marine engineering, aerospace engineering, vehicle engineering, and sound and vibrations. It originated in the field of civil engineering where in the nineteenth century attention was paid to the dynamic amplification effects of bridges due to vehicle loading [178, 179, 415]. Abundant research can be found concerning the interactions of, for example, train-track, train-bridge, train-track-bridge, vehicle-bridge, wind-bridge-vehicle, earthquake-bridge-vehicle, and aircraft/taxiway–bridge systems. Due to the increase in train traffic density, vehicle speed and heavier train axle loads, together with restricted maintenance time and limited budgets for bridges, tracks and roads, the knowledge concerning vehicles-structure interactions became increasingly important. In particular for the design of bridges, the high speed of running vehicles implies a large amount of kinetic energy that is transferred into the bridge structure, resulting in resonance of the bridge under certain circumstances. Also, large amplitude vibrations of a bridge affect the ride comfort, stability, service life, and safety of vehicles and passengers that are located on the bridge.

In vehicle engineering, the ride comfort and handling stability are of great importance, and they are determined by vehicle mechanisms, i.e., the vehicle suspension system. The prediction of service loads for durability studies is another important application of vehicle dynamics simulation in the automotive industry. Such studies require accurate predictions of, for example, the loads that tires transmit to the spindle rather than accurate predictions of the detailed tire deformation and stress fields [416].

J. Jia, *Essentials of Applied Dynamic Analysis*, Risk Engineering,
DOI: 10.1007/978-3-642-37003-8_19, © Springer-Verlag Berlin Heidelberg 2014

Fig. 19.1 Lashed cars secured on ship decks

In the field of marine engineering, increasing demands on RO/RO ships (vessels designed to carry wheeled cargo, such as cars, trucks, semi-trailer trucks, trailers and railroad cars that are driven on and off the ship on their own wheels. This is in contrast to lo-lo (lift-on/lift-off) vessels that use a crane to load and unload cargo) with less damaged vehicles during transport require a further understanding of the responses of deck structures under vehicle loads. Figure 19.1 shows that vehicles are secured on decks with lashings during transportation. To obtain more efficient and reliable vehicle securing and handling, studies of the reaction forces between vehicles and decks during ship motions are of great interest. Due to the special dynamic characteristics of lightweight deck structures, they may be more sensitive to interactions with the loaded vehicles compared to conventional deck structures.

There are similarities between applications that are concerned with vehicle-structure interactions modeling: they all consist of a vehicle model and a structure model that supports the vehicle, where the vehicle is represented by mass-

19.1 Introduction to the Topic

suspension systems. It can also be seen that the modeling techniques and considerations for different types of interactions share common characteristics between applications. Modeling of the interactions between vehicles and their supporting structures is a complicated task because of the modeling of the complex vehicle-structure mechanisms: one moving or standing vehicle system interacting with its supporting structures, or even moving structures.

Ahlbeck [417] studied railway wagon-track interactions and identified some typical problems corresponding to each frequency bandwidth. If stability and ride comfort are of primary interest, a frequency bandwidth of about 100 Hz is sufficient. Wheel-rail impact loads and rail corrugations require a frequency bandwidth of over 1,000 Hz, which leads to additional degrees-of-freedom in the numerical analysis such as axle and rail bending modes. That is, the frequency range of interest affects the degrees-of-freedom and the distributed flexural modes necessary to account for in the model. Since vehicle-track interaction forces are well below the natural frequency of track structures, the simplified spring-damper-track models are sufficient to describe the wheel-rail contact forces and the response of track components.

In addition, Table 19.1 [411] presents an overview of various engineering applications of vehicle-structure interaction problems in different frequency ranges. It should be noted that the vehicle's dynamic characteristics (natural frequencies and mode shapes) are normally not influenced by the interactions (but the vehicle's vibration amplitude is influenced by the interactions). However, the supporting structure's dynamic characteristics are influenced by interactions from vehicles. This is because the forces applied on the vehicles (interaction forces, dead weight, wind loads, etc.) do not significantly influence the natural vibrations of vehicles. One can utilize this phenomenon if the structures are rigid enough. Then it is only necessary to apply the rigid motions of the structures on the vehicles, without considering the small amplitude of structural vibrations.

In the field of marine engineering, various mathematical models for trailer-lashing deck systems have been studied, such as that by Dallinga [418], who stated that the target trailer he studied will overturn instead of sliding if the friction (friction is hereafter referred to as the fraction between the resultant tangential force and the normal force, see [179, 418]) is larger than 0.5. By modeling a trailer deck system at different levels of detail, Turnbull [419–421] found that the reduction of suspension flexibility (i.e., tire flexibility + spring flexibility) leads to a decrease of maximum lashing loads and the maximum loads can also be reduced by increasing the friction between the trailer tire and the deck surface. It was also stated that the distribution of loads among lashings is not uniform, and when one lashing yields, it starts to shift its loads onto other lashings that have to carry more loads until they yield. By studying the dynamic behavior of flexible semi-trailers on board RO/RO ships, Turnbull concluded that the chassis stiffness has only a secondary effect on the lashing force.

Car-deck interactions were studied by Mora [422] who presented the phenomenon that a decrease in lashing stiffness increases the friction between the tire and the ship deck, which is in agreement with Turnbull's conclusions on the relation between the

Table 19.1 Related research topics in different frequency ranges concerning vehicle-structure interactions, by Jia and Ringsberg [411]

Frequency range (Hz)	Vehicle-bridge interactions	Train-track interactions	Vehicle engineering	Marine engineering
0–3	Tire-deck interaction due to the sprung mass bouncing and pitching motion	Ride comfort; stability of vehicles; suspension loads and deflections; elevated guide-way beam responses due to bending	Vibrations of suspensions; slight human responses: whole body vibration aversion (mainly in the horizontal plane)	Vehicle vibrations and vehicle securing due to wave-induced ship motions; hull girder vibrations
3–15			Human responses: motion sickness	Deck's structural vibrations during propeller or slamming excitations; motion sickness
15–100			Acoustics; human responses	The deck structural responses due to water jet, main engine or auxiliary machinery excitations; human responses
0–1,000		Tie-fastener load and deflections; ballast and sub-grade pressures; truck frame loads; lading damage	Acoustics; human responses; radial vibrations of tires; sidewall vibrations (400–600 Hz); engine shaking	Acoustics; human responses; engine shaking
50–2,000		Wheel-rail contact; rail accelerations and stresses; fastener and tie loads and accelerations; wheel bearing loads; gearbox; wheel; axle loads and accelerations; wheel-rail contact noise	Acoustics	Acoustics
100–10,000		Wheel-rail flanging; structural noise; wheel-rail contact noise	Acoustics; stick–slip	

19.1 Introduction to the Topic 383

lashing stiffness and the friction. In order to study the fatigue behavior of a ship deck using trapezoidal-shaped stiffeners during the rolling of a truck on the deck, Eylmann et al. [423] performed a test that simulated a truck tire (isolated from the sprung system) rolling (rolling length: 1.5 m) on a deck plate right above the stiffener. It was observed in the test that the first and the second visible crack occurred at the web frame intersection on two opposite sides of the tire.

Sternsson [424] measured lashing forces of a car parked on one of the worst places for cargo securing on a PCTC (Pure Car Truck Carrier) vessel operating in the North Atlantic. He concluded that even if the head sea caused the greatest lashing force amplitude, the influence of wave directions on the forces was small. The lashing forces and the required friction coefficient were also found to be small and the largest measured friction occurred in the lateral direction of the car as the lashings in this direction gave little support.

In Jia and Ulfvarson [178], a review of systematic studies on various levels of problems concerning the dynamics of vehicle-ship deck interactions was presented. It incorporated the investigation of the dynamic structural behavior of vehicle-deck systems, vehicle vibrations, damping effects of vehicles on structural systems, dynamic interactions between tires and deck surfaces, and vehicle securing on decks during ship motions. The results and observations from this investigation identified some interesting topics that were investigated systematically in detail in the references by Jia [179], Jia and Ulfvarson [68, 177, 178, 412–414], and Jia and Ringsberg [411], using numerical analyses (analytical and the finite element (FE) method) and experiments:

- Vehicle modeling.

  - Numerical modeling of a vehicle using masses and coupled spring-damper systems.
  - Numerical modeling of tires.
  - Simplified numerical modeling of the contact between the tire and the ship deck.

- Structure/ship deck modeling.

  - Parametric study of lightweight structure design.
  - Comparison between different deck designs, conventional and lightweight.

- Vehicle-deck interactions.

  - Modeling of contact forces between tire and ship deck structure.
  - Dynamic responses of a deck structure depending on where on a ship deck vehicles are parked.
  - Dynamic responses of a deck structure depending on the number of cars parked on a ship deck.
  - Securing of vehicles on ship decks using lashings or without lashings.

The following sections give an overview of the methodologies used and results achieved in comparison to research work published in the literature. Section 19.2

describes the detailed physical modeling of vehicles, supporting structures, lashings and tires, followed by interaction models for vehicles and supporting structures, and ship motions. Results from some numerical analyses and experiments are presented and discussed in Sect. 19.3. By utilizing the knowledge presented, Sect. 19.4 presents the application of the method on the evaluation of vehicle securing onboard a ship.

19.2 Physical Modeling

19.2.1 General

This section describes mechanisms for vehicles and supporting structures (decks, beams, bridges, rails, etc.) needed for performing a dynamic vehicle-structure interaction analysis. The mechanisms are described in detail, since the interactions between them are highly influenced by the characteristics of one single part of the system.

19.2.2 Vehicle, Lashing and Tire Models

The loads applied from vehicles on structures mainly include the vehicles' gravity loads and the inertia loads. Based on the vehicle load effects, the vehicle interaction models can be distinguished as the following types:

- *The sprung mass model* reduces the vehicle body, suspension and wheel to a discrete system by modeling them as masses connected by springs and dampers. Most of the vehicle-structure interaction research is based on this type of model, with various degrees-of-freedoms to represent vibrations of a vehicle's body-suspension-tire system. The vertical direction normal to supporting structures is generally the most important direction with respect to the degrees-of-freedom. Thus, most of the sprung mass models are created based only on the consideration of vertical dynamics. In addition, when studying the vehicle-structure interactions instead of the structural behaviors of vehicles, compared to the contributions from the suspension systems and tires, the displacements and motions of vehicle frame structures may be neglected if resonance does not occur.
- *The mass model* is a simplification of the sprung mass model, since the model keeps the inertia effects (mass) of the sprung mass model but it neglects the spring and damper modeling. The model can account for the inertia effects in all directions. One of the drawbacks of the mass model is that it ignores the bouncing actions that are rather significant in the presence of rail irregularities or pavement roughness, or when vehicles are moving at rather high speeds. Occasionally, the "jumping" action (when vehicles jump from structures and re-contact with the

19.2 Physical Modeling 385

structures) may occur due to, for example, bad road conditions or large excitations through the movement or vibration of structures, and under those situations the bouncing effects are important to consider (see Yang et al. [425]).

- *The dead weight (force) model* is a further simplification of the sprung mass model. When the inertia forces of vehicles are much smaller than the dead weight of the vehicles, the coupling of vehicles and structures can be neglected, which means that the inertia and elastic effects of vehicles are not taken into account. In this case, if only the responses of structures in vertical directions are of interest, the vehicles can then be described as vertical forces applied on the supporting structures. The majority of the design codes for designing supporting structures are based on this assumption, and, as a merit, it can guarantee a closed form of solution of the equations of equilibrium. The factors contributing to the inertia effects mainly comprise the supporting structure motions (e.g., accelerations of decks due to ship motions or accelerations of bridge decks due to strong wind loads and earthquake excitations), deformations, and irregularities (when vehicles are moving on structures, etc.). The detailed cases for the inertia effects to be considered include:

  - Large excitation accelerations to supporting deck/beam structures, e.g., under slamming and springing, the accelerations of ships may be quite significant to make the inertia effects of parked vehicles more obvious.
  - Flexible (slender) structures.
  - Large vehicle mass.
  - Low structural mass.
  - Stiff suspensions.
  - Large structural irregularities (for moving vehicles).
  - High vehicle speed (for moving vehicles).

Since the vehicle mass inertia effects are not included in the modeling of the dead weight model, the fundamental frequency of the vehicle-structure system is higher than that of the moving mass model and it is generally higher than that of the sprung mass model.

Note that the mass model and the dead weight model can be seen as simplifications of the sprung mass model. In the following, the sprung mass model is used for the description of different types of vehicles.

19.2.2.1 Car Models

It is assumed that the influence from the vibrations of continuous structural elements on the vehicle frame is normally small and can therefore be neglected [426]. It is also suggested by Cebon [427] from vehicle interaction simulations that the sprung masses can be assumed to be rigid. The vehicle body is then assumed to be connected to the axles through the suspension system. The vehicle body is referred to as the sprung mass, whereas the vehicle's axles are the unsprung masses. A parallel connection of springs and dampers is used to represent the suspension and

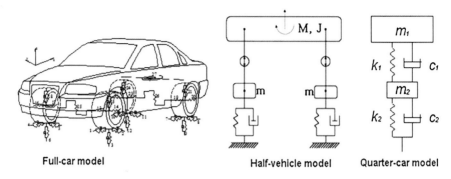

Fig. 19.2 Full-car, half-vehicle, and quarter-car models with various degrees-of-freedom [410]

vehicle tire. In the present section, three types of car models with different levels of detailed modeling are presented (see Fig. 19.2): a full-car model, a half-vehicle model, and a quarter-car model. Detailed descriptions of these three types of sprung-mass models with various degrees-of-freedoms can be found in Jia and Ulfvarson [178].

There are no unified selection criteria for which type of vehicle model should be adopted. Generally, the full-car model is the most general and accurate model to study. However, in order to simplify the problem, one needs to select as simple a model as possible without losing the necessary accuracy. This selection is mainly based on the problems being studied. For example, if one just needs to perform a dynamic analysis to obtain the overall structural behavior of decks, such as eigenpairs of a ship deck under vehicle loads, a quarter-car model is generally sufficient for modeling the vehicle. However, when the study is extended to obtain the local detailed dynamic structural behavior, at least a half-car model may be adopted, since the coupling between roll and bounce motions of the vehicles may be relevant in influencing the local structural behavior at different positions. Furthermore, if the detailed information concerning interactions (along both vertical and horizontal directions) between the tire and the supporting structure as well as the lashing forces (in case cars are lashed to decks) are required, a full-car model is highly recommended for use due to the fact that the complexity of the interaction is far more than a quarter-car model can represent. In addition, the selection criteria also depend on the requirement for accuracy.

Note that if the suspension's vertical stiffness is much lower than the tire's vertical stiffness, the vertical motions of the vehicle body and the tires are almost uncoupled. Hence, a change in the stiffness of the suspension has only a secondary effect on the vehicle body's vertical motion, and may therefore be neglected.

In some specific cases, the vehicle suspension has "critical damping" (relative damping $\xi = 1$), which means that the suspension reaches a limiting case between oscillatory and non-oscillatory motions and does not give any vibrations, i.e., the damped fundamental frequency is close to zero. In those cases, only the stiffness and the damping for the tire need to be considered.

19.2 Physical Modeling

It should also be noticed that, for vehicles in which partially filled tanks with free surface liquid are equipped, eigenfrequencies of the vehicle may shift to lower values.

19.2.2.2 Lorry and Trailer Models

The shifting of trailers has been found to contribute to a number of RO/RO ship casualties. Some research efforts have been devoted to the securing of trailers; see Turnbull [419–421] and Andersson [428] for the numerical modeling of trailers. In addition, a trailer that is lashed on board a ship constitutes a complicated system, which can be simplified to a system of girders and springs with different stiffness and damping coefficients. A typical model representation of a trailer is shown in Fig. 19.3. On the deck of a ship, the trailer usually rests on a trestle close to the "fifth wheel" (i.e., the part of the trailer that rests on the motor unit) and its own suspension. It is normally lashed to the deck using chain lashings or web lashings (Fig. 19.5), and, if possible, these lashings are attached to anchor points on the deck and special anchor points on the trailer's chassis. In practice, however, very few trailers have anchor points fitted to their chassis, and the deck crew have to find any possible place they can to attach the lashings to the trailers. Due to the high center of trailer mass and the relative low positions of lashings, large lashing loads and large movements of the trailer on decks may occur [179].

Figure 19.4 shows the trailer models developed by Turnbull [419] with various levels of detail. The main differences between each model are characterized by the modeling of bending and torsion stiffness of chassis, and the mass distributions, which are shown in Table 19.2.

Note that suspensions on vehicles are either air suspension or steel leaf suspension. Air suspension gains stiffness through the action of a pressurized

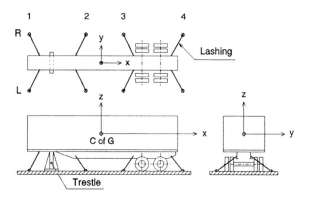

Fig. 19.3 A typical illustration of a semi-trailer from Turnbull [419]: *L* and *R* refer to the *left* and the *right hand side* of the trailer; the *numbers* refer to the positions of the lashings along the trailer axis

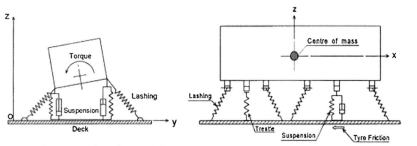

A two-dimensional trailer model.   A trailer model with uniform mass distribution.

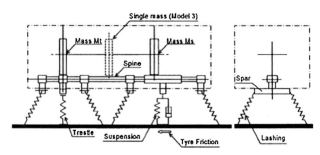

A trailer model with two concentrated masses.

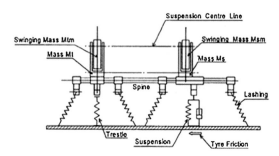

A trailer model with two concentrated masses plus two swing masses (with ±10° free swing).

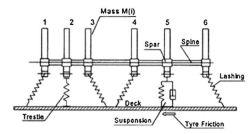

A trailer model with six concentrated masses.

Fig. 19.4 Semi-trailer models from Turnbull [419–421]

19.2 Physical Modeling

Table 19.2 Differences between chassis models and trailer models from Turnbull [419]

	Types of chassis model		
	Bending stiffness assumption	Torsional stiffness assumption	Mass distributions
Two-dimensional model	Rigid	Flexible	Uniform
Uniform-mass-distribution model	Rigid	Rigid	Uniform
Two-concentrated-mass model	Rigid	Flexible	Two concentrated masses
Two-concentrated-mass-plus-two-swing-mass model	Rigid	Flexible	Two concentrated masses plus two swinging masses with ± 10° of free swing
Six-concentrated-mass model	Flexible in both vertical and horizontal directions	Flexible	Six concentrated masses

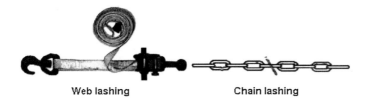

Web lashing   Chain lashing

Fig. 19.5 Typical lashings for vehicles: web and chain lashings

elastometric bag used to provide the lift force, and uses hydraulic shock absorbers to obtain necessary damping. Stiffness in steel leaf suspension is gained through the size, geometry, and alignment of steel strips, which bend elastically during loading, while the damping effects are due to the dry ("Coulomb") friction between the steel strips at a number of contact points. Thus, leaf-sprung vehicles normally induce higher supporting structure responses than air-sprung vehicles do [426, 429].

19.2.2.3 Modeling of Lashings

During the transportation of cars and trailers, lashings may be used to secure the vehicles. For cars, they are normally attached to the front and the back of the cars and fixed to the lashing holes on the deck; for trailers, see Figs. 19.3 and 19.4. Lashings should be applied very tightly and firmly. For example, in the shipping industry, the International Maritime Organization (IMO) [430] states that if vehicles are to be secured on ship decks during transportation, lashings should be of high strength and have elongation characteristics that are at least equivalent to steel chains or wires. The vehicle should be stowed in a fore-and-aft direction rather than athwart ships to prevent shifting. The main types of lashings for securing vehicles are web and chain lashings as shown in Fig. 19.5. Conventionally, the lashing can be modeled as a component with axial linear stiffness and damping that is active during elongation but zero during shortening. The lashing can even be simplified as an elastic component with only axial stiffness. It should be noted that pretension (in the magnitude of 0.2 kN to several kNs, or around 10 % of the breaking load) of lashings may have to be included in the modeling. Moreover, in some severe cases, yielding of the lashing may have to be taken into consideration in the modeling and analysis.

19.2.2.4 Tire Models

For cars, trucks and trailers, tire characteristics are of crucial importance for the dynamic behavior. The component of an air-filled tire includes the tread, carcass, air-filled volume belt (only with radial ply tires) and rim. For studying vehicle-structure interactions, most spring-damper systems can represent the characteristics of the tire for the calculations of vehicle-structure interactions with sufficient accuracy, provided that the deformations of the tires are small. However, because of the continuous nature of deformation and contact, tires actually exhibit significant nonlinear behavior due to contacts (in reality, not point contacts or contact with uniform contact pressure) and large tire deformations. When the detailed information about tire vibrations or noise (during the driving of a car on structures) needs to be calculated, the tire must be modeled in more detail. These detailed models are based on diverse theories of shell vibrations [431].

Tire pressure has an influence on the structural behavior of supporting structures. It is presented in Jia and Ulfvarson [68] that during the same static vehicle loading, the higher the tire pressure is (corresponding to a small contact area with the same tire-loading value), the higher the deformation and local stress the supporting deck structure will be. In addition, when the vehicle ride and durability loads are of interest, the nonlinear behavior of a tire can be essential. Thus, in reality the dynamic responses of vehicles may be simulated by coupling the calculation from the multi-body system program to simulate vehicles without tires and the responses from the nonlinear FE analysis to simulate tires.

Even though most of the pavement analysis is based on the assumptions that the contact pressure between tires and road surfaces is equal to the inflation pressure and uniform, experiments show that the pressure in the shoulder area (around the edge of the contact areas) is higher. It should be noted that, due to the difference in the stiffness of the side walls and the tread, even in a straight-line steady-speed motion of vehicles, the contact pressure between tires and road surfaces contains significant lateral and longitudinal shear tractions as well as vertical pressure [427].

19.2 Physical Modeling

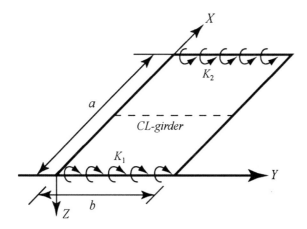

Fig. 19.6 Description of an analytical model of a typical deck

19.2.3 Modeling of Supporting Structures

Before calculating structural responses, a number of assumptions (idealizations) must be made in order to simplify the modeling without losing its accuracy. The main assumptions should be made after considering the following items: support conditions, material constitutive relations (elastic or elastic–plastic), geometric linearity or nonlinearity, shear deformations, rotary inertia effects, initial imperfections, structural details modeling (welds, joints, brackets, etc.), damping, inclusion of orders of vibration modes, geometric symmetry, etc. It should be noted that when geometric symmetry/asymmetry is adopted, one may lose asymmetrical/symmetrical vibration modes as discussed in Sect. 4.3. For calculating the structural responses during vehicle-deck interactions, it is normally assumed that the contribution from higher-order vibration modes of the supporting structure can be neglected. However, if acoustic characteristics are to be studied during vehicle-deck interactions, the higher order of vibration modes must be included.

In cases in which the supporting structures are of a type of deck structures, there are different levels for describing the structural models. One of the most sophisticated methods is to model the structure by using the orthotropic plate theory [413]. An analytical model that uses an orthotropic plate to describe a typical deck is shown in Fig. 19.6. In two opposite edges, the deck is fixed for translation in the global X, Y, and Z directions with two torsion stiffnesses K_1 and K_2 around the Y direction. The other two edges are either free to resemble bridge edges or fixed for translation in the Y direction to resemble the in-plane constraints of decks. In the middle of the deck, a centerline (CL) girder is installed that can resemble either the centerline girder of ship decks or the middle support for continuous bridge decks.

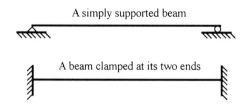

Fig. 19.7 A beam with different support conditions to resemble supporting structures [411]

Using the orthotropic thin plate theory [432], the governing differential equation can be expressed as:

$$D_X \frac{\partial^4 \omega(X,Y,t)}{\partial X^4} + 2H \frac{\partial^4 \omega(X,Y,t)}{\partial X^2 \partial Y^2} + D_Y \frac{\partial^4 \omega(X,Y,t)}{\partial Y^4} + \cdots \\ C \frac{\partial \omega(X,Y,t)}{\partial t} + \rho t_z \frac{\partial^2 \omega(X,Y,t)}{\partial t^2} = f(X,Y,t) \qquad (19.1)$$

$$H = (D_X v_Y + D_Y v_X)/2 + 2 D_{XY} \qquad (19.2)$$

where D_X and D_Y are the flexural stiffnesses in the X and Y directions, respectively; D_{XY} is the torsion stiffness; $f(X, Y, t)$ is the lateral moving load at time t and position (X, Y); ρ and t_z are the density and the thickness of the plate, respectively; and ω is the vertical deflection at point (X, Y). v_X and v_Y are the Poisson's ratios associated with the X and Y directions, respectively, along the plate. For a surface panel with isotropic material, v_X and v_Y are equal to the Poisson's ratio v. H is defined as the effective torsional rigidity of the orthotropic plate.

Provided that a deck structure is supported by evenly distributed transverse beams, and only the structural responses in the vertical direction are of interest, it can be simplified as a beam with appropriate support conditions. Figure 19.7 shows two types of supported beams where the beam clamped at its two ends is often used to resemble a deck with transverse beams connected to stiff side-girders, while the beam simply supported at its two ends can generally resemble the case of decks with relatively thin cross-sections.

For the two cases presented in Fig. 19.7, a beam model can be expressed using the Euler–Bernoulli differential equation (classical beam theory):

$$E I_z \frac{\partial^4 \omega(x,t)}{\partial x^4} + \rho A \frac{\partial^2 \omega(x,t)}{\partial t^2} + c A \frac{\partial \omega(x,t)}{\partial t} = f(x,t) \qquad (19.3)$$

where E is Young's modulus of the beam material; I_z is the second moment of area of the beam cross-section, A; $\omega(x, t)$ is the vertical deflection of a beam at point x and time t; ρ is the density of the beam material; c is the viscous damping coefficient; and $f(x, t)$ represents the force applied at position x and time t, for example due to the interactions with vehicles.

In addition, for high-frequency structural vibrations, which may be of interest in the study of noise, the cross-sectional deformation becomes significant. Hence, the

19.2 Physical Modeling 393

Table 19.3 Coupling relations for various types of vehicle and supporting structure models

Vehicle models	Rigid surface with or without irregularities	One-dimensional beam model	Plate model
Dead weight (force) model		Uncoupled	Uncoupled
Mass model	Uncoupled	Coupled	Coupled
Spring-damper mass model	Uncoupled	Coupled	Coupled

Euler–Bernoulli beam model may be replaced by the Timoshenko beam model. Differences in displacement amplitude can then be observed, since the Timoshenko beam theory accounts for transverse shear and rotary inertia-induced flexibility. In other words, the Euler–Bernoulli beam model is not preferred in analyses of short and thin-webbed beams and beams where higher modes are excited. From a wave point of view, the Euler-Bernoulli beam theory predicts the unrealistic wave speed because it approaches infinity for very high frequencies [433].

Selection and design of a deck model depends on output requirements and loading conditions. For example, if the reaction forces or the motions of a car are of interest, given that the excitation frequency is well below the natural frequency of the flexural vibration modes of a relatively stiff deck structure, the deformation of the deck can then be neglected. Hence, the deck structure can be assumed to be a rigid surface that simplifies the vehicle-deck interaction problem to a situation where the vehicle (car) is attached to/interacting with a rigid surface.

In reality, for a vehicle moving on a deck structure, the initial surface profile and the dynamic deflection of the structure contribute to the excitation of the vehicle and the tire forces/contact forces between the tires and the deck, which vary in time. Large vibration displacement amplitudes of the deck structure may occur when the excitation frequency of the vehicle is close to the natural frequency of the supporting structure.

Table 19.3 illustrates various types of modeling for vehicles and supporting structures and their coupling relations. If a dead weight (force) model is used for modeling the vehicle, see Ref. [68] for simulation of vehicles running on decks, or if the supporting structure (deck) is modeled as a rigid surface, see references [177, 179]. There is no dynamic coupling between the vehicle and the supporting structure. Consequently, the solution procedure of these cases can be simplified to a great extent, since only the responses of the structure or the vehicle need to be analyzed.

19.2.4 Interaction Models for Vehicle and Supporting Structures

The interaction between a vehicle and its supporting structure (e.g. ship deck or rail track) can be calculated by coupling the responses from the vehicle and the supporting structure:

1. Prescribe an initial excitation from the structure to the vehicle; this excitation may represent, for example, an initial surface irregularity of the structure surface, or a motion or deformation of the structure.
2. The prescribed excitation gives rise to the contact forces between the vehicle tires and the supporting structure. It will also give information about the dynamic characteristics (vibrations) of the vehicle for the given excitation.
3. The calculated contact (reaction) forces are applied to the supporting structure, which, in turn, result in a dynamic response of the structure.
4. Iteratively, the registered dynamic response in step 3 is used as an updated excitation and steps 1–3 are repeated until the response of the vehicle in the contact with the structure agrees with the structural response in the same position.

Numerically, two equations of motions for the vehicle and the supporting structure are defined, respectively. A coupling between them is formulated as an equation system with matrices through the contact forces. In the condition when the vehicle is moving, the matrices are time-dependent, which then requires that the equations are updated for every time step [425].

19.3 Finite Element Simulations

For simplified structural layouts and idealized support conditions, analytical models can be used to analyze vehicle-structure interaction problems. However, the finite element (FE) method is more convenient to use as a tool in investigations with complex structures with many degrees-of-freedom.

There are several ways of utilizing the FE method for calculating or analyzing dynamic interactions. One of the most simplified ways is to use structure elements, such as beam or shell elements, for representation of the supporting structure, and lumped masses connected by spring and dashpot elements for representation of the vehicle. These types of models are suitable in the evaluation of interaction responses when the vehicle is no longer moving in relation to its supporting structure; see [178, 411] for a thorough description of vehicle models.

Figure 19.8 shows an example from an FE calculation where a truck is running on a concrete deck bridge [435]. In this calculation, which requires a higher order of the finite elements used and also a more complex FE model, the concrete part of the bridge is represented using solid elements, and the steel rebars and strands are modeled using a one-dimensional bar element, the nodes of which coincide with the corresponding nodes of the solid elements. The truck is modeled with a three-dimensional suspension system as well as pneumatic and rotating wheels, and appropriate contact algorithms are used to simulate the contacts between the tires and the bridge deck. Fig. 19.8 shows the FE mesh (top) and parts of the time history of the rear suspension resultant forces (bottom) when the truck is running on the bridge.

19.3 Finite Element Simulations

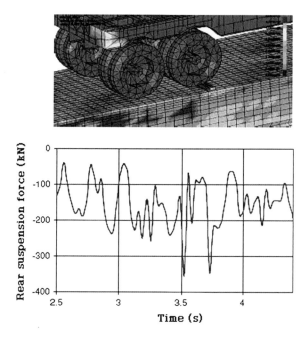

Fig. 19.8 Finite element mesh (*top*) and time history of rear suspension resultant forces (*bottom*) when the truck is running on a concrete bridge deck at a speed of 80 km/h [435]

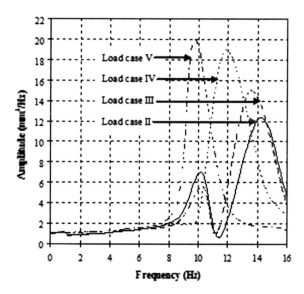

Fig. 19.9 Response spectra of the vertical displacement at the center of the deck for four different load cases; the horizontal axis represents the excitation frequency. See the text for explanation of the different load cases [411]

Jia and Ulfvarson [68] modeled the vehicles as spring-mass systems and the lightweight decks with thin shell elements. The results in Fig. 19.9 show the vertical displacement of a lightweight high-tensile steel deck structure under different vehicle load cases. The load cases are: (II) deck loaded with six cars parked on the right half of the deck; (III) deck loaded with one car at the center of the deck above the CL-girder; (IV) deck loaded with one car on the right side of the deck close to the side-shell; (V) deck loaded with a rigid cargo load on the right half of the deck. The value of the cargo load in case (V) is equal to the load of six cars parked on the deck. It is found that the cars act as a dynamic absorber of the vibrations. As such, the cars reduce the structural responses when the excitation frequency is below 11 Hz, while above 11 Hz the carloads increase the structural response. Therefore, it is suggested that carloads have a similar mechanism as that of mass dampers, i.e., through the momentum exchange between the spring-mass system and the deck structure, the system will absorb the energy initially provided through the vibrations of the deck. The above observations are specific to the deck structure; similar analyses may give different results for another structure. However, it is reasonable to point out that parked vehicles can reduce at least one mode shape response. It is also shown that the locations of the cars parked on the deck have an influence on the dynamic response of the deck. When a car is parked at the center of the deck (i.e., an anti-node location), the maximum dynamic response is lower than that for a case with the car parked on the panel close to the side-shell (i.e., node location). Obviously, if the car/cars are placed on the locations of the structure where vibration amplitude is lowest, less influence of the parked car/cars on the natural frequencies and eigenmodes would be seen.

19.4 Analysis of Vehicle Securing

A practical problem addressed in dynamic vehicle-ship deck interaction research is schematically described in Fig. 19.10. Vehicles such as cars, train wagons, trailers and trucks, are parked on RO/RO vessels for transportation. During the ship motions and other types of load/motion excitations, the vehicles may vibrate together with the deck structures. In order to prevent the vehicles from moving on the deck when the vessel is at sea, all vehicles are nowadays fixed to the deck structure using different types of lashings. If the lashings are for any reason disconnected/fractured during transportation, the cargo may shift and unstable seafaring can result, which poses great risk for the entire vessel.

Mechanical testing experiments were carried out in order to investigate, during dynamic excitations, the friction coefficients between a tire and two different types of deck surfaces. From the testing, it was shown that the friction coefficients were not sensitive to the variation of the deck excitation frequencies, but they were quite sensitive to the normal forces caused by the wheel loads (see Jia and Ulfvarson [177]).

19.4 Analysis of Vehicle Securing

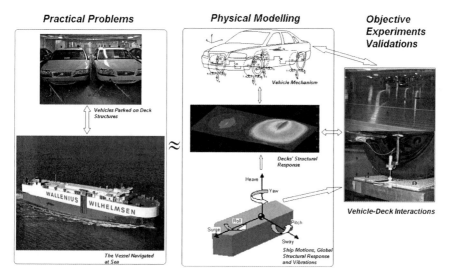

Fig. 19.10 Schematic description of a practical problem addressed in dynamic vehicle-ship deck interaction research [410]

In the mechanical friction testing, it was assumed that the tire had the same friction characteristics regardless of which direction the resultant tangential force is applied along the deck surface. However, due to the variation of the tire surface pattern, it may yield a different friction coefficient when the resultant tangential horizontal force is applied in different directions along the deck.

In Jia and Ulfvarson [177], a car was modeled numerically as a spring-damping-mass system. Harmonic roll and pitch motions were applied on it, and it was concluded that the value of maximum required friction coefficients between tires and decks to prevent sliding was highly relevant to the roll and pitch amplitude, the pitch period, the orientation of vehicles on decks, and the vertical location of vehicles from the base line of the vessel.

A "lashing-free" concept for cargo securing in ships was investigated in Jia [179]. A computational code was developed for the calculation of vehicle-deck interactions under ship motions. It was found that, for a target ship of a 6,000-car unit RO/RO vessel, vehicle securing was mainly influenced by the ship's roll motions, which were highly dependent on the wave height and the loading condition. It was suggested that, based on the analyses, vehicles can be secured without being lashed in a large area of the ship on some voyage routes in specific weather conditions with less adverse sea states. However, it was still suggested that conventional lashing holes on the deck should be constructed to cope with severe sea states.

Contrary to the suggestions from cargo-securing codes by IMO [430] and experience, Jia and Ulfvarson [177] found that the fore-and-aft direction for securing vehicle cargoes may not be the most optimal direction to prevent sliding,

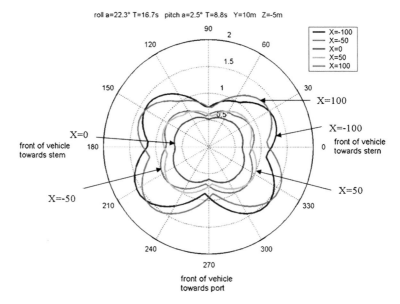

Fig. 19.11 The maximum required friction coefficient to resist sliding of vehicles with the variation of orientation and location of a car parked on decks (*X* the distance from the center of a ship along the surge direction; the phase angle between roll and pitch motions: 0°)

as shown in Fig. 19.11. The influence of the phase angle between the roll and the pitch motions and the effects of the car locations on the friction are important parameters that can be elaborated with.

The criterion of vehicle securing in the references [177, 179] was based on the assumption that shifting of vehicles occurs when any of the tires reaches the maximum required friction coefficient between the tire and the deck surface. Realistic contact, which involves dynamic contact effects, is more complex than that treated in those publications, and should therefore be addressed in future work. Additionally, Jia and Ulfvarson [177] found that, theoretically, the maximum required friction coefficient normally occurs when the phase angle between the roll and the pitch is 40°, even though the phase angle will normally not exceed 30° [439]. In order to get a better understanding of the relation between the phase angle and the vehicle-deck interaction, it is still worth investigating in a future study why this happens.

References

1. Amr S Elnashai and Luigi Di Sarno, Fundamentals of Earthquake Engineering, John Wiley and Sons, UK, 2008.
2. BM Broderick, AS Elnashai, NN Ambraseys, JM Barr, RG Goodfellow and EM Higazy, The Northridge (California) earthquake of 17 January 1994: Observations, strong motion and correlative response analysis, Engineering Seismology and Earthquake Engineering, Research Report No ESEE 94/4, Imperial College, London, 1994.
3. ISO 2631-1, Mechanical vibration and shock – evaluation of human exposure to whole body vibration – Part 1: General requirements, International Organization for Standardization, 2nd ed., Geneva, 1997.
4. Tianjian Ji and Adrian Bell, Seeing and Touching Structural Concepts, Taylor and Francis, Oxon, 2008.
5. PK Berg and O Bråfel, Noise and vibrations on board, Joint Industrial Safety Council, Stockhom, 1991.
6. Odd M Faltinsen and Alexander N Timokha, On sloshing modes in a circular tank, Journal of Fluid Mechanics, Vol 695: 467-477, 2012.
7. JA Romero, R Hildebrand, M Martinez, O Ramirez and JA Fortanell, Natural sloshing frequencies of liquid cargo in road tankers, International Journal of Heavy Vehicle System, Vol 2, No 2: 121-138, 2005.
8. Ken Hatayama, Lessons from the 2003 Tokachi-oki, Japan, earthquake for prediction of long-period strong ground motions and sloshing damage to oil storage tanks, J Seismol, Vol 12: 255-263, 2008.
9. Crazy "wave pool" aboard Sun Princess, http://www.youtube.com/watch?v=AJCurMmkNTY, 2007.
10. Pekka Ruponen, Jerzy Matusiak, Janne Luukkonen and Mikko Ilus, Experimental study on the behavior of a swimming pool onboard a large passenger ship, Marine Technology, Vol 46, No 1: 27–33, 2009.
11. Kathryn Brown, Tsunami! At Lake Tahoe? 2008.
12. The Joint Accident Investigation Commission of Estonia, Finland and Sweden, Final Report on the Capsizing on 28 September 1994 in the Baltic Sea of the RoRo passenger vessel MV ESTONIA, 1997.
13. Torgeir Moan, Safety of offshore structures, No 2005-04, Centre for Offshore Research and Engineering, National University of Singapore, 2005.
14. Matthys Levy and Mario Salvadori, Why Buildings Fall Down, WW Norton and Company, New York, NY, 2002.
15. RN White, P Gergely and RG Sexsmith, Structural Engineering, Volume 1, Introduction to Design Concepts and Analysis, John Wiley and Sons, New York, NY, 1972.

J. Jia, *Essentials of Applied Dynamic Analysis*, Risk Engineering,
DOI: 10.1007/978-3-642-37003-8, © Springer-Verlag Berlin Heidelberg 2014

16. A Larsen, S Esdahl, JE Andersen and T Vejrum, Storebaelt suspension bridge – vortex shedding excitation and mitigation by guide vanes, Journal of Wind Engineering and Industrial Aerodynamics, Vol 88: 283–296, 2000.
17. Junbo Jia, Wind and structural modeling for an accurate fatigue life assessment of tubular structures, Engineering Structures, Vol 33, No 2: 477–491, 2011.
18. William T Thomson, Vibration theory and applications, George Allen and Unwin, London, 1966.
19. www.taipei-101.com.tw.
20. David Lee and Martin Ng, Application of tuned liquid dampers for the efficient structural design of slender tall buildings, CTBUH Journal, Issue 4: 30–36, 2010.
21. PA Hitchcock, KCS Kwok and RD Watkins, Characteristics of liquid column vibration absorbers (LCVA)-I., Engineering Structures, Vol 19, No 2: 126–134, 1997.
22. PA Hitchcock, KCS Kwok and RD Watkins, Characteristics of liquid column vibration absorbers (LCVA)-II, Engineering Structures, Vol 19, No 2: 135–144, 1997.
23. Junbo Jia, Modern Earthquake Engineering for Offshore and Onland Structures, Springer, Heidelberg, 2014.
24. Arvid Naess and Torgeir Moan, Stochastic Dynamics of Marine Structures, Cambridge University Press, Cambridge, 2012.
25. Petr Krysl, A Pragmatic Introduction to the Finite Element Method for Thermal and Stress Analysis, World Scientific, London, 2006.
26. Alain Curnier, Computational Methods in Solid Mechanics, Kluwer Academic Publishers, Dordrecht, Netherlands, 1994.
27. Herbert Goldstein, Charles P Poole Jr., and John L Safko, Classical Mechanics, 3rd ed., Addison-Wesley, Boston, MA, 2001.
28. ET Whittaker, A Treatise on the Analytical Dynamics of Particles and Rigid Bodies, Cambridge Mathematical Library, Cambridge, 1988.
29. S Timoshenko and DH Young, Advanced Dynamics, 1st ed., McGraw-Hill Book Company, New York, NY, 1948.
30. JP den Hartog, Forced vibrations with combined Coulomb and viscous damping, Translations of the ASME, 53: 107–115, 1930.
31. JL Synge and BA Griffith, Principles of Mechanics, McGraw-Hill, New York, NY, 1959.
32. AH Nayfeh, Perturbation Methods, Wiley, New York, NY, 1973.
33. SH Crandall and WD Mark, Random Vibration in Mechanical Systems, Academic Press, New York, NY, 1963.
34. JD Robson, Random Vibration, Edinburgh University Press, Edinburgh, 1964.
35. Olek C Zienkiewicz, Robert L Taylor, JZ Zhu, The Finite Element Method: Its Basis and Fundamentals, 6th ed., Butterworth-Heinemann, Oxford, 2005.
36. Roger F Gans, Engineering Dynamics—from the Lagrangian to simulation, Springer, Heidelberg, 2013.
37. Meirovitch L, Elements of Vibration Analysis, McGraw-Hill Kogakusha, Tokyo, 1975.
38. Kerson Huang, Fundamental Forces of Nature—the story of gauge fields, World Scientific, London, 2007.
39. Olof Friberg, Course notes of structural dynamics, Chalmers University of Technology, Gothenburg, 2000.
40. Jianlian Cheng and Hui Xu, Inner mass impact damper for attenuating structure vibration, International Journal of Solids and Structures, Vol 43: 5355–5369, 2006.
41. De Morgan, Budget of Paradoxes, Longmans, Green, London, 1872.
42. Mohamed S Gadala, Finite Element Applications in Dynamics, in Vibration and Shock Handbook, Taylor and Francis, Oxon, 2005.
43. JS Bendat and AG Piersol, Engineering Applications of Correlation and Spectral Analysis, A Wiley-Interscience Publication, New York, NY, 1980.
44. Kihong Shin and Joseph K Hammond, Fundamentals of Signal Processing for Sound and Vibration Engineers, John Wiley and Sons, West Sussex, 2008.

References

45. HJ Pradlwarter and GI Schuëller, Stochastic structural dynamics—a primer for practical applications, in Bilal M. Ayyub, Ardeshir Guran and Achintya Haldar, Uncertainty Modeling in Vibration, Control and Fuzzy Analysis of Structural Systems, World Scientific, London, 1997.
46. Paul H Wirsching, Thomas L Paez and Keith Ortiz, Random Vibrations—theory and practice, John Wiley and Sons, New York, NY, 1995.
47. Arthur Schuster, On the Investigation of Hidden Periodicities with Application to a Supposed 26 Day Period of Meterological Phenomena, Terrestr. Magnet., Vol 3: 13–41, 1898.
48. Arthur Schuster, The periodogram of magnetic declination as obtained from the records of Greenwich Observatory during the years 1871–1895, Trans. Camb. Phil. Soc., Vol 18: 107–135, 1900.
49. Arthur Schuster, The periodogram and its optical analogy, Proc. Roy. Soc. Lond. Ser. A, Vol 77: 136–140, 1906.
50. Dag Myrhaug, Probabilistic theory of sealoads, Department of Marine Hydrodynamics, Compendium No UK-93-52, Norwegian Institute of Technology, Trondheim, 1993.
51. John Rice, Mathematical Statistics and Data Analysis, 2nd ed., Duxbury Press, CA, 1995.
52. A Naess, Prediction of extremes of Morison-type loading, an example of general method, Ocean Engineering, Vol 10: 313–324, 1983.
53. Junbo Jia, An efficient nonlinear dynamic approach for calculating wave induced fatigue damage of offshore structures and its industrial applications for lifetime extension, Applied Ocean Research, Vol 30, No 3: 189–198, 2008.
54. M Isaacson and J Baldwin, Random wave forces near free surface, ASCE J. Waterways, Port, Coastal Ocean Eng. Vol 116, No 2: 232–251, 1990.
55. CC Tung, Effects of free surface fluctuation on total wave forces on cylinder, ASCE J. Eng. Mech. Vol 121, No 2: 274–280,1995.
56. CY Liaw and XY Zheng, Inundation effect of wave forces on jack-up platforms, Int. J. Offshore Polar Eng. Vol 11, No 2, 2001.
57. XY Zheng and CY Liaw, Non-linear frequency-domain analysis of jackup platforms, Int. J. Non-Linear Mech. Vol 39, No 9: 1519–1534, 2004.
58. DE Cartwright and MS Longuet-Higgins, The statistical distribution of the maxima of a random function, Proc. Roy. Soc. Ser. A, Vol 237: 212–232, 1956.
59. MS Longuet-Higgins, On the statistical distributions of the heights of sea waves, Journal of Marine Research, Vol 9, No 3: 245–266, 1952.
60. A Almar-Næss, Fatigue Handbook, TAPIR Publishers, Trondheim, 1985.
61. W Waloddi Weibull, A statistical distribution function of wide applicability, J. Appl. Mech.-Trans. ASME, Vol 18, No 3: 293–297, 1951.
62. Ulf Björkenstam, Reliability Based Design, 1st ed., Department of Naval Architecture and Ocean Engineering, Chalmers University of Technology, 1998.
63. Milutin Srbulov, Geotechnical Earthquake Engineering—simplified analyses with case studies and examples, Springer, Heidelberg, 2008.
64. S Winterstein, T Ude, CA Cornell, P Bjerager and S Haver, Environmental parameters for extreme response: Inverse FORM with omission factors. Proceedings of the 6th International Conference on Structural Safety and Reliability, Balkema, Innsbruck, 1993.
65. S Haver, G Sagli and TM Gran, Long term response analysis of fixed and floating structures, Ocean Wave Kinematics, Dynamics and Loads on Structures (OTRC), April-May: 240–250, 1998.
66. EH Vanmarcke, Properties of spectral moments with applications to random vibration, Journal of Engineering Mechanics Division, ASCE 98: 425–426, 1972.
67. WG Price and RED Bishop, Probability Theory of Ship Dynamics, Chapman and Hall, Dordrecht, 1974.

68. Junbo Jia and Anders Ulfvarson, Structural behaviour of a high tensile steel deck using trapezoidal stiffeners and dynamics of vehicle–deck interactions, Marine Structures, Vol 18: 1–24, 2005.
69. Jan Van Der Tempel, Design of Support Structures for Offshore Wind Turbines, PhD thesis, Technische Universiteit Delft, 2006.
70. ISO 19902, Petroleum and natural gas industries – fixed steel offshore structures, International Standard Orgnization, Geneva, 2007.
71. NORSOK N-003, Actions and Action Effects, 2nd ed, Standard Norway, Lysaker, September, 2007.
72. Typical Metocean Reports, Statoil ASA, Stavanger, Norway.
73. Takewaki Izuru, Critical Excitation Methods in Earthquake Engineering, 1st ed., Elsevier Science, Oxford, 2007.
74. JN Brune, Tectonic stress and the spectra of seismic shear waves from earthquakes, Journal of Geophysical Research, Vol 75: 4997–5009, 1970.
75. JN Brune, Correction, Journal of Geophysical Research, Vol 76: 5002, 1971.
76. TC Hanks, Bulletin of the Seismological Society of America, Vol 72: 1867–1879, 1982.
77. AS Papageorgiou and K Aki, A specific barrier for the quantitative description of inhomogeneous faulting and the prediction of strong ground motion. II. Applications of the model, Bulletin of the Seismological Society of America, Vol 73: 953–978, 1983.
78. Eric W Weisstein, Maximum entropy method, MathWorld—A Wolfram Web Resource, 2011.
79. JR Morison, MP O'Brian, JW Johnson and SA Schaaf, The force exerted by surface waves on piles, Petroleum Transactions, AIME, Vol 189: 149–154, 1950.
80. O Mo, Stochastic time domain analysis of slender offshore structures, Report UR 83-33, Norwegian University of Science and Technology, Trondheim, 1983.
81. NDP Barltrop and AJ Adams, Dynamics of Fixed Marine Structures, 3rd ed., MTD, London, 1991.
82. Yoshimi Goda, Random Seas and Design of Maritime Structures, 3rd ed., World Scientific, London, 2010.
83. Turgut Sarpkaya, Mechanics of Wave Forces on Offshore Structures, Van Nostrand Reinhold Co, New York, NY, 1981.
84. A Gelb and WE Vander Velde, Multiple-input Describing Functions and Nonlinear System Design, McGraw-Hill, New York, NY, 1968.
85. James F Wilson, Dynamics of Offshore Structures, John Wiley and Sons, New Jersey, 2003.
86. Sverre Haver and Torgeir Moan, On some uncertainties related to short term stochastic modeling of ocean waves, Applied Ocean Research, Vol 5, No 2: 93–108, 1983.
87. OM Faltinsen, Sea Loads on Ships and Offshore Structures, Cambridge University Press, Cambridge, 1990.
88. MK Ochi and WE Bolton, Statistics for prediction of ship performance in a seaway, Parts 1-3. ISP Nos 222, 224 and 229, 1973.
89. DE Cartwright and MS Longuet-Higgins, The statistical distribution of the maxima of a random function, PRS (Series A): 237, 1956.
90. WC Price and RED Bishop, Probabilistic Theory of Ship Dynamics, Chapman and Hall, London, 1974.
91. ARJM Lloyd, Seakeeping: Ship behaviour in rough weather, Hampshire, 1998.
92. J Ewing, Some Results from the Joint North Sea Wave Project of Interest to Engineers, Joint ONR/RINA symposium on The Dynamics of Marine Vehicles and Structures in Waves, UCL, London April 1974.
93. K Hasselmann, TP Barnett, E Bouws, H Carlson, DE Cartwright, K Eake, JA Euring, A Gicnapp, DE Hasselmann, P Kruseman, A Meerburg, P Mullen, DJ Olbers, K Richren, W Sell and H Walden, Measurement of wind wave growth and swell decay during the Joint North Sea Wave Project (JONSWAP), Deutschen Hydrographischen Zeitschrift Reihe A, Vol 8, No 12, DEUTSCHES HYDROGRAPHISCHES INSTITUT, Hamburg, 1973.

References

94. K Torsethaugen and S Haver, Simplified double peak spectral model for ocean waves. Proceedings of the 14th International Offshore and Polar Engineering Conference, Toulon, France, 2004.
95. G Najafian, R Burrows and RG Tickell, A review of the probabilistic description of Morison wave loading and response of fixed offshore structures, Journal of Fluids and Structures, Vol 5: 585–616, 1995.
96. K Anastasiou, RG Tickell and JR Chaplin, The non-linear properties of random wave kinematics. Proceedings of the 3rd International Conference on the Behaviour of Offshore Structures (493–515), MIT, Cambridge MA, 1982.
97. Osborne Reynolds, An experimental investigation of the circumstances which determine whether the motion of water shall be direct or sinuous, and of the law of resistance in parallel channels, Philosophical Transactions of the Royal Society, Vol 174: 935–982, 1883.
98. I van der Hoven, Power spectrum of horizontal wind speed in the frequency range of 0.0007 to 900 cycles per hour, Journal of Meterology, Vol 14: 160–164, 1957.
99. Notes for meeting, Comment during IEA expert meeting on offshore wind energy, Ringkøbing, 2004.
100. Einar N Strømmen, Theory of Bridge Aerodynamics, Springer, Heidelberg, 2010.
101. AR Collins, The modern design of wind-sensitive structures, Proceedings of the Seminar held on 18 June 1970 at The Institution of Civil Engineers, Great George Street London, 1970.
102. AG Davenport, The spectrum of horizontal gustiness near the ground of high winds, Quarterly Journal Royal Meteorological Society, Vol 87: 194–211, 1961.
103. OJ Andersen and J Løvseth, The Frøya database and boundary layer wind description, Marine Structures, Vol 19: 173–192, 2006.
104. NS-EN ISO 19901-1, Petroleum and natural gas industries – Specific requirements for offshore structures – Part 1: Metocean design and operating conditions, International Organization for Standardization, 2005.
105. NS 3491-4:2002, Design of structures—Design actions—Part 4: Wind loads, 2002.
106. Det Norske Veritas, Recommended practice Dnv-Rp-C205—Environmental conditions and environmental loads, April 2007.
107. O Ditlevsen, and HO Madsen, Structural Reliability Methods, TECHNICAL UNIVERSITY OF DENMARK, Lyngby, 2005.
108. C Dyrbye and SO Hansen, Wind Loads on Structures, John Wiley and Sons, Chichester, 1997.
109. E Simiu and RU Scanlan, Wind Effects on Structures—An Introduction to Wind Engineering, John Wiley, New York, NY, 1978.
110. MK Ochi and YS Shin, Wind turbulent spectra for the design consideration of offshore structures, OTC paper 5736, Houston, TX, 1988.
111. ESDU 86010. Characteristics of atmospheric turbulence near ground, Part II and Part III, Engineering and Science Data Unit, London, 1985.
112. K Aas-Jakobsen and E Strømmen, Time domain buffeting response calculations of slender structures, Journal of Wind Engineering and Industrial Aerodynamics, Vol 89, No 5: 341–364, 2001.
113. TJO Sanderson, Ice Mechanics—Risks to Offshore Structures, Graham and Trotman, London, 1988.
114. ISO 19906, Petroleum and natural gas industries—Arctic offshore structures, International Standard Orgnization, Geneva, 2010.
115. ME Johnston, Ken R Croasdale and Ian J Jordaan, Localized pressures during ice–structure interaction: relevance to design criteria, Cold Regions Science and Technology, Vol 27, Iss 2: 105–117, April 1998.
116. Ning Xu, Qianjin Yue, Fengwei Guo, Mitigating ice-induced vibration by adding ice-breaking cone, The Proceedings of the 20th (2010) International Offshore and Polar Engineering Conference, Beijing, China 2010.

404 References

117. T Kärnä, K Kamesaki and H Tsukuda, A numerical model for dynamic ice structure interaction, Computers and Structures, Vol 72: 645–658, 1999.
118. T Kärnä, Y Qu and W Kühnlein, New spectral method for modeling dynamic ice actions, Proceedings of 23rd International Conference on Offshore Mechanics and Arctic Engineering, OMAE 2004, New York: ASME, 2004.
119. H Gravesen, SL Sorensen, P Volund, A Barker and G Timco, Ice loading on Danish wind turbines: Part 2. Analyses of dynamic model test results, Cold Regions Science and Technology, Vol 41, No 1: 25–47, 2005.
120. X Liu, G Li, R Oberlies and Q Yue, Research on short term dynamic ice cases for dynamic analysis of ice-resonant jacket platform in the Bohai Gulf, Marine Structures, Vol 22, No 3: 457–479, 2009.
121. QJ Yue, Y Qu, XJ Bi and T Kärnä, Ice force spectrum on narrow conical structures, Cold Regions Science and Technology, Vol 49, No 2: 161–169, 2007.
122. G Neumann, On wind generated ocean waves with special reference to the problem of wave forecasting, NYU, Coll. of Eng., Res. Div, Dept. of Metcor. and Ocean-ogr: 136–141, 1952.
123. AM Cornett and GW Timco, Ice loads on an elastic model of the Molikpaq, Applied Ocean Research, Vol 20: 105–118, 1998.
124. Anil K Chopra, Dynamics of Structures – Theory and application to earthquake engineering, 2nd ed., Prentice Hall, NJ, 2000.
125. Steven Kramer, Geotechnical Earthquake Engineering, Prentice Hall, London, 1996.
126. K Kanai, Semi-empirical formula for the seismic characteristics of the ground, Bulletin of the Earthquake Research Institute, University of Tokyo, Vol 35: 309–325, 1967.
127. H Tajimi, A statistical method of determining the maximum response of a building structure during an earthquake, Proceedings of the 2nd World Conference on Earthquake Engineering, Vol 2: 781–798, 1960.
128. Nam Hoang, Yozo Fujino, Pennung Warnitchai, Optimal tuned mass damper for seismic applications and practical design formulas, Engineering Structures, Vol 30, No 3: 707–715, 2008.
129. T Kubo and J Penzien, Simulation of three-dimensional strong ground motions along principal axes, San Fernando Earthquake, Earthquake Engineering and Structural Dynamics, Vol 7: 265–278, 1979.
130. Abbas Moustafa, Identification of resonant earthquake ground motion, Sãdhanã, Indian Academy of Science, Vol 5, Part 3: 355–371, 2010.
131. Det Norske Veritas (DnV), Theory Manual of Framework Version 2.1, Høvik, 1993.
132. H Kanaji, N Hamada, T Naganuma, Seismic retrofit of a cantilever truss bridge in the Hanshin expressway, Proceedings of the International Symposium on Earthquake Engineering, 2005.
133. G Chen and J Wu, Optimal placement of multiple tuned mass dampers for seismic structures, Journal of Structural Engineering ASCE, Vol 127, No 9: 1054–1062, 2001.
134. C Li and Y Liu, Ground motion dominant frequency effect on the design of multiple tuned mass dampers, Journal of Earthquake Engineering, Vol 8, No 1: 89–105, 2004.
135. Bjarni Bessason, Assessment of Earthquake Loading and Response of Seismically Isolated Bridges, PhD thesis, MTA-rapport 1991: 88, Norges Tekniske Høgskole, Trondheim, 1992.
136. TC Hanks and RK McGuire, The character of high-frequency strong ground motion in space and time, Engineering Mechanics, Vol 112: 154–174, 1986.
137. DM Boor, Stochastic simulation of high-frequency ground motions based on seismological models of the radiated spectra, Bull. Seism. Soc. Am, Vol 73, No 6: 1865–1894, 1983.
138. RB Herman, Letter to the editor—an extension of random vibration theory estimates of strong ground motion to large distances, ull. Seism. Soc. Am, Vol 75, No 5: 1447–1453, 1985.
139. JE Luco and HL Wong, Response of a rigid foundation to a spatially random ground motion, J. Earthquake Eng. and Struct. Dyn., Vol 14: 891–908, 1986.

References 405

140. JC Wilson and PC Jennings, Spatial variation of ground motion determined from accelerograms recorded on a highway bridge, Bulletin of the Seism. Soc. of America, Vol 75, No 6: 1515–1533, 1985.

141. RS Harichandran and EH Vanmarcke, Stochastic variation of earthquake ground motion in space and time, J. Eng. Mech., ASCE, Vol 112: 154–175, 1986.

142. CH Loh and YT Yeh, Spatial variation and stochastic modeling of seismic differential ground movement, Earthquake Eng. Struct. Dyn., Vol 16: 583–596, 1988.

143. QM Feng and YX Hu, Mathematical model of spatial correlative ground motion, Earthquake Eng. Eng. Vib., Vol 1: 1–8, 1981.

144. CS Oliveira, H Hao and J Penjien, Ground motion modeling for multiple-input structural analysis, Struct. Saf., Vol 10: 79–93, 1991.

145. TJ Qu, JJ Wang and QX Wang, A practical power spectrum model for spatially varying seismic motion, J. Seismol., Vol 18: 55–62, 1996.

146. William F Stokey, Vibration of systems having distributed mass and elasticity, in Cyril M Harris and Allan Piersol (eds), Harris' Shock and Vibration Handbook, 5th ed., McGraw-Hill, New York, 2002.

147. S Graham Kelly, Theory and Problem of Mechanical Vibrations, McGraw-Hill, New York, 1996.

148. RED Bishop and DC Johnson, The Mechanics of Vibration, Reissue edition, Cambridge University Press, Cambridge, 2011.

149. M Moore, B Phares, B Graybeal, D Rolander and G Washer, Reliability of visual inspection for highway bridges, Federal Highway Administration Report, No FHWA-RD-01-020 and FHWA-RD-01-021, 2001.

150. Kinemetrics, System for SHM, Pasadena, CA, 2001.

151. Branko Glisic and Daniele Inaudi, Fibre Optic Methods for Structural Health Monitoring, John Wiley and Sons, Chichester, 2007.

152. Guidelines for structural health monitoring, ISIS Canada, 2001.

153. Hongna Li, Dongsheng Li and Gangbing Song, Recent applications of fiber optic sensors to health monitoring in civil engineering, Engineering Structures, Vol 26, No 11: 1647–1657, 2004.

154. J Zhang, J Prader, KA Grimmelsman, F Moon, AE Aktan, A Shama, Challenges in Experimental Vibration Analysis for Structural Identification and Corresponding Engineering Strategies, Third Experimental Vibration Analysis for Civil Engineering Structures conference, 3rd ed., Wroclaw, 2009.

155. Daniel Balageas, Claus-Peter Fritzen, Alfredo Güemes, Structural Health Monitoring (ISTE), Wiley-ISTE, 2006.

156. Srini-vasan Gopalakrishnan, Massimo Ruzzene, Sathyanaraya Hanagud, Computational Techniques for Structural Health Monitoring, Springer, Heidelberg, 2011.

157. Bharat Bhushan Prasad, Advanced Soil Dynamics and Earthquake Engineering, Prentice-Hall, Upper Saddle River, NJ, 2010.

158. Leonard Meirovitch, Elements of Vibration Analysis, McGraw-Hill Kongakusha, Tokyo, 1975.

159. WC Hurty and MF Rubinstein, Dynamics of Structures, Prentice-Hall, Englewood Cliffs, NJ, 1964.

160. S Timoshenko, DH Young and W Weawer Jr., Vibration Problem in Engineering, 4th ed., Van Nostrand, New York, NY, 1974.

161. Ray W Clough and Joseph Penzien, Dynamics of Structures, 3rd ed., Computers and Structures, Berkeley, CA, 2003.

162. S Graham Kelly, Fundamentals of Mechanical Vibrations, 2nd ed., McGraw-Hill, Boston, 2000.

163. Youn-Ju Jeong, Young-Jun You, Du-Ho Lee and Min-Su Park, Experimental study on water damping effects of hybrid floating structure, Proceedings of the ASME 32nd International Conference on Ocean, Offshore and Arctic Engineering, Nantes, 2013.

164. A Alexia, SR Wendy, R Dominique, F Patri, G Wayne, Feasibility and design of the clubstead: a cable-stayed floating structure for offshore dwellings, Proceedings of the 29th International Conference on Ocean, Offshore and Arctic Engineering, Shanghai, 2010.
165. de Silva and Clarence W, Vibration: Fundamentals and Practice, CRC Press, Boca Raton, FL, 2000.
166. Cheung Hun Kim, Nonlinear Waves and Offshore Structures, World Scientific, London, 2008.
167. Paul H Wirsching, Thomas L Paez and Keith Ortiz, Random Vibrations—theory and practice, John Wiley and Sons, New York, 1995.
168. Multi-Semester Interwoven Project for Teaching Basic Core Science, Technology, Engineering, Mathematics Material Critical for Solving Dynamic System Problems, UMass Lowell, 2006.
169. John Melvin Biggs, Introduction to Structural Dynamics, 1st ed., McGraw-Hill, New York, NY, 1964.
170. Maurice Petyt, Introduction to Finite Element Vibration Analysis, 2nd ed., Cambridge University Press, Cambridge, 2010.
171. Edward L Wilson, Three-Dimensional Static and Dynamic Analysis of Structures, Chapter 13, Dynamic Analysis Using Mode Superposition, Computers and Structures, 3rd ed., Berkeley, 2002.
172. Junbo Jia, Seismic analysis for offshore industry: Promoting state of the practice toward state of the art, Proceedings of the 22nd International Ocean and Polar Engineering Conference (ISOPE), Rhodes, 2012.
173. KJ Bathe, Finite Element Procedures in Engineering Analysis. Prentice-Hall, Englewood Cliffs, NJ, 1982.
174. PE Kloeden and E Platen, Numerical Solution of Stochastic Differential Equations, Springer, Heidelberg, 1999.
175. Meg Chesshyre, Offshore Engineer, October: 34–35, 2011.
176. http://www.offshore-mag.com/articles/2012/04/siri-platform-repair-cost-soars.html.
177. Junbo Jia and Anders Ulfvarson, The friction between car tires and decks under ship motions, Marine Technology(SNAME) Vol 43, No 1: 27–39, 2006.
178. Junbo Jia and Anders Ulfvarson, Dynamic analysis of vehicle-deck interactions, Ocean Engineering, Vol 33, No 13: 1765–1795, 2006.
179. Junbo Jia, Investigations of vehicle securing without lashings for Ro-Ro ships, Journal of Marine Science and Technology, Vol 12, No 1: 43–57, 2007.
180. LN Virgin, Vibration of axially loaded structures, Cambridge University Press, Cambridge, 2007.
181. Paolo L Gatti and Vittorio Ferrari, Applied Structural and Mechanical Vibrations: Theory, methods and measuring instrumentation, Taylor and Francis, Oxon, 2003.
182. Michiel Zaaijer, Loads, Lecture notes: dynamics and structural design, Offshore Windfarm Design (OE 5662), Delft University of Technology, Delft, 2007.
183. R Nataraja and CL Kirk, Dynamic response of a gravity platform under random wave forces, OTC-2904, Proceedings of the Offshore Technology Conference, 1977.
184. Peter Kocsis, The equivalent length of a pile or caisson in soil, Civil Engineering, Vol 46, No 12: 63,1976.
185. K Terzaghi, Evaluation of coefficients of subgrade reactions, Geotechnique, Vol 5, No 4: 297–326, 1955.
186. LN Virgin and RH Plaut, Postbuckling and vibration of linearly elastic and softening columns under self-weight, International Journal of Solids and Structures, Vol 41: 4989–5001, 2004.
187. AE Galef, Bending frequencies of compressed beams, Journal of the Acoustical Society of America, Vol 44 No 8: 643, 1968.
188. Olga Lebed and Igor A Karnovsky, Nonclassical Vibrations of Arches and Beams: Eigenvalues and eigenfunctions, McGraw-Hill, New York, NY, 2004.

References

189. A Bokaian, Natural frequency of beams under compressive axial loads, Journal of Sound and Vibration, Vol 126, No 1: 49–65, 1988.
190. A Bokaian, Natural frequencies of beams under tensile axial loads, Journal of Sound and Vibration Vol 142 No 3: 481–498, 1990.
191. FJ Shaker, Effect of axial load on mode shapes and frequencies of beams, NASA TN D-8109, 1975.
192. LN Virgin, ST Santillan and DB Holland, Effect of gravity on the vibration of vertical cantilevers, Journal of Mechanics Research Communications, Vol 34: 312–317, 2007.
193. LN Virgin and RH Plaut, Postbuckling and vibration of linearly elastic and softening columns under self-weight, International Journal of Solids and Structures, Vol 41: 4989–5001, 2004.
194. RH Plaut and LN Virgin, Use of frequency data to predict buckling, Journal of Engineering Mechanics, Vol 116, No 10: 2330–2335, 1990.
195. J Singer, Vibration correlation techniques for improved buckling predictions of imperfect stiffened shells, in JE Harding, PJ Dowling and N Agelidis (eds), Buckling of Shells in Offshore Structures (285–329), Granada Publishing, London, 1982.
196. J Singer, Vibration and buckling of imperfect stiffened shells—Recent developments, in JMT Thompson and GW Hunt (eds), Collapse: The buckling of structures in theory and practice (443–479), Cambridge University Press, Cambridge, 1983.
197. EN Carlos, César T Mazzilli, Odulpho GP Sanches, Neto Baracho, Wiercigroch Marian, Keber Marko, Non-linear modal analysis for beams subjected to axial loads: Analytical and finite-element solutions, International Journal of Non-Linear Mechanics, Vol 43, No 6: 551–561, 2008.
198. E Rabinowicz, The intrinsic variables affecting the stick-slip process, Proceedings of the Royal Society of London A, Vol 71: 668–675, 1959.
199. D Tabor, Friction: The present state of our understanding, Journal of Lubrication Technology, Vol 103: 169–179, 1981.
200. Robert F Steidel, An Introduction to Mechanical Vibrations, John Wiley& Sons, 3rd ed, The Atrium, West Sussex, 1989.
201. JR Morison, MP O'Brian, JW Johnson and SA Schaaf, The force exerted by surface waves on piles, Petroleum Transactions, AIME, Vol 189: 149–154, 1950.
202. PC Jennings, Equivalent viscous damping for yielding structures, Journal of Engineering Mechanics Division, ASCE, Vol 94: 131–165, 1968.
203. AJ King, The measurement and suppression of noise, with special reference to electrical machines, Chapman and Hall, London, 1965.
204. LN Virgin, Introduction to Experimental Nonlinear Dynamics, Cambridge University Press, Cambridge, 2000.
205. SS Rao, Mechanical Vibrations, 4th ed., Prentice Hall, Singapore, 2005.
206. TK Caughey, Classical normal modes in damped linear dynamic systems, J. of Applied Mechanics, Vol 27: 269–271, 1960.
207. DJ Ewins, Modal Testing: Theory, practice and application, 2nd ed., Research Studies Press, New York, NY,2000.
208. Quan Qin and Lei Lou, Effects of nonproportional damping on the seismic responses of suspension bridges, World Conference on Earthquake Engineering, Auckland, 2000.
209. AM Claret and F Venancio-Filho, A model superposition pseudo-forces method for dynamic analysis of structural systems with non-proportional damping, EESD, Vol 20, No 4: 303–315, 1991.
210. H Bachmann, WJ Ammann, F Deischl, J Eisenmann, I Floegl, GH Hirsch, GK Klein, GJ Lande, O Mahrenholtz, HG Natke, H Nussbaumer, AJ Pretlove, JH Rainer, EU Saemann and L Steinbeisser, Vibration Problems in Structures—Practical Guidelines, Springer, Heidelberg, 1995.
211. Benjamin Joseph Lazan, Damping of Materials and Members in Structural Mechanisms, 1st ed., Pergamon Press, Oxford, 1968.

212. Cyril M Harris and Allan G Piersol, Harris' Shock and Vibration Handbook, 5th ed., McGraw-Hill, New York, 2002.
213. JK Vandiver, The Significance of Dynamic Response in the Estimation of Fatigue Life, MIT, Cambridge, MA, 1980.
214. HE Krogstad, Analysis of wave spectra from the Norwegian continental shelf, Proceedings of the POAC, Trondheim, 1979.
215. Overall Damping for Piled Offshore Support Structures, Germanischer Lloyd Industrial Services GmbH, Hamburg, 2005.
216. Recommendations for revision of seismic damping values in regulatory guide 1.61, Brookhaven National Laboratory, US Nuclear Regulatory Commision Office of Nuclear Regulatory Research, Washington DC, 2006.
217. ASME, BPVC Section III—Rules for Construction of Nuclear Facility Components-Division 1-Appendices, 2007.
218. SM Wong, Computational Methods in Physics and Engineering, 2nd ed., World Scientific, London, 2003.
219. Romulus Militaru and Ioan Popa, On the numerical solving of complex linear systems, International Journal of Pure and Applied Mathematics, Vol 76, No1: 113–122, 2012.
220. Paul Manneville, Instabilities, Chaos and Turbulence: An introduction to nonlinear dynamics and complex systems, World Scientific, London, 2004.
221. Seyed Habibollah, Hashemi Kachapi and Davood Domairry Ganji, Dynamics and Vibrations, Springer, Heidelberg, 2014.
222. Subrata K Chakrabarti, The Theory and Practice of Hydrodynamics and Vibration, World Scientific, Singapore, 2002.
223. Robert D Cook, David S Malkus and Michael E Plesha, Concepts and Applications of Finite Element Analysis, 3rd ed., John Wiley and Sons, New York, NY, 1989.
224. Ali H Nayfeh and Balakumar Balachandran, Applied Nonlinear Dynamics: Analytical, computational, and experimental methods, Wiley-Vch Verlag GmbH and Co KGaA, Weinheim, 2004.
225. Junbo Jia, Investigation of sea wave and structure modeling for fatigue assessment of offshore structures, Journal of Ship Mechanics, Vol 15, No 9: 1005–1021, 2011.
226. James M Gere and Stephen P Timoshenko, Mechanics of Materials, 3rd ed., Chapman and Hall, London, 1991.
227. PV Lade, Three-parameter failure criterion for concrete, J. Eng. Mech. Div., ASCE, Vol 8, No 108, EM4: 850–863, 1982.
228. B Bresler and KS Pister, Strength of concrete under combined stresses, J. ACI, Vol 55: 321–345, 1958.
229. NS Ottosen, A failure criterion for concrete, J. Eng. Mech. Div., ASCE, Vol 103, EM4: 527–534, 1977.
230. W Prager, A new method of analyzing stresses and strains in work-hardening plastic solids, J Appl Mech, ASME, Vol 23: 493–496, 1956.
231. Akhtar S Khan and Sujian Huang, Continuum Theory of Plasticity, Wiley, Hoboken, NJ, 1995.
232. J Besson, G Cailletaud, JL Chaboche and S Forest, Non-Linear Mechanics of Materials, Springer, Heidelberg, 2010.
233. H Kauderer, Nonlinear mechanics, Izd. Inostr. Lit. Moscow, translated from Nichtlineare Mechanik, Berlin, 1958.
234. EE Khachian and VA Ambartsumyan, Dynamical Models of Structures in the Seismic Stability Theory, Nauka, Moscow, 1981.
235. NG Bondar, Non-linear Problems of Elastic Systems, Budivel'nik, Kiev, 1971.
236. IA Karnovsky and O Lebed, Free Vibrations of Beams and Frames: Eigenvalues and eigenfunctions, McGraw-Hill, New York, NY, 2003.
237. WF Chen and EM Lui, Stability Design of Steel Frame, CRC Press, Boca Raton, FL, 1991.

References

238. Y Goto and WF Chen, Second order analysis for frame design, Journal of Structural Engineering, ASCE, Vol 113, No 7, 1987.
239. D Nixon and D Beaulieu, Simplified second order frame analysis, Canadian Journal of Civil Engineering, Vol 2, No 4: 602–605, 1975.
240. A Rutenberg, A direct P-Delta analysis using standard plane frame computer programs, Computer and Structures, Vol 14: 1–2, 1987.
241. AS Monghadam and A Aziminejad, Interaction of torsional and P-Delta effects in tall buildings, Proceedings of the 13th World Conference on Earthquake Engineering, Vancouver, 2004.
242. YV Zakharov, KG Okhotkin and AD Skorobogatov, Bending of bars under a follower load, Journal of Applied Mechanics and Technical Physics, Vol 45 No 5: 756–763, 2004.
243. Norman E Dowling, Mechanical Behavior of Materials, Prentice-Hall, London, 1999.
244. JBP, Common structural rules for double hull oil tankers, Technical report, Joint Tanker Project, London, UK, 2005.
245. Inge Lotsberg, Assessment of fatigue capacity in the new bulk carrier and tanker rules, Marine Structures, Vol 19, No 1: 83–96, 2006.
246. Kjell Magne Mathisen, Solution methods for nonlinear finite element analysis, Norwegian University of Science and Technology, January, 2012.
247. Junbo Jia, The load sequence effects on structures' ultimate limit strength evaluation, Journal of Constructional Steel Research, Vol 67, No 2: 255–260, 2011.
248. J Case, L Chilver and CTF Ross, Strength of Materials and Structures, Elsevier, Oxford, 1999.
249. Nonlinear Finite Element Methods, Lecture notes, University of Colorado Boulder, 2012.
250. Zhibin Jia, Anders Ulfvarson, Jonas Ringsberg and Junbo Jia, A return period based plastic design approach for ice loaded side-shell/bow structures, Marine Structures, Vol 22, No 3: 438–456, 2009.
251. Sofia Leon, A unified library of nonlinear solution schemes: An excursion into nonlinear computational mechanics, University of Illinois at Urbana-Champaign, 2010.
252. DP Mondkar and GH Powell, Evaluation of solution schemes for nonlinear structures, Computers and Structures, Vol 9, No 3: 223–236, 1978.
253. YB Yang and SR Kuo. Theory and Analysis of Nonlinear Framed Structures, Prentice Hall, London, 1994.
254. Bjørn Skallerud and Jørgen Amdahl, Nonlinear Analysis of Offshore Structures, Research Studies Press, Baldock, England, 2002.
255. Carlos A. Felippa, Nonlinear finite element methods, Course material, University of Colorado Boulder, 1999.
256. Bashir Ahmed Memon and Xiaozu Su, Arc-length technique for nonlinear finite element analysis, Journal of Zhejiang University Science, Vol 5, No 5: 618–628, 2004.
257. Joseph Padovan and Surapong Tovichakchaikul, Self-adaptive predictor-corrector algorithms for static nonlinear structural analysis, Computers & Structures, Vol 15, No 4:365–377, 1982.
258. G Powell and J Simons, Improved iteration strategy for nonlinear structures, International Journal for Numerical Methods in Engineering, Vol 17, No 10: 1455–1467, 1981.
259. J Simons and PG Bergan, Hyperplane displacement control methods in nonlinear analysis, in WK Liu, T Belytschko and KC Park (eds), Innovative Methods for Nonlinear Problems (345–364), Pineridge Press, Swansea, 1984.
260. F Fujii, K Choong and S Gong, Variable displacement control to overcome turning points of nonlinear elastic frames, Computers & Structures, Vol 44, No 1–2: 133–136, 1992.
261. M Crisfield, An incremental iterative algorithm that handles snap through, Computers and Structures, Vol 13: 55–62, 1981.
262. E Riks, The application of Newton's method to the problem of elastic stability, Journal of Applied Mechanics, Vol 39: 1060–1065, 1972.

410 References

263. Norwegian Geotechnical Institute, Presentation notes of seismic analyses for Aker Solutions, Bergen, 16 April, 2009.
264. Y Yamamoto and JW Baker, Stochastic model for earthquake ground motion using wavelet packets, Department of Civil and Environmental Engineering, Stanford University, 2011.
265. R Dobry, I Oweis and A Urzua, Simplified procedures for estimating the fundamental period of a soil profile, Bull. Seismol. Soc. Am., Vol 66: 1293–1321, 1976.
266. S Kumar and JP Narayan, Importance of quantification of local site effects based on wave propagation in seismic microzonation, Journal of Earth System Science, Vol 117, No 2: 731–748, 2008.
267. ISO 19901-2, Petroleum and natural gas industries—Specific requirements for offshore structures—Part 2: Seismic design procedures and criteria, International Standard Orgnization, Geneva, 2004.
268. Eurocode 8, Design of Structures for Earthquake Resistance, British Standard, London, 2004.
269. Stephen A Nelson, Earthquake Prediction, Control and Mitigation (lecture notes), Tulane University, September, 2011.
270. Francie Diep, Fast facts about the Japan earthquake and tsunami, Scientific American, March 2011.
271. Ronald S Harichandran, Spatial Variation of Earthquake Ground Motion, Department of Civil and Environmental Engineering, Michigan State University, 1999.
272. A Lupoi, P Franchin, PE Pinto and G Monti, Seismic design of bridges accounting for spatial variability of ground motion, Earthquake Eng. Struct. Dyn., Vol 34: 327–348, 2005.
273. F Romanelli, GF Panzal and F Vaccari, Realistic modeling of the effects of asynchronous motion at the base of bridge piers, Journal of Seismology and Earthquake Engineering, Vol 6, No 2: 19–28, Summer 2004.
274. N Tzanetos, AS Elnashai, FH Hamdan and S Antoniou, Inelastic dynamic response of RC bridges subjected to spatial non-synchronous earthquake motion, Adv. Struct. Eng., Vol 3, No 3: 191–214, 2000.
275. Nicholas J Burdette, Amr S. Elnashai, Alessio Lupoi, and Anastasios G Sextos, Effect of asynchronous earthquake motion on complex bridges. I: Methodology and input motion. Journal of Bridge Engineering, Vol 13, No 2: 166–172, 2008.
276. Aspasia Zerva, Spatial Variation of Seismic Ground Motion – Modeling and engineering applications, CRC Press, Boca Raton, FL, 2008.
277. TJ Zhu, AC Heidebrecht and WK Tso, Effects of peak ground acceleration to velocity ratio on the ductility demand of inelastic systems, Earthquake Engineering and Structural Dynamics, Vol 16: 63–79, 1988.
278. Manolis Papadrakakis, Nikolaos D Lagaros and Michalis Fragiadakis, Seismic Design Procedures in the Framework of Envolutionary Based Structural Optimization, Computational Mechanics—Solids, structures and coupled problems, Springer, Netherlands, 2006.
279. Najib Bouaanani, Patrick Paultre and Jean Proulx, Two-dimensional modeling of ice cover effects for the dynamic analysis of concrete gravity dams, Earthquake Engineering and Structural Dynamics, Vol 31, No 12: 2083–2102, 2002.
280. Gregory P Tsinker, Chapter 2: Marine structures in cold regions in Marine Structures Engineering: Specialized Applications, Chapman and Hall, New York, 1995.
281. Ahmed Ghobarah, Review Article: Performance-based design in earthquake engineering: State of development, Engineering Structures, Vol 23, No 8: 878–884, 2001.
282. D Vamvatsikos and CA Cornell, Seismic performance, capacity and reliability of structures as seen through incremental dynamic analysis, Report No RMS-55, RMS Program, Stanford University, Stanford, MA, 2000.
283. M Razavi, Evaluation of endurance time method compared to spectral dynamic analysis method in seismic analysis of steel frames. Department of Civil Engineering, Sharif University of Technology, Tehran, 2007.

References

284. Weicheng Cui, Review Article: A state-of-the-art review on fatigue life prediction methods for metal structures, Journal of Marine Science and Technology, Vol 7, No 1: 43–56, 2002.
285. D Radaj, CM Sonsino and W Fricke, Fatigue Assessment of Welded Joints by Local Approaches, Woodhead, Cambridge, 2006.
286. Zhiyuan Li, Fatigue assessment of containerships—a contribution to direct caculation procedures, Report No 3329, Chalmers University of Technology, Gothenburg, 2013.
287. A Fatemi and L Yang, Cumulative fatigue damage and life prediction theories: A survey of the state of the art for homogeneous materials, International Journal of Fatigue, Vol 20: 9–34, 1998.
288. D Krajcinovic, Damage mechanics, Mechanics of Materials, Vol 8, No 2-3: 117–197, 1989.
289. HO Madsen, S Krenk and NC Lind, Methods of Structural Safety, Prentice-Hall, Englewood Cliffs, NJ, 1986.
290. EJ Niemi, P Tanskanen, Hot spot stress determination for welded edge gussets, The International Institute of Welding—IIW Doc. XIII-1781-1799, 1999.
291. O Doerk, W Fricke and C Weissenborn, Comparison of different calculation methods for structural stresses at welded joints, International Journal of Fatigue, Vol 25, No 5: 359–369, 2003.
292. W Fricke and A Kahl, Comparison of different structural stress approaches for fatigue assessment of welded ship structures, Marine Structures, Vol 18: 473–488, 2005.
293. Inge Lotsberg and Gudfinnur Sigurdsson, Hot spot stress s-n curve for fatigue analysis of plated structures, Journal of Offshore Mechanics and Arctic Engineering, Vol 128: 330–336, 2006.
294. Jae-Myung Lee, Jung-Kwan Seo, Myung-Hyun Kim, Sang-Beom Shin, Myung-Soo Han, June-Soo Park and Mahen Mahendran, Comparison of hot spot stress evaluation methods for welded structures, International Journal of Naval Architecture and Ocean Engineering, Vol 2: 200–210, 2010.
295. W Fricke, Recommended hot spot analysis procedure for structural details of FPSO's and ships based on round-robin FE analyses, International Journal of Offshore and Polar Engineering, Vol 12: 1–8, 2001.
296. Inge Lotsberg, Recommended methodology for analysis of structural stress for fatigue assessment of plated structures, in OMAE Specialty Symposium on FPSO Integrity (1–14), Houston, TX, 2004.
297. W Fricke and A Kahl, Numerical and experimental investigation of weld root fatigue in fillet-welded structures, International Shipbuilding Progress, Vol 55: 29–45, 2008.
298. Mustafa Aygül, Fatigue analysis of welded structures using the finite element method, Lic. No. 2012:4, Chalmers University of Technology, Gothenburg, 2012.
299. CG Schilling, KH Klippstein, JB Barsom and GT Blake, Fatigue of welded steel bridge members under variable-amplitude loadings, NCHRP—Project 12-12, Transportation Research Board, National Reach Council, Washington DC, 1975.
300. DnV-OS-C201, Structural Design of Offshore Units (WSD Method), Det Norske Veritas, Høvik, 2005.
301. PW Marshall and WH Luyties, Allowable stress for fatigue design, Proceedings of the 3rd International Conference on Behavior of Offshore Structures (BOSS 82), Boston, MA, 1982.
302. AK Williams and JE Rinne, Fatigue analysis of steel offshore structures, Proc Instn. Civ. Engnrs., Part 1, Vol 60, No 4: 635–654,1976.
303. Fatigue Assessment of Offshore Structures, American Bureau of Shipping (ABS), Houston, TX, 2003.
304. K Sobczyk and BF Spencer, Random Fatigue—from Data to Theory, Academic Press, Boston, MA, 1992.
305. FJ Zwerneman, Fatigue damage accumulation under varying-amplitude loads, in R Narayanan and TM Roberts (eds), Structures Subjected to Repeated Loading—stability and strength (25–53), Elsevier, Oxford, 1991.

412 References

306. N Karlsson and PH Lenander, Analysis of Fatigue Life in Two Weld Class Systems, LITH, UniTryck, Linköping, 2005.
307. H Tada, PC Paris and GR Irwin, The Stress Analysis of Cracks Handbook, Del Research Corporation, Hellertown, PA, 1973.
308. DP Rooke and DJ Cartwright, Compendium of Stress Intensity Factors, HM Stationery Office, London, 1975.
309. GCM Sih, Handbook of Stress Intensity Factors, Lehigh University, Bethlehem, PA, 1973.
310. RJ Donahue, HM Clark, P Atanmo et al., Crack opening displacement and the rate of fatigue crack growth, International Journal of Fracture Mechanics, Vol 8: 209–291, 1972.
311. Julie A Bannantine, Jess J Comer and James L Handrock, Fundamentals of Metal Fatigue Analysis, Prentice Hall, London, 1990.
312. Ted L Anderson, Fracture Mechanics: Fundamentals and Applications, 3rd ed., CRC Press, Boca Raton, FL, 2004.
313. PC Paris, MP Gomez and WP Anderson, A rational analytical theory of fatigue, Rends Engineering, Vol 13: 9–14, 1961.
314. PC Paris and F Erdogan, A critical analysis of crack propagation laws, Journal of Basic Engineering, Vol 85: 528–534, 1963.
315. J Martinsson, Fatigue Assessment of Complex Welded Steel Structures, Royal Institute of Technology, Stockholm, 2005.
316. A Hobbacher (ed), Recommendations for fatigue design of welded joints and components, IIW document, XIII-1823-07, Welding Research Council Bulletin 520, NY, 2009.
317. Z Barsoum, Residual Stress Analysis and Fatigue of Welded Structures, Royal Institute of Technology, Stockholm, 2006.
318. M Byggnevi, LEFM Analysis and Fatigue Testing of Welded Structures, Royal Institute of Technology, Stockholm, 2005.
319. NORSOK standard N-001, Structural Design, Oslo, Norway: Norwegian Technology Standards Institution, 2000.
320. Inge Lotsberg, and Gudfinnur Sigurdsson, Use of Probabilistic Methods for Planning of Inspection for Fatigue Cracks in Offshore Structures (JIP), Report No 2011-0666, DnV, Høvik, 2012.
321. HJ Wessel, Norwegian University of Science and Technology, Report No UR-86-49, 1986.
322. Plåthandboken, SSAB Tunnplåt AB, 1990.
323. AF Blom, Overload retardation during fatigue crack propagation in steels of different strength, Swedish Journal of Metallurgy, Vol 18, 1989.
324. RG Forman, VE Kearney and RM Engle, Numerical analysis of crack propagation in cyclic-loaded structures, Journal of Basic Engineering, Vol 89, No3: 459–463, 1967.
325. NORSOK standard N-001, Integrity of Offshore Structures, 5th ed., August 2008.
326. NORSOK standard N-004, Design of Steel Structures, rev. 2, October 2004.
327. BS 7910, Guidance on Methods for Assessing the Acceptability of Flaws in Fusion Welded Structures, British Standard, London, 1999.
328. JG Kuang, AB Potvin and RD Leick, Stress Concentrations in Tubular Joints, Offshore Technology Conference (OTC) No 2205, Texas, 1975.
329. MB Gibstein, Fatigue strength of welded tubular joints, Steel in Marine Structures Conference, Paris, October 1981.
330. Fatigue Design of Offshore Steel Structures, Recommended Practice, DnV-RP-C203, Det Norske Veritas, Høvik, August 2008.
331. Eurocode 3: Design of steel structures—Part 1–9: Fatigue, 1993.
332. ASME Boiler and Pressure Vessel Code, Section VIII, Division 2, Alternative Rules, 2007.
333. BS7608:1993, Code of Practice for Fatigue Design and Assessment of Steel Structures, 1993.
334. American Bureau of Shipping, Guide for Fatigue Strength Assessment of Tankers, Part 3: Steel Vessel Rules, ABS, NY, 1992.

References

335. Eurocode 9, Design of Aluminum Structures – Part 2: Structures Susceptible to Fatigue, EN 1999-2:1998 E, British Standard, London, 1998.
336. JL Fayard, A Bignonnet and Van K Dang, Fatigue design criterion for welded structures, Fatigue and Fracture of Engineering Materials and Structures, Vol 19, No 6: 723–729, 1996.
337. Zhiyuan Li, Jonas Ringsberg and G Storhaug, Time-domain fatigue assessment of ship side-shell structures, International Journal of Fatigue, Vol 55, No 1: 276–290, 2013.
338. A Hobbacher, Recommendations for fatigue design of welded joints and components, IIW document IIW-1823-07 ex XIII-2151r4-07/XV-1254r4-07, International Institute of Welding, 2008.
339. Classification Note No 30.7: Fatigue Assessment of Ship Structure, Det Norske Veritas, Høvik, 2010.
340. ZG Xiao and KA Yamada, A method of determining geometric stress for fatigue strength evaluation of steel welded joints, International Journal of Fatigue, Vol 26: 1277–1293, 2004.
341. B Noh, J Song and S Bae, Fatigue strength evaluation of the load-carrying cruciform fillet welded joints using hot-spot stress, Key Engineering Materials, Vols 324–325: 1281–1284, 2006.
342. P Dong, A structural stress definition and numerical implementation for fatigue analysis of welded joints, International Journal of Fatigue, Vol 23: 865–876, 2001.
343. P Dong, JK Hong and Z Cao, Structural stress based master S-N curve for welded joints, IIW Doc. XIII -1930-02/XV-1119-02: 24, 2002.
344. P Dong and JK Hong, Analysis of hot spot stress and alternative structural stress methods, The 22nd International Conference on Offshore Mechanics and Artic Engineering, Vol 3: 213–224, 2003.
345. I Poutiainen, P Tanskanen, G Marquis, Finite element methods for structural hot spot stress determination – A comparison of procedures, International Journal of Fatigue, Vol 26: 1147–1157, 2004.
346. D Radaj, Design and analysis of fatigue resistant welded structures, Woodhead, Cambridge, 1990.
347. CM Sonsino, New local concept for the design of welded joints—industrial application, UTMIS autumn course, 2010, Södertälje.
348. Kristin Nielsen, Crack propagation in cruciform welded joints, UPTEC F11 008, Uppsala University, Uppsala, 2011.
349. W Fricke, Round-Robin Study on Stress Analysis for the Effective Notch Stress Approach, IIW document, XIII-2129-06/XV-1223-06, 2006.
350. JL Fayard, A Bignonnet, KD Van, G Marquis and J Solin, Fatigue design of welded thin sheet structures, in European Structural Integrity Society (145–152), Elsevier, Oxford, 1997.
351. EJ Niemi and GB Marquis, Structural hot spot stress method for fatigue analysis of welded components, International Conference on Metal Structures (39–44), Miskolc, Hungary, 2003.
352. EJ Niemi, Stress determination for fatigue analysis of welded components, IIW documentation, IIS/IIW-1221-93, The International Institute of Welding, 1995.
353. Å Eriksson, AM Lignell, C Olsson and H Spennare, Weld Evaluation Using FEM—A guide to fatigue-loaded structures, Industrilitteratur AB, Gothenburg, 2003.
354. Alewyn Petrus Grové, Development of a Finite Element Based Nominal Stress Extraction Procedure for Fatigue Analysis of Welded Structures, University of Pretoria, 2006.
355. M Fermér, M Andréasson and B Frodin, Fatigue Life Prediction of MAG-welded Thin Sheet Structures, Volvo Car Corporation, SAE 982311, 1998.
356. Tom Lassen and Naman Récho, Fatigue Life Analyses of Welded Structures: Flaws, 1st ed., Wiley-ISTE, Hoboken, NJ, 2013.
357. Wikipedia, Fatigue (material), 2013.
358. T Moan, G Hovde and A Blanker, Reliability-based Fatigue Design Criteria for Offshore Structures Considering the Effect of Inspection and Repair. In Proc. 25th OTC, Vol 2: 591–599, OTC 7189, Houston, TX, 1993.

359. E Ayala-Uraga and T Moan, Time-variant reliability assessment of FPSO hull girder with long cracks, Journal of Offshore Mechanics and Arctic Engineering, Vol 129, No 1: 81–89, 2007.

360. AK Vasudevan and K Sadananda, Short crack growth and internal stresses, International Journal of Fatigue, Vol 19, No 1: 99–108, 1997.

361. WC Cui, RG Bian and XC Liu, Application of the two-parameter unified approach for fatigue life prediction of marine structures, The 10th International Symposium on Practical Design of Ships and Other floating Structures (PRADS), Houston, TX, Vol 2: 392–398, 2007.

362. MA Miner, Cumulative damage in fatigue. J Appl Mech, Vol 12: 159–164, 1945.

363. A Palmgren, Die lebensdauer von kugellagern. Z Vereins Deut Ingenieure, Vol 68: 339–341, 1924.

364. OM Faltinsen, Sea loads on ships and offshore strucutres, Cambridge University Press, Cambridge, 1990.

365. Junbo Jia, An efficient nonlinear dynamic approach for calculating the wave induced fatigue damage of offshore structures and its industry applications for the lifetime extension purpose. Applied Ocean Research, Vol 30, No 3: 189–198, 2008.

366. WJ O'Donnell and BF Langer, Fatigue Design Basis for Zircaloy Components, Nuclear Science and Engineering, Vol 20, No 1: 1–12, 1964.

367. ASTM Designation E 1049-85, Standard practices for cycle counting in fatigue analysis. American Society for Testing and Materials, West Conshohocken, PA, 1985.

368. DV Nelson, Cumulative Fatigue Damage in Metals, PhD thesis, Stanford University, Stanford, CA, 1978.

369. M Matsuishi and T Endo, Fatigue of metals subject to varying stress, paper presented to Japan Soc Mech Engs, Jukvoka, Japan, 1968.

370. NE Dowling, Fatigue failure predictions of complicated stress-strain histories, ASTM J. Mater., Vol 7, No 1: 71–87, 1972.

371. PH Wirsching and MC Light, Fatigue under wide band random loading, J Struct. Div., ASCE: 1593–1607, 1980.

372. JCP Kam and WD Dover, Advanced tool for fast assessment of fatigue under offshore random wave stress history, Proc. Instn. Civ Engrs., Part 2, 87: 539–556, 1988.

373. JW Hancock and DS Gall, Fatigue under narrow and broad band stationary loading, Final report of the Cohesive Programme of Research and Development into the Fatigue of Offshore Structures, Marine Technology Directorate Ltd., 1985.

374. T Dirlik, Application of Computers in Fatigue Analysis, University of Warwick, PhD thesis, 1985.

375. GK Chaudhury and WD Dover, Fatigue analysis of offshore platforms subject to sea wave loading, International Journal of Fatigue, Vol 7, No 1: 13–19, 1985.

376. Igor Rychlik, A new definition of the rainflow cycle counting method, International Journal of Fatigue, Vol 9, No 2: 119–121, 1987.

377. Igor Rychlik, Note on cycle counts in irregular loads. Fatigue Fract Eng Mater Structure, Vol 16: 377–390, 1993.

378. M Frendhal and I Rychlik, Rainflow analysis: Markov method. Int.J. Fatigue, Vol 15: 265–273, 1993.

379. Zheng Gao and Torgeir Moan, Accuracy of the narrow-band approximation of stationary wide-band Gaussian processes for extreme value and fatigue analysis, in H Furuta, D Frangopol and M Shinozuka (eds), Proceedings of 10th International Conference on Structural Safety and Reliability (997–1004), CRC Press, Boca Raton, 2009.

380. NWM Bishop and F Sherratt, A theoretical solution for the estimation of rainflow ranges from power spectral density data, Fatigue Fract. Engng. Mater. Struct., Vol 13, No 4:311–326, 1990.

381. NWM Bishop, Spectral methods for estimating the integrity of structural components subjected to random loading., in A Carpinteri (ed), Handbook of Fatigue Crack Propagation in Metallic Structures (1685–1720), Elsevier, Oxford, 1994.

References

382. A Halfpenny, A frequency domain approach for fatigue life estimation from finite element analysis, paper presented at International Conference on Damage Assessment of Structure (DAMAS 99), Dublin, 1999.
383. D Benasciutti and R Tovo, Spectral methods for lifetime prediction under wide-band stationary random processes, International Journal of Fatigue, Vol 27, No 8: 867–877, 2005.
384. Zheng Gao and Torgeir Moan, Frequency-domain fatigue analysis of wide-band stationary Gaussian processes using a trimodal spectral formulation, International Journal of Fatigue, Vol 30, No 10–11: 1944–1955, 2008.
385. Zheng Gao, Stochastic Responses Analysis of Mooring Systems with Emphasis on Frequency-domain Analysis of Fatigue due to Wind-band Response Process, PhD thesis, Norwegian University of Science and Technology, February 2008.
386. D Benasciutti and R Tovo, On fatigue damage assessment in bimodal random processes, International Journal of Fatigue, Vol 29, No 2: 232–244, 2007.
387. TT Fu and Cebon, Predicting fatigue lives for bi-modal stress spectral densities International Journal of Fatigue, Vol 22, No 1: 11–21, 2000.
388. S Sakai and H Okamura, On the distribution of rainflow range for Gaussian random processes with bimodal PSD. JSME Int J Ser A 38: 440–445, 1995.
389. G Jiao and T Moan, Probabilistic analysis of fatigue due to Gaussian load process, Probabilistic Engineering Mechanics, Vol 5, No 2: 76–83, 1990.
390. ISO 19901-7, ISO, Petroleum and natural gas industries – Specific requirements for offshore structures – Part 7: Stationkeeping systems for floating offshore structures and mobile offshore units, International Standard Orgnization, Geneva, 2005.
391. API, Recommended practice for design and analysis of stationkeeping systems for floating structures. API RP 2SK, 2005.
392. DnV Offshore Standard – Position mooring, DnV-OS-E301, Det Norske Veritas, Høvik, 2010.
393. JH Nieuwenhuijsen, Experimental Investigations on Seasickness, Drukkerij Schotanus and Jens, Utrecht, 1958.
394. Arthur Bolton, Structural Dynamics in Practice – A guide for professional engineers. McGraw-Hill Europe, London, 1993.
395. Mai Tongt, Guo-Quan Wang and George C Lee, Time derivative of earthquake acceleration, Earthquake Engineering and Engineering Vibration, Vol 4, No 1: 1–16, 2005.
396. Per-Åke Berge and Olle Bråfelt, Noise and Vibration On Board, Joint Industrial Safety Council, Stockholm, 1991.
397. Larson Davis Inc. Industrial Hygiene Vibration Monitor – About human exposure to vibration in the workplace, www.LarsonDavis.com, 2003.
398. MJ Griffin, Handbook of Human Vibration, Academic Press, London, 1996.
399. NORSOK S-002, Working environment, www.standard.no, rev.04, 2004.
400. ISO 6954-4, Mechanical Vibration—Guidelines for the measurement, reporting and evaluation of vibration with regard to habitability on passenger and merchant ships, 2000.
401. ISO 4867, Code for the Measurement and Reporting of Shipboard Vibration Data, 1st ed., Geneva, 1984.
402. NATO STANAG 4154, Comman Procedures for Seakeeping in the Ship Design Process, NATO, 2000.
403. Priyan Mendis and Tuan Ngo, Vibration and shock problems of civil engineering structures, in Clarence W de Silva (ed.), Vibration and Shock Handbook, CRC Press, Boca Raton, FL, 2011.
404. EL Wilson, A Der Kiureghian and ER Bayo, A replacement for SRSS method in seismic analysis, Earthquake Engineering and Structural Dynamics, Vol 9: 187–192, 1981.
405. A Der Kiureghian, A response spectrum method for random vibration analysis of MDF systems, Earthquake Engineering and Structural Dynamics, Vol 9: 419–435, 1981.
406. E Rosenblueth and J Elorduy, Responses of linear systems to certain transient disturbances, Proceedings of the 4th World Conference on Earthquake Engineering, Santiago, Vol I: 185–196, 1969.

407. MB Zaaijer, Sensitivity analysis for foundations of offshore wind turbines, Section Wind Energy, WE 02181, Delft, 2000.
408. TH Søreide, J Amdahl, E Eberg, T Holmås and Ø Hellan Ø, USFOS – A computer program for progressive collapse analysis of steel offshore structures, Theory Manual. Report No F88038, SINTEF, Trondheim, 1993.
409. AF Dier and M Lalani, New code formulations for tubular joint static strength, Proceedings of the 8th International Symposium on Tubular Structures, edited by US Choo and GJ van Der Vegte, Balkema, Innsbruck, 1998.
410. Junbo Jia, Dynamics of Vehicle-Deck Interaction, Chalmers University of Technology, 2006.
411. Junbo Jia and Jonas W Ringsberg, Numerical and experimental investigation of dynamics of vehicle/ship-deck interactions, Marine Technology, Vol 45, No 1: 28–41, 2008.
412. Junbo Jia and Anders Ulfvarson, Modal testing and finite element calculations for lightweight aluminum panels in car carriers, Marine Technology, Vol 43, No 1: 11–21, 2006.
413. Junbo Jia and Anders Ulfvarson, A systematic approach toward the structural behavior of a lightweight deck-side shell system, Journal of Thin-Walled Structures, Vol 43, No 1: 83–105, 2005.
414. Junbo Jia and Anders Ulfvarson, A parametric study for the structural behavior of a lightweight deck, Engineering Structures, Vol 26, No 7: 963–977, 2004.
415. GG Stokes, Discussion of a differential equation relating to the braking of railway bridges, Transactions of the Cambridge Philosophical Society Vol 8, No 5: 707–735, 1849.
416. CW Mousseau and GM Hulbert, An efficient tire model for the analysis of spindle forces produced by a tire impacting large obstacles, Computer Methods in Applied Mechanics and Engineering, Vol 135, No 1–2: 15–34, 1996.
417. DR Ahlbeck, Effects of track dynamic impedance on vehicle-track interactions: Interaction of railway vehicles with the track and its substructure, Proceedings of the 3rd Herbertov Workshop on Interaction of Railway Vehicles with the Track and its Substructure, September 1994(58–71), edited by K Knothe, SL Grassie, JA Elkins, Swets and Zeitlinger, Herbertov, 1994.
418. RP Dallinga, Safe securing of trailers and deck cargo, Proceedings of the 12th International Conference and Exhibition on Marine Transport using Roll-on/Roll-off Methods, Gothenburg, 1994.
419. SR Turnbull, Roll-on/roll-off semi-trailer models: A comparison of results, Journal of Marine Science and Technology, Vol 5, No 3: 101–106, 2001.
420. SR Turnbull and D Dawson, The dynamic behavior of flexible semi-trailers on board Ro-Ro ships, International Journal of Mechanical Science, Vol 41, No 12: 1447–1460, 1999.
421. SR Turnbull and D Dawson, The securing of rigid semi-trailers on roll-on/roll-off ships, International Journal of Mechanical Science, Vol 39, No 1: 1–14, 1997.
422. V Mora, Modeling and analysis of a car lashed on a ship deck, Report No X-99/112, Department of Naval Architecture and Ocean Engineering, Chalmers University of Technology, Gothenburg, 1999.
423. S Eylmann, H Paetzold and B Bohlmann, Fatigue behavior of car decks made of VHTS 690, 9th Symosium on Practical Design of Ships and Other Floating Structures, Luebeck-Travemuende, Vol 1: 449–456, 2004.
424. M Sternsson, Statistical Study of Lashing Forces and Encountered Wave Height. Licentiate thesis, Department of Naval Architecture and Ocean Engineering, Chalmers University of Technology, Sweden, 2001.
425. YB Yang, JD Yau and YS Wu, Vehicle-Bridge Interaction Dynamics with Applications to High-Speed Railways. World Scientific, Beijing, 2004.
426. D Cebon, Interaction between heavy vehicles and roads. SP-951, L Ray Buckendale Lecture, SAE, 1993.

References

427. D Cebon, Handbook of Vehicle-Road Interaction, Swets and Zeitlinger, Lisse, Abingdon, Exton, Tokyo, 1999.
428. P Andersson, Securing of road trailers on board Ro-Ro ships, Research report by MariTerm AB, Gothenburg, 1983.
429. C MacDougall, MF Green and S Shillinglaw, Fatigue damage of steel bridges due to dynamic vehicle load. Journal of Bridge Engineering, Vol 11, No 3: 320–328, 2006.
430. International Maritime Organization (IMO), Code of safe practice for cargo stowage and securing, IMO: 27, 2003.
431. AW Leissa, Vibration of Shells, The Acoustical Society of America, New York, NY, 1993.
432. NJ Huffington and WH Hoppmann II, On the transverse vibrations of rectangular orthotropic plates, Journal of Applied Mechanics ASME, Vol 25: 389–395, 1958.
433. MA Foda and BA Albassam, Vibration confinement in a general beam structure during harmonic excitations, Journal of Sound and Vibration, Vol 295, No 3–5: 491–517, 2006.
434. ARJM Lloyd, Seakeeping—Ship behavior in rough weather, Sussex 1998.
435. L Kwasniewski, HY Li, J Wekezer and J Malachowski, Finite element analysis of vehicle-bridge interaction, Journal of Finite Elements in Analysis and Design, Vol 42, No 11: 950–959, 2006.
436. BT Wang, Structural modal testing with various actuators and sensors, Journal of Mechanical Systems and Signal Processing, Vol 12, No 5: 627–639, 1998.
437. EV Lewis, Principles of Naval Architecture, 2nd rev, Vol II: Resistance, propulsion and vibration, The Society of Naval Architects and Marine Engineers: 306–315, 1988.
438. ISO 6954, Mechanical vibration: Guidelines for the measurement, reporting and evaluation of vibration with regard to habitability on passenger and merchant ships, 2000.
439. P Andersson, B Allenström and M. Niilekselä, Safe stowage and securing of cargo on board ships, Research report by MariTerm AB, Gothenburg, 1982.

Index

A
a/v ratio, 315, 316
Acceleration resonance frequency, 53
Accelerometer, 76
Across wind, 193
Added mass, 46, 84, 93, 174, 291
Air suspension, 387
Aliasing, 115
Allowable stress range, 352
Along wind, 187, 194
Amplification-factor, 249
Anti-symmetry, 72, 73
Apparent velocity, 201
Approximate solution, 27, 61, 68, 79, 262
Arc-length method. *See* Load-displacement
    control
Atmospheric boundary layer, 185
Atomic bonds, 322
Auto-spectra density. *See* Power spectra
    density

B
Bandwidth, 123, 124, 254, 256, 362, 363, 365,
    366, 368, 370, 381
Bandwidth parameter, 123, 124, 365, 368, 370
Base excitations, 152–154, 161, 221
Bauschinger effect, 274
Benasciutti and Tovo's method, 368–369
Bi-modal, 366, 368–370
Broad band. *See* Wide band
Buckling, 28, 85, 87–89, 91, 92, 94, 276, 278,
    294, 299, 321
Buffeting effects, 187

C
Cartesian deflection, 215
Caughey damping, 246
Central difference method, 220, 221, 224–226
Characteristic ground frequency, 199
Characteristic value. *See* Eigenvalue
Cholesky decomposition, 195
Coefficient vector, 215
Coffin–Manson relationship, 341
Coherence function, 170–172, 189, 190, 193,
    195, 200, 201, 309
Complementary solution, 142, 155–157
Complex periodical excitation, 155
Conditionally stable. *See* Numerical stability
Consistent mass, 215, 226, 230, 231
Continuous system, 33, 79, 83, 86
Convolution integral, 158, 160, 161, 216
Corner frequency, 114
Corrosion, 257, 323, 326
Coupled equations, 27, 211, 214, 220, 248
Courant number, 225
CQC method, 313
Crack growth rate, 342–345
Crack growth threshold, 342, 344
Crack initiation, 322, 323, 329, 340, 341
Crack size, 344, 345
Creep, 275, 276, 321, 324, 341
Critical damping, 48–51, 161, 256
Critical viscous damping, 256
Cross-correlation function. *See* Cross-
    covariance
Cross-covariance, 170, 171, 194, 200
Cross-spectra density, 167, 170, 171
Cumulative damage, 347

J. Jia, *Essentials of Applied Dynamic Analysis*, Risk Engineering,
DOI: 10.1007/978-3-642-37003-8, © Springer-Verlag Berlin Heidelberg 2014

420 Index

Cumulative probability distribution function, 126, 129, 136
Cutoff frequency, 114
Cycle counting, 282, 348, 363, 364

D

Damping force, 32, 47, 63, 143, 145, 157, 235, 237–240, 243, 244, 247
Damping matrix, 204, 218, 231, 245, 247, 248
Davenport spectrum, 190
Dead weight (force) model, 385, 393
Depth attenuation, 177
Design S–N curves, 325, 327, 340
Design wave height, 349, 354
Deterministic, 25, 26, 99, 100, 106, 164, 201, 220, 309–311, 349–351, 353, 354, 356, 360, 361
Deterministic fatigue analysis, 349, 341, 353, 360
Deviator stress, 270–272, 275
Differential equation, 24, 26, 27, 35, 47, 58, 79, 86, 88–90, 141, 155, 205, 220, 262, 283, 285, 288, 392
Dirac delta function. *See* Unit impulse
Direct integration, 203, 220, 221, 226, 230, 291
Directional wave energy spectrum. *See* Short crest waves
Dirlik's method, 367
Discontinuous system, 44
Displacement boundary nonlinearity, 279
Displacement control, 294, 299, 300
Displacement limit point, 293, 294, 300
Displacement resonance frequency, 53
Distributed mass, 58, 59, 61, 63, 64, 68
Drag coefficient, 174, 183, 184, 240
Drag force, 122, 174, 183, 184, 240, 279, 358, 363
Dunkerley method, 68, 72, 79, 83
Dynamic amplification factor, 25, 146
Dynamic equilibrium, 31–34, 218
Dynamic stiffness, 89

E

Earthquake, 2–6, 18, 19, 21, 24, 40, 44, 45, 75, 102, 103, 111–114, 134, 152, 156, 172, 173, 199–201, 226, 238, 248, 303–311, 314, 317, 318, 374, 379, 385
Effective radius, 336, 337
Eigen-analysis, 72, 88

Eigenfrequency, 46, 51, 55, 56, 64, 87, 91, 92, 193, 195, 203, 209, 211, 212, 238, 243, 249, 253, 254, 358
Eigenpairs, 26, 56, 58, 94, 212, 213, 230, 386
Eigenperiod, 92, 152, 168, 180, 216, 250, 285, 289, 290, 310
Eigen-problem, 55, 87, 88, 212
Eigenvalue, 56, 82, 87, 89, 212, 213
Eigenvector, 87, 212
Elastic linear-hardening, 267
Elastic-perfectly plasticity, 266
Elastic power-hardening, 266, 270
Endurance limit. *See* Threshold stress
Energy dissipation, 40, 233, 234, 239, 241, 242, 244, 247, 249, 252, 256–258
Engineering strain, 264–268
Engineering stress, 264–267
Environmental contour line, 138
Equations of motions, 26, 27, 31, 33, 34, 36, 37, 40, 203, 214, 220, 224, 225, 240, 246–248, 394
Equilibrium equations, 26, 58
Equilibrium path, 94, 293–295, 299, 300, 302
Equivalent viscous damping, 239, 241, 243, 244, 256
Ergodic, 106
Erosion, 323
Euler-Bernoulli beam, 58, 393
Euler's equation, 41
Euler buckling, 85, 91
Euler's equation, 41
Exact solution, 26, 27, 56, 58, 61, 81, 82, 262
Explicit integration, 224–226

F

Fatigue, 3, 7–9, 14, 22, 23, 25, 29, 119, 148, 173, 181, 190, 192, 193, 198, 258, 259, 282, 283, 373, 383
Fatigue design factor, 348
Fatigue ductility coefficient, 341
Fatigue ductility exponent, 341
Fatigue limit. *See* Threshold stress
Finite difference, 27, 220, 261
Finite element, 27, 33, 36, 47, 58, 73, 195, 206, 224–226, 229, 230, 248, 261, 262, 266, 270, 295, 298, 383, 394, 395
Flow rule, 270, 275
Fluid damping, 233, 258
Follower loads, 276, 279
Force boundary nonlinerity, 279
Forced vibration, 26, 77, 190, 216, 253, 254

Index 421

Forman's equation, 343
Fourier integral, 156, 162
Fourier spectrum, 111, 112, 114, 115
Fracture, 196, 263, 292
Fracture mechanics, 323, 324, 326, 336, 341–343, 345, 346, 348
Fracture toughness, 343, 345
Free decay, 17, 239, 249–251
Free vibration, 25, 81, 100, 180, 193, 203, 208, 209, 212, 216, 228, 235, 238, 249–252
Frequency domain, 101, 106, 107, 109, 111, 112, 174, 178, 191, 195, 200, 201, 211, 243, 249, 253, 255
Frequency response function, 148
Friction coefficient, 236, 237, 383, 396–398
Friction/Coulomb damping, 236, 240, 243
Frøya (NPD) spectral density, 190
Froude Krylov force, 174
Full-car model, 386

G
Galef's formula, 91, 92, 94
Gaussian distribution, 121, 122, 125, 133, 176, 183
GBS, 20, 21, 97, 217
General solution of motions, 49, 51
Generalized coordinate, 79, 215, 216
Generalized damping, 215
Generalized excitation force, 215
Generalized mass. *See* Modal mass
Generalized stiffness, 215
Geometric stress. *See* Hot-spot stress
Geometrical nonlinearity, 276, 278, 285, 287, 288, 292
Global vibrations, 117
Ground motion duration, 306
Ground motions, 5, 21, 102, 103, 110–114, 116, 117, 198–201, 303–311, 313–317, 319

H
Half-vehicle model, 386
Hamilton's principle, 31, 33–36, 86
Hardening nonlinearity, 283, 285
Hardening rule, 266, 270, 273–275
Harmonic excitations, 313, 314
Harris spectrum, 189, 190
Homogeneous equation, 55
Hot-spot stress, 323, 327–333, 336, 338, 372
Houbolt method, 220, 227
Human body vibration, 29

Hydraulic shock absorber, 233, 389
Hysteresis loop, 76, 239
Hysteretic damping, 238–241, 243, 256

I
Ice breaking length, 197
Ice load, 196–198, 379
Ice velocity, 197
Impact damper, 39, 40
Implicit integration, 226, 228
Impulsive force, 234
Impulsive responses, 158–160, 163
Incremental iterative, 295
Inertia dominant, 143
Inhomogeneous material, 340
Instantaneous wind speed, 185
Integration point, 331
Interpolation, 220, 221
Irregular wave, 176, 182
Irregularity factor, 124
Isotropic hardening, 273, 274

J
Jacket, 8, 10, 22–24, 58, 59, 92, 93, 174, 179, 180, 198, 227, 229, 247, 289, 291, 313, 317, 318
Jack-up, 174, 228–230
Jerk, 33
Joint probability distribution, 136
JONSWAP spectrum, 137, 179

K
Kanai–Tajimi model, 199
Kinematic hardening, 273, 274
Kinetic energy, 35, 63, 64, 79, 80, 84, 196, 211, 242, 379
Kurtosis, 122, 126

L
Lagrange's equation, 31, 34, 36, 37
Lashing, 381, 383, 384, 386, 387, 389, 390, 397
Lift coefficient, 183
Lightly damped, 226, 233, 239, 247, 252–254
Linear iteration, 27, 261, 263
Linearized (airy) wave theory, 175
Load control, 294, 295, 299, 300
Load-displacement control, 300, 301
Load imbalance. *See* Residual force

422 Index

Load limit point, 292–294, 299, 300
Load sequence effects, 280–283
Load step, 295–297
Local maximum, 120, 123, 294
Local vibrations, 117
Logarithmic decrement, 243, 250
Lo-lo, 380
Loss factor, 241, 242, 253, 254, 256
Luco and wong coherency function, 201
Lumped mass, 79, 83, 206, 208, 215, 226, 231, 394

M

Magnification factor, 243, 244, 253
Mass matrix, 206, 207, 211, 213, 215
Mass model, 384, 385, 393
Mass proportional, 245, 246
Mass scaling, 254–256
Material damping, 263, 275, 283
Material nonlinearity, 261, 273, 281
MDOF. *See* Multi-degrees-of-freedom
Mean period of motions, 123
Mean S–N curve, 324, 326
Mean wind speed, 130, 135, 184–190, 194, 219
Membrane stress, 278
Meteorological cycle, 184
Microstructural length, 336, 337
Miner rule, 282, 283
Miner summation. *See* Miner rule
Miner's accumulation rule. *See* Miner rule
Mixed hardening, 274
Modal analysis, 27, 29, 211
Modal combination, 312, 313
Modal damping. *See* Generalized damping
Modal force. *See* Generalized excitation force
Modal mass, 209, 213, 215–217
Modal matrix, 215, 247
Modal stiffness. *See* Generalized stiffness
Modal superposition, 27, 33, 211, 214, 218, 220, 229, 230, 243, 244, 248, 261, 262
Modal testing, 77, 85, 94, 246
Mode shape, 26, 55–57, 59, 61, 68, 74, 75, 77, 80–87, 89, 92, 95, 203, 209–218, 243–247, 262, 283–285, 305, 381, 396
Mode shape orthogonality, 213, 214, 246
Modified Newton–Raphson method, 297, 299
Modulus of transfer function, 148, 166, 168, 355
Moment of inertia, 206, 283
Momentum equilibrium, 40

Morison's equation, 122, 173
Motion induced load effects, 187
Multi-degrees-of-freedom, 56, 203
Multi-modal frequency, 359, 362

N

Narrow band, 112, 178, 179, 197, 199
Narrow conical structure, 196, 197
Natural frequency, 15, 16, 19, 24–26, 58, 61, 68, 72, 79–83, 85, 88, 91, 95, 199, 233, 234, 254, 283–285, 309, 381, 393
Natural period, 1, 14, 15, 19, 83–85, 230, 285–289, 304, 306, 310–315
Negative damped, 51
Neumann spectrum, 197
Newmark method, 27, 220, 221, 227, 296
Newton–Raphson method, 27, 297–299, 301
Nominal stress, 323, 327–333, 336, 338, 340, 342
Non-damaging stress. *See* Threshold stress
Non-dimensional gain function, 146, 154
Non-homogeneous equation, 140
Nonlinear, 25, 27, 87, 94, 122, 125, 174, 183, 184, 193, 220, 221, 238, 240, 244, 249, 310, 390
Nonlinear elasticity, 275
Nonlinear equilibrium, 292, 294
Non-periodical excitations, 156, 162
Nonproportional damping, 244, 248
Non-stationary, 103, 105, 106, 310, 319
Non-trivial solution, 55, 212
Non-uniform system, 33, 79
Normal distribution, 122, 130
Normality rule, 275
Normalized eigenvectors, 212
Normal value. *See* Eigenvalue
Notch stress, 323, 327, 329, 331, 336–339, 349
Numerical integration, 220, 229
Numerical stability, 227
Nyquist frequency, 115
Nyquist theorem, 115

O

Off-diagonal term, 247, 248, 313
Over-damped, 50

P

Palmgren–Miner rule. *See* Miner rule
Paris law, 343–345

Index 423

Particle velocity, 34, 174, 177, 183, 240
Particular solution, 142, 157, 160
PCTC, 383
P-delta effects, 288–291
Peak time, 252
Performance Based Design, 319
Period of peaks, 123
Period of zero crossing, 123
Phase angle, 106, 110, 195, 256, 398
Plastic potential, 272, 275
Plasticity, 28, 190, 198, 239
Plastic strain increment, 275
PM spectrum, 178, 179
Poisson distribution, 119, 133, 134
Potential energy, 34–36, 61, 64, 79, 80, 84, 87, 92, 233
Power hardening, 266, 269, 270
Power spectra density, 200
Prandtl–Reuss relation, 275
Principal stress, 272–274, 327, 331, 332, 346, 355
Probability density function, 121, 125, 136, 138

Q
Quarter-car model, 386
Quasi-static, 173, 192, 193, 228, 243, 253, 254

R
Radiation damping, 252, 257
Rain-flow counting, 193
Ramberg–Osgood rule 266 270
Random, 99, 102–104, 106, 111, 116, 119–121, 123, 125, 126, 129, 133, 136, 173, 175, 176, 195, 303, 313
Random excitation, 164
Random process, 99, 102, 103, 106, 111, 119, 121, 175, 367, 368
Rayleigh distribution, 125, 126, 129, 130
Regular wave, 174, 175
Residual force, 298
Residual stress, 328, 339, 340, 371
Resonance, 1–4, 7, 11, 14–16, 22, 23, 26, 111, 119, 173, 190, 197, 199, 214, 234, 249, 254, 256, 304, 309, 379, 384
Resonance frequency, 16, 51, 53
Resonance period, 22, 304
Response spectrum, 201, 305, 309–313
Reynolds number, 184
Rice distribution, 122, 126
Rigid-body vibrations, 55
Riks method. *See* Load–displacement control

RO/RO ship, 380, 381, 387

S
SDOF. *See* Single-degree-of-freedom
Seismic analysis, 309, 310, 319
Secant stiffness, 296
Seismic wave, 2, 6, 201, 303, 304, 306
Seismic wave passage effect, 201
Semi-stochastic fatigue analysis method, 370, 372
Shallow foundation, 97, 304
Shannon theorem, 115
Shape function, 231
Shear wave velocity, 201, 304, 305
Short crest waves, 180
Short term distribution, 125
Significant wave height, 130, 135, 137, 138, 178, 180
Simplified fatigue analysis, 349, 351, 352
Single-degree-of-freedom, 27, 34
Singularity, 346
Site period, 304, 305
Skewness, 121, 122, 126
Sloshing, 4–6, 19
Snap back, 292–294, 300
Snap through, 292–294, 299
S–N curve, 282, 323–331, 333, 334, 338–340, 347, 352, 358, 363, 366, 367, 370
Softening nonlinearity, 284–286
Soil-structure interactions, 303
Spatial variation, 194, 200, 201, 308, 319
Spectral gap, 185
Spectral moment, 123, 125
Spherical stress, 271
Spring-damper, 381, 383, 390
Sprung mass, 382, 384, 385
SRSS method, 312, 313
Stability point, 294
Stable crack growth, 321
Standard deviation, 103, 121, 187–189, 193, 313
Stationary process, 103–106, 303
Statistics, 119, 130, 164, 350, 357
Steady-state solution. *See* Particular solution
Steel leaf suspension, 387, 389
Step excitations, 159
Step response, 249
Stiffness matrix, 207, 208, 211, 231, 245
Stiffness proportional, 245, 246
Stochastic, 25, 26, 99, 102, 106, 116, 176, 199, 201, 303, 304, 310, 311
Stochastic fatigue analysis, 353, 354, 360, 370, 371

424 Index

Strain based approach, 323, 324, 341
Strain hardening, 263, 264, 266, 269, 270, 273, 274, 283, 284
Stress amplitude, 324, 326, 348, 359, 365
Stress based approach, 323, 324, 327
Stress concentration factor, 328–331, 370
Stress intensity factor, 342–345
Stress interaction, 347
Stress invariant, 271, 272, 275
Stress range, 324, 326, 331–333, 340, 343, 347, 349–354, 356–358, 361, 363, 366, 367, 370, 371
Stress stiffening/softening, 87
Structural discontinuities, 45
Structural health monitoring, 74, 75
Structural stress. *See* Hot-spot stress
Structural/slip damping, 257
Sub-modeling, 333, 334
Support factor, 336, 337
Suspension system, 379, 384, 385, 394
Swell, 183
Symmetry, 72, 73, 221, 391
System damping, 248, 251, 258

T

Tangent stiffness, 292, 296–298
Taylor series expansion, 224
Threshold stress, 324, 326
Through thickness crack, 340, 341
Time domain, 99, 173, 174, 176, 200, 221, 223, 249, 319
Time lag, 259
Timoshenko beam, 393
TMD. *See* Tuned mass damper
Trailer model, 387, 389
Train-track interaction, 382
Transfer function, 76, 211, 319
Transient responses, 238, 256
Tresca theory, 272
Tri-modal, 365, 368
Tuned mass damper, 18, 19, 40, 200
Turbulence intensity, 184, 188, 189
Tire, 8, 33, 221, 223, 236, 237, 379, 381, 383, 384, 386, 390, 393, 397

U

Ultimate ductile failure, 321
Unconditionally stable. *See* Numerical stability
Uncoupled/independent equations, 27, 211
Uniform system, 79, 147
Unit impulse, 156, 158

Unstable crack-growth, 321
Upper stability limit, 187

V

Variance, 102, 105, 106, 177, 182
Variation of accelerations, 220
Vehicle–bridge interaction, 51
Vehicle securing, 380, 382–384, 396–398
Vehicle–structure interactions, 382, 384, 390
Velocity resonant frequency, 53
Vibration perception, 375
Virtual displacement, 33, 34
Viscoelasticity, 260
Viscous damping, 17, 29, 63, 231, 235, 238–241, 243, 244, 248–250, 252, 256
VIV. *See* Vortex induced vibrations
von Mises criterion, 272, 273
Vortex induced vibrations, 14, 51, 368

W

Wave elevation, 175–177, 182
Wave energy, 175, 177, 178, 180, 181
Wave kinematics, 175, 177, 183
Wave load, 8, 10, 21, 22, 24, 103, 173, 174, 177, 180, 227, 229, 258, 259
Wave propagation speed, 224
Wave spreading, 180
Wear, 233, 257
Weibull distribution, 129, 130, 138, 351, 352
Weibull shape parameter, 352
Weld toe, 328–339, 344
Whipping effects, 317
White noise, 199
Wide band, 199
Wilson–method, 220, 221, 227
Wind load, 14, 183, 184, 190, 219, 282, 349, 381, 385
Wind turbulence, 103, 184, 185, 196
Wind velocity fields, 184

Y

Yield criterion, 266, 270, 272–275
Yield surface, 273–275
Youngs' modulus, 65, 283

Z

Zero crossing period. *See* Period of zero crossing
Zero upcrossing, 123, 369

Printed by Publishers' Graphics LLC
LMO140122.15.17.45